Lecture Notes in Computer Science

Lecture Notes in Artificial Intelligence 14993

Founding Editor

Jörg Siekmann

Series Editors

Randy Goebel, *University of Alberta, Edmonton, Canada*
Wolfgang Wahlster, *DFKI, Berlin, Germany*
Zhi-Hua Zhou, *Nanjing University, Nanjing, China*

The series Lecture Notes in Artificial Intelligence (LNAI) was established in 1988 as a topical subseries of LNCS devoted to artificial intelligence.

The series publishes state-of-the-art research results at a high level. As with the LNCS mother series, the mission of the series is to serve the international R & D community by providing an invaluable service, mainly focused on the publication of conference and workshop proceedings and postproceedings.

Oliver Brock · Jeffrey Krichmar
Editors

From Animals to Animats 17

17th International Conference
on Simulation of Adaptive Behavior, SAB 2024
Irvine, CA, USA, September 9–12, 2024
Proceedings

Editors
Oliver Brock
Technical University Berlin
Berlin, Berlin, Germany

Jeffrey Krichmar
University of California
Irvine, CA, USA

ISSN 0302-9743 ISSN 1611-3349 (electronic)
Lecture Notes in Artificial Intelligence
ISBN 978-3-031-71532-7 ISBN 978-3-031-71533-4 (eBook)
https://doi.org/10.1007/978-3-031-71533-4

LNCS Sublibrary: SL7 – Artificial Intelligence

© The Editor(s) (if applicable) and The Author(s), under exclusive license
to Springer Nature Switzerland AG 2025

This work is subject to copyright. All rights are solely and exclusively licensed by the Publisher, whether the whole or part of the material is concerned, specifically the rights of translation, reprinting, reuse of illustrations, recitation, broadcasting, reproduction on microfilms or in any other physical way, and transmission or information storage and retrieval, electronic adaptation, computer software, or by similar or dissimilar methodology now known or hereafter developed.
The use of general descriptive names, registered names, trademarks, service marks, etc. in this publication does not imply, even in the absence of a specific statement, that such names are exempt from the relevant protective laws and regulations and therefore free for general use.
The publisher, the authors and the editors are safe to assume that the advice and information in this book are believed to be true and accurate at the date of publication. Neither the publisher nor the authors or the editors give a warranty, expressed or implied, with respect to the material contained herein or for any errors or omissions that may have been made. The publisher remains neutral with regard to jurisdictional claims in published maps and institutional affiliations.

This Springer imprint is published by the registered company Springer Nature Switzerland AG
The registered company address is: Gewerbestrasse 11, 6330 Cham, Switzerland

If disposing of this product, please recycle the paper.

Preface

These proceedings contain the articles accepted and presented at *From Animals to Animats 17: The 17th International Conference on the Simulation of Adaptive Behavior (SAB 2024)*, held at the University of California, Irvine, on September 9–12, 2024.

The Simulation of Adaptive Behavior (SAB) conference started in September 1990 and is held every two years. The SAB interdisciplinary conference brings together researchers in artificial intelligence, artificial life, computer science, cybernetics, ethology, evolutionary biology, neuroscience, robotics, and many other fields, to further our understanding of the behaviors and underlying mechanisms that allow natural animals and artificial agents to adapt and survive in complex, dynamic and uncertain environments. Animats denote the connection between animals and synthetic agents. The conference focuses on models and experiments designed to help characterize and compare various organizational principles and architectures underlying adaptive behavior in real animals and in animats.

Thirty papers were submitted to the conference. Each paper was reviewed by at least two anonymous reviewers and a meta-reviewer. In the cases where there were discrepancies among the reviewers, additional reviewers were recruited to assist in the decision process. Of these, 26 papers were accepted for publication in the proceedings.

The 26 articles that make up these proceedings follow the theme of SAB by presenting research that connects biology and natural behavior to synthetic agents. Although the scope of SAB 2024 was broad, all papers followed this tradition. However, it was hard to pigeonhole the collection of papers into specific sections. With that disclaimer, in this volume we have organized the papers into the following:

- **Bio-Inspired Navigation:** Navigation is a critical behavior for all organisms. Animals ranging from insects to humans have developed successful behaviors that suggest solutions for navigating robots and self-driving vehicles. Models of navigation can also help us understand behavioral patterns, as well as neural correlates of navigation. The papers in this section take inspiration from ant route following, recent findings in place cell neurophysiology, bat foraging, and human navigation strategies.
- **Biomimetic Robots:** The wide variety of biological organisms can provide inspiration for unique robot designs. Some of these robots focus on morphology, while others focus on behavior. The papers in this section present soft robot morphologies, a robot capable of camouflaging, and a robot interacting with people through touch.
- **Collective Behavior:** Sometimes referred to as swarm intelligence, simulations of collective behavior can solve complex problems by relatively simple agents that interact. Population dynamics can lead to unique solutions that would be difficult to achieve individually. Simulations of collective behavior can assist with the study of social behaviors. The papers in this section investigate collective behavior for controlling of simulated agent groups, teams of robots, and as a tool for studying animal populations.

- **Evolutionary Approaches to Adaptive Behavior:** Evolutionary computation, which is inspired by natural evolution, can solve problems where no known solution exists. Using an objective function to guide the evolution, agents mutate and crossover until the population fitness reaches a desired level. Since the process is agnostic to the type of solution, interesting and counterintuitive results can emerge. The papers in this section apply the evolutionary approach to reaching tasks, population dynamics, and theories of consciousness.
- **Motor Learning:** The range of motions observed in biological organisms surpasses that of artificial agents. In most cases these skills are learned through trial and error, as well as reinforcement. The papers in this section take inspiration from biology to learn skills and behavioral repertoires in robots and other agents.
- **Problem Solving and Decision-Making:** Simulated agents solve problems with different machine learning and artificial intelligence methods that have roots in biological decision-making and adaptive behavior. The papers in this section solve a range of problems such as exploration strategies, theory of mind with socioeconomic games, and reversal learning with biologically plausible neural networks.

September 2024

Oliver Brock
Jeffrey Krichmar

Organization

General Chairs

Oliver Brock — Technische Universität Berlin, Germany
Jeffrey L. Krichmar — University of California, Irvine, USA

Program Committee

Lola Cañamero — ETIS, CY Cergy Paris Université, France
Jens Krause — Leibniz Institute, Berlin, Germany
Praveen Pilly — HRL Laboratories, USA
Jun Tani — Okinawa Institute of Science and Technology, Japan
Myra Wilson — Aberystwyth University, UK

Organizing Committee

Derrik Asher — DEVCOM Army Research Laboratory, USA
Michael Beyeler — University of California, Santa Barbara, USA
Kexin Chen — Meta Platforms, Inc., USA
Ting Shuo-Chou — ABL Space Systems, USA
Tiffany Hwu — Riot Games, USA
Hirak Kashyap — Samsung Research America, USA
Kenneth Stewart — US Naval Research Laboratory, USA
Thomas Trappenberg — Dalhousie University, Canada
Jinwei Xing — Google Cloud AI, USA
Alina Yuan — University of California, Irvine, USA

Additional Reviewers

Andy Alexander — UCSB
Nikhil Bhattasali — NYU
Nicholas Cheney — UVM
Yoonsuck Choe — Texas A&M
Anurag Daram — UTSA

Emmanuel Daucé	Marseille
Harrison Espino	UC Irvine
Shane Forbrigger	Dalhousie
Guillermo Gallego	TU Berlin
Martin Gillis	Dalhousie
Jennifer Hallinan	BioThink
Susu He	Meta
Dietmar Heinke	Birmingham
Olaf Hellwich	TU Berlin
Andrew Howe	HRL
Alexander Koenig	TU Berlin
Si-Yuan Kong	Activision
Oliver Layton	Colby
Christopher Lewis	Meta
Louis L'Haridon	ETIS
Xing Li	TU Berlin
Vito Mengers	TU Berlin
Alican Mertan	U. Vermont
Sohum Misra	Dusty
Bhavishya Mittal	Google
Mohammad Nadji-Tehrani	Neuraville
Duc Nguyen	TU Berlin
Uroš Petković	TU Berlin
Andrew Philippides	U. Sussex
Pawel Romanczuk	HU Berlin
Baljinder Singh Bal	Cergy
Antoine Sion	U. Namur
Steven Skorheim	HRL
Nigel Stepp	HRL
Masami Tatsuno	U. Lethbridge
Elio Tuci	U. Namur
Barbara Webb	U. Edinburgh
Isaac Xu	Dalhousie
Brian Yik Tak Tsui	UCSD
Magdalena Yordanova	HU Berlin
Oussama Zenkri	TU Berlin
Veronica Zhang	Neuraville

Contents

Bio-inspired Navigation

Efficient Visual Navigation with Bio-inspired Route Learning Algorithms 3
    Efstathios Kagioulis, James C. Knight, Andrew Philippides,
    Anindya Ghosh, Amany Amin, Paul Graham, and Thomas Nowotny

Insect-Based Navigation Model Allows Efficient Long-Range Visual
Homing in a 3D Simulated Environment 15
    Thomas Misiek, Andrew Philippides, and James Knight

Vector-Based Navigation Inspired by Directional Place Cells 27
    Harrison Espino and Jeffrey L. Krichmar

The Cost of Behavioral Flexibility: Reversal Learning Driven by a Spiking
Neural Network 39
    Behnam Ghazinouri and Sen Cheng

A Behavior-Based Model of Foraging Nectarivorous Echolocating Bats 51
    Thinh H. Nguyen and Dieter Vanderelst

Benefit of Varying Navigation Strategies in Robot Teams 63
    Seyed A. Mohaddesi, Mary Hegarty, Elizabeth R. Chrastil,
    and Jeffrey L. Krichmar

Biomimetic Robots

No-brainer: Morphological Computation Driven Adaptive Behavior
in Soft Robots 81
    Alican Mertan and Nick Cheney

CuttleBot: Emulating Cuttlefish Behavior and Intelligence in a Novel
Robot Design 93
    Michael A. Pfeiffer, Sriskandha Kandimalla, Jiahe Liu, Katherine Hsu,
    Eleanore J. Kirshner, Alina Yuan, Hin Wai Lui, and Jeffrey L. Krichmar

The Emergence of a Complex Representation of Touch Through
Interaction with a Robot 106
    Louis L'Haridon, Raphaël Bergoin, Baljinder Singh Bal,
    Mehdi Abdelwahed, and Lola Cañamero

Collective Behavior

Analyzing Multi-robot Leader-Follower Formations in Obstacle-Laden
Environments 121
   Zachary Hinnen and Alfredo Weitzenfeld

Spatio-Temporal Dynamics of Social Contagion in Bio-inspired
Interaction Networks 133
   *Yunus Sevinchan, Carla Vollmoeller, Korbinian Pacher,
David Bierbach, Lenin Arias-Rodriguez, Jens Krause,
and Pawel Romanczuk*

Behavioural Contagion in Human and Artificial Multi-agent Systems:
A Computational Modeling Approach 145
   *Maryam Karimian, Fabio Reeh, Asieh Daneshi, Marcel Brass,
and Pawel Romanczuk*

Transient Milling Dynamics in Collective Motion with Visual Occlusions 157
   Palina Bartashevich, Lars Knopf, and Pawel Romanczuk

Extended Swarming with Embodied Neural Computation for Human
Control over Swarms 169
   Jonas D. Rockbach and Maren Bennewitz

Bio-Inspired Agent-Based Model for Collective Shepherding 182
   Yating Zheng and Pawel Romanczuk

DaNCES: A Framework for Data-inspired Agent-Based Models
of Collective Escape 194
   Marina Papadopoulou, Hanno Hildenbrandt, and Charlotte K. Hemelrijk

Evolutionary Approaches to Adaptive Behavior

The Role of Energy Constraints on the Evolution of Predictive Behavior 211
   William Kang, Christopher Anand, and Yoonsuck Choe

Influence of the Costs of Acquisition of Private and Social Information
on Animal Dispersal 223
   Antoine Sion, Matteo Marcantonio, and Elio Tuci

Integrated Information in Genetically Evolved Braitenberg Vehicles 236
   Hongju Pae and Jeffrey L. Krichmar

Motor Learning

Neural Chaotic Dynamics for Adaptive Exploration Control
of an Autonomous Flying Robot 251
    Vatsanai Jaiton and Poramate Manoonpong

Non-instructed Motor Skill Learning in Monkeys: Insights from Deep
Reinforcement Learning Models 263
    Laurène Carminatti, Lucio Condro, Alexa Riehle, Sonja Grün,
    Thomas Brochier, and Emmanuel Daucé

Memory-Feedback Controllers for Lifelong Sensorimotor Learning
in Humanoid Robots 275
    Magdalena Yordanova and Verena V. Hafner

Problem Solving and Decision-Making

Extracting Principles of Exploration Strategies with a Complex Ecological
Task 289
    Oussama Zenkri, Florian Bolenz, Thorsten Pachur, and Oliver Brock

"Value" Emerges from Imperfect Memory 301
    Jorge Ramírez-Ruiz and R. Becket Ebitz

The Role of Theory of Mind in Finding Predator-Prey Nash Equilibria 314
    Tiffany Hwu, Chase McDonald, Simon Haxby, Flávio Teixeira,
    Israel Knight, and Albert Wang

Nonverbal Immediacy Analysis in Education: A Multimodal
Computational Model 326
    Uroš Petković, Jonas Frenkel, Olaf Hellwich, and Rebecca Lazarides

Author Index 339

Bio-inspired Navigation

Efficient Visual Navigation with Bio-inspired Route Learning Algorithms

Efstathios Kagioulis[✉], James C. Knight, Andrew Philippides, Anindya Ghosh, Amany Amin, Paul Graham, and Thomas Nowotny

Centre for Computational Neuroscience and Robotics, University of Sussex, Brighton BN1 9QJ, UK
e.kagioulis@sussex.ac.uk
https://www.sussex.ac.uk/ccnr/

Abstract. Visual navigation through complex environments is a challenging task, yet ants navigate in them easily and accurately with low-resolution vision and limited neural resources. Inspired by ants, we have developed a series of visual familiarity-based navigation algorithms for teach-and-repeat style navigation. These algorithms learn the egocentric visual appearance of the world on a training route and, during subsequent navigation, move in the direction that leads to the best match of the current view with one of the scenes encountered during training. Because they do not depend on accurate feature extraction or map building these algorithms use relatively unprocessed low-resolution panoramic views, making them computationally efficient. However, the computational cost of comparing the current view with all training images is still quite large and the algorithm can get confused if the path crosses itself. Here we develop and test novel algorithms where the agent uses sequence information to adaptively select a window of route memories to navigate with. This algorithm is shown to successfully navigate real-world routes through ant-like habitats, including a figure of 8-route, as well as a long route along a corridor, with all computation performed onboard the robot.

Keywords: Bio-inspired Robotics · Visual Navigation · Teach and Repeat · Autonomous Robotics

1 Introduction

Even though they have fewer than one million neurons [10], desert ants habitually navigate routes tens of thousands of times their body length [16]. After only one traversal of a route [13,24] ants can repeat it purely guided by learned visual information, thus solving a problem akin to teach-and-repeat route navigation [8,11,28]. Inspired by this, a model of route navigation has been developed that enables visual routes to be learned rapidly and robustly by using images as so-called 'visual compasses'. In this paradigm, each image defines an *action* (the direction to move in) rather than specifying a *location* [2,35]. In visual compass models (also called familiarity-based algorithms), an agent stores panoramic views of the environment while moving along the training route. To navigate along the learned route subsequently, the agent visually scans the world at each step and compares all orientations of the current panoramic view to all the stored images. The stored image that is most familiar, i.e. most similar to the current view at a given orientation, is selected and the ant moves in the direction specified by that best orientation of the current view. The underlying rationale is that ants and many robots are constrained to move in the direction they are facing and therefore, if a current view is familiar to a stored image it is very likely that the agent is on the route and facing in the right direction to follow it.

In previous work, we have demonstrated the plausibility of navigation in a variety of environments using familiarity-based algorithms with low-computation, memory and visual resolution, suggesting this way of using information to be the basis of ant route navigation [1,2,17]. The original algorithms do not, however, use any information about the sequence in which the training images are acquired, meaning any training image can be matched to any point on the route, which can result in both increased compute time and a risk of visual aliasing when trying to navigate complex routes [15]. While the computational cost can be reduced by implementing the memory as a simple neural network [1,2,17], visual aliasing remains problematic. For instance, considering a route where the path crosses itself, there are two points on the route where the agent will be at the same physical location but two different actions are required, depending on when in the route sequence the agent encounters the crossing point. Recent work has shown that ants do have some useful knowledge of visual sequence during route navigation [30]. Further, incorporating route sequence information into familiarity-based algorithms tested in simulations of ant habitats was found to reduce problems associated with visual aliasing while also significantly reducing compute time [15]. As this suggests the utility of sequence information for familiarity-based navigation, it is natural to next ask whether benefits transfer from simulated ant habitats to the noise and uncertainty of real-world dynamic indoor environments.

The benefits of incorporating sequence information for navigation have been highlighted previously in a number of forms. We and others [4,22,31,33] demonstrated homing to a series of way-points which uses sequence information. This can have very impressive results [22] though we found the use of a strict sequence of images as attractors to a location to be a little brittle [31] and moved on to consider familiarity-based algorithms in which waypoints are stored continually [3]. A second branch of work has used sequences of images to improve the robustness of visual place recognition (VPR), starting with the SeqSLAM algorithm [25], with further models from the same lab demonstrating the utility of sequence information for VPR [7,14,27]. While VPR does not directly provide the guidance information our robots need, other robotic teach-and-repeat route navigation algorithms also have exploited sequence information [6,36].

[36] use a panoramic image for visual heading correction and, like ours, compares it to a (very restricted) subset of the training routes, around the current best estimate of position. However, unlike our algorithm, the estimate of position comes from odometric information. This paper has inspired several others with the most recent paper showing particularly impressive performance again through the integration of odometric information with sequence [6]. In a different vein, [20] extracted and stored SURF features from monocular route views and selected which features to try and match based on an estimate of the distance travelled by the robot obtained using odometry. [19] extended this approach to use stereo cameras and AGAST and BRIEF image features, while in the most recent iteration [29], sequence information is included by using a particle filter to improve state estimation, demonstrating impressive performance in a range of environments.

Given the improvements seen by adding sequence information into different navigation algorithms operating in the real world, and the promise shown in [15], we here evaluate how visual familiarity-based algorithms perform with and without sequence information. Our algorithms are tested on a small wheeled robot performing all computations autonomously on board, using a monocular omnidirectional camera with no feature detection. We first use a test environment inspired by ant habitats and containing only complex proximal landmarks (plastic plants) before challenging the algorithms with a longer corridor route. We focus on this class of algorithm as we want our results to be relevant to the ants that inspired them. For this reason we also do not include explicit odometric information, setting our algorithm apart from many others. A key part of the robust navigation system of ants is the independence of odometry and visual guidance [18]. While navigating ants do have odometry, they are very able to perform accurate visual navigation in the absence of odometric information (see [24]). A second point of difference is that we use an adaptive temporal window size which changes continuously based on the familiarity of the current view, addressing the challenge that the optimal length of sequence appears to be environment-dependent [15].

2 Methods

2.1 Robot

We used a small robot (25 cm × 22 cm × 12 cm) with mecanum wheels [9] equipped with an NVIDIA Jetson TX2 module (256-core Pascal GPU, Dual-Core NVIDIA Denver2 64-Bit CPU, Quad-Core ARM Cortex-A57 MPCore, 8 GB 128-bit LPDDR4 Memory, 32 GB eMMC storage, 15 W maximum power). Images were acquired with an on-board Kodak PIXPRO SP360 panoramic camera at 1440 × 1440 pixel resolution, unwrapped and down-sampled to 180 × 50 resolution (Fig. 1).

2.2 Test Environments

We test the navigation algorithms in two real-world environments. The first is a 6 m × 6 m robot arena that is inspired by the challenging environments within which ants navigate. The arena contains artificial plants of complex shape and texture but very limited information from distant objects (Figs. 1, 2) as the arena is enclosed in a uniformly white plastic sheet and we restrict the field of view of the robot to not extend above it. In this arena we use a VICON positioning system [34] to determine the accurate ground truth robot position in order to evaluate the performance of the different algorithms.

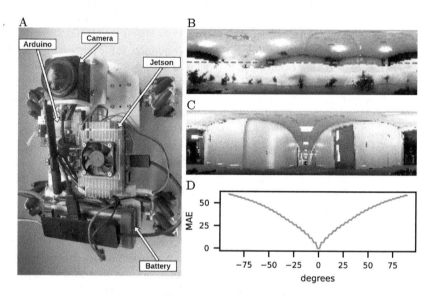

Fig. 1. **A)** Robot used in the experiments. **B)** and **C)** Robot's eye views. The visual input to the algorithms is an insect-like resolution of 2° per pixel. For navigation in the robot arena, (**B**) the views were cropped by 50% vertically (shown here by the red dashed line) to reduce the effect of the ceiling as a visual cue. In the corridor environment (**C**) we used the whole image. **D)** illustration of an RIDF for the view **C** with itself.

We then challenged the algorithms to navigate a route of 21 m along a corridor (Fig. 3) where we tracked the robot using mobile phone cameras and reconstructed the trajectories with custom tracking code based on the OpenCV [5] implementation of the CSR-DCF tracking algorithm [23].

2.3 Route Navigation

Following the teach-and-repeat paradigm, the robot was first driven along a route by a human operator and images were stored at 10 Hz unless the robot was stationary or turning rapidly. Subsequently, the robot repeated the route by comparing rotated versions of its current view – generated at steps of 1° through $\theta \in [-90°, 90°]$ – with the stored images captured along the training route. While the rotated views *could* be generated by physically rotating the robot, to speed up navigation, we instead perform this rotation 'in silica', i.e. we simulate rotation in unwrapped panoramic images (such as the one shown in Fig. 1) by moving image pixel matrix columns from the left-hand side to the right-hand side for a rotation to the right and vice versa for a rotation to the left.

The generated views are then compared to training images using an *Image Difference Function (IDF)*. While there are many possible IDFs, here we use the simplest, Mean Absolute Error (MAE),

$$\text{IDF}(C^\theta, S) = \frac{\sum_{i=1}^{M} \sum_{j=1}^{N} |(C_{ij}^\theta - S_{ij})|}{M \times N}, \quad (1)$$

where C^θ is the current $M \times N$ panoramic view rotated by θ relative to the current heading of the agent with greyscale values C_{ij}^θ and S is a $M \times N$ panoramic view stored during training. We calculate the IDF for all rotations of the current view (in steps of 1°) to generate the values of the so-called Rotational Image Difference Function (RIDF) (Fig. 1D),

$$\text{RIDF}(C, S, \theta) = \text{IDF}(C^\theta, S) \quad (2)$$

By determining the minimum of the RIDF function with respect to $S \in I$ and $\theta \in [-90°, 90°]$ for a given current view C_t, we determine the angle to turn given a route memory represented by the set of images I:

$$\theta_t, S_t = \underset{\theta \in [-90°, 90°], S \in I}{\arg\min} \text{RIDF}(C, S, \theta), \quad (3)$$

where t denotes the current timestep. The angle θ_t is then used as the error signal for the robot's turn controller, i.e. the difference between the desired and current directions. Theoretically, a full PID controller could be used to translate θ_t into robot motor commands. However, the robot moves quite slowly and the speed of the control loop running on the Jetson is limited by the panoramic camera's 15fps frame rate meaning that the effect of the I and D terms would be negligible. Therefore, here we use a simple P controller to calculate the turn speed u_t:

$$u_t = K_p \theta_t. \quad (4)$$

where K_p is a free gain parameter. If $|\theta_t| > 9°$, the robot turns on the spot at speed u_t. Otherwise, it moves forward at a fixed speed while also turning with speed u_t.

In this work, we compare two navigation algorithms, called 'Perfect Memory' (PM) and 'Adaptive Sliding Memory Window' (ASMW), which differ in what part of the available route memory they use at any given timestep. In the **Perfect Memory (PM)** algorithm the agent compares *all* images from the training route to the rotated versions of the current view [2] i.e. the set I contains all known route images. Notably, PM does not take into account the order in which the images were collected but simply determines the global minimum across all views and orientations. In practice, this can be trivially parallelised across the 6 core CPU of the Jetson. The direction of travel for the next movement is then the orientation of the current view that provided the best match across all compared stored views.

The **Adaptive Sliding Memory Window Algorithm ASMW** algorithm builds on [15] in which training images from within a 'window' of size w around an estimated current position, the *Memory Pointer* S_{MP_t}, within the route memory are included in I, i.e. $I_t^{\mathrm{win}} = \{S_{\mathrm{MP}_t - \frac{w}{2} - 1}, \ldots, S_{\mathrm{MP}_t + \frac{w}{2}}\}$. At $t = 0$, all route images are used and MP_t is set to the index of S_t obtained by evaluating Eq. 3. At each step a new window is formed around the MP_t as described above. From that moment onwards only images within a given window are considered. Subsequently, and for each new query image, it is updated to the index of S_t obtained from Eq. 3 used with I_t^{win}. The position of the S_{MP_t} and thus the window are calculated and defined using the same RIDF method described above.

In this work, we introduce an *adaptive* window size w_t that can change at every timestep t. The intuition behind this change is that with an adaptive memory window, the agent can use more information if the circumstances require it and less if not needed. For example, if the agent is navigating an occluded and/or visually indistinct part of the environment, the window can grow to include more informative memories. In contrast, if the views are of high quality the window can shrink. Based on this intuition, we grow or shrink the window depending on whether the value of the RIDF minimum at subsequent timesteps increases or decreases. That is, we calculate the change in the RIDF minimum, $\beta_t - \beta_{t-1}$, where

$$\beta_t = \min_{\theta \in [-90°, 90°], i \in I_t^{\mathrm{win}}} (\mathrm{RIDF}(C_t, S_i, \theta)). \tag{5}$$

We then update the window size w_t as follows:

$$w_{t+1} = \begin{cases} w_t + \gamma & \text{if } \beta_t - \beta_{t-1} \geq 0 \\ \max\{w_t - \gamma, w^{\min}\} & \text{otherwise.} \end{cases} \tag{6}$$

Here, $w^{\min} = 10$ denotes a fixed minimum window size and γ is the window size update step. We have used $\gamma = 5$ for the experiments in this work. MP is then set to the best match index i_t (see Eq. 3), $\mathrm{MP}_{t+1} = i_t$, and thus the indices of images in the window at time t are $W_t = \{\mathrm{MP}_t - \frac{w_t}{2} - 1, \ldots, \mathrm{MP}_t + \frac{w_t}{2}\}$.

3 Results

3.1 Robot Arena

We start by comparing the algorithms in the robot arena using three routes of increasing difficulty. On the first route we find that both the PM and the

ASMW algorithm work effectively (Fig. 2A-B). From the successful paths of each algorithm, we see the characteristics of this class of algorithms: like the ants that the algorithms are inspired by [24], route recapitulations often run in parallel to the original training route (solid blue line in Fig. 2A-F) forming a narrow corridor. As the algorithms recover movement direction from best matches to stored views from the route, the movement is parallel to the training route rather than converging back to it when there are small deviations from the route.

We next tested both algorithms on a figure-of-eight route (Fig. 2C-D). Here, the PM algorithm struggles in one of the tests (*pm1* green dashed line in Fig. 2C) with the route intersection, as expected. The robot repeats the lower loop of the figure-of-eight rather than completing the route. Analysis of the RIDF data along the completed route illustrates why this is the case. Figure 2G shows the local minima of the RIDF between each image from the test route (along the y-axis) and every image from the training route (along the x-axis), together with the location of the overall minimum for each given test route image when using PM (black) or ASMW (blue). The green lines indicate the boundary of the memory window that was used. We see that once the robot comes to the crossing point of the figure-of-eight for the second time, the PM algorithm matches the current view with an earlier memory (the black line jumps to the left) and instead of carrying on to complete the route, the robot retraces the lower loop of the figure-of-eight as the erroneously matched memory commands. This is consistent with results from [15], obtained in simulation, that showed that routes that cross themselves need sequence information (or other task-relevant information) for robust navigation.

However, as the robot only considers headings that are forward from its current pose (i.e. from -90° to 90° around its current heading), it will not always fail even if it encounters a redundant view at the crossing point. Therefore, most trials pass through the first crossing point without any issues. Nonetheless, PM is still only successful in one of the other trials. The failures arise because the testing routes happened to all start a little bit on the outside of the initial starting point of the training route and the robot is pushed a bit wider because the shape of the route means that the best-matching position specifies a heading which points slightly away from the route. As there is no mechanism to bring the robot back towards the training route, this results in the robot moving too far outside the route corridor and in turn the current view becoming sufficiently different from the nearby training images such that better matches are found with a view at an unrelated location in the training route. This leads to a random heading and the robot either runs into a wall (orange trace) or a plant. The robot is not equipped with any collision avoidance to avoid simplifying navigation by canalising routes into successful corridors between objects.

The ASMW algorithm suffers from the same issue of route shape but performs better on this route. Firstly, because the window keeps track of the approximate position in the route sequence, the crossing point is not an issue (see window positions and widths in Fig. 2F). Secondly, although the route dynamics again guide the robot wide, they do not overshoot as much because the lower latency

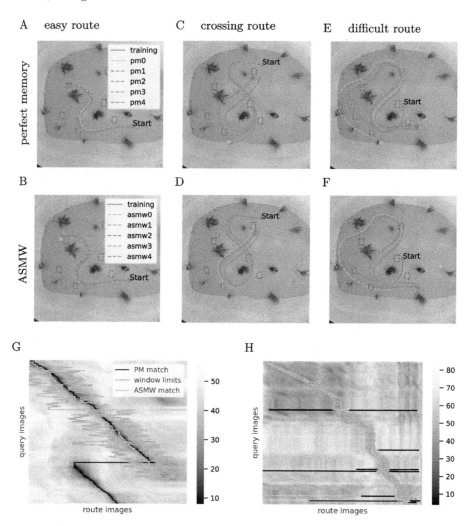

Fig. 2. Navigation in the robot arena. (**A**) PM trials on route 1, and (**B**) ASMW trials on route 1. Both algorithms completed 5/5 trials. (**C**) PM trials on the figure-of-eight route (completed 1/5 trials), and (**D**) ASMW trials on the figure-of-eight route (completed 3/5 trials). (**E**) PM trials on route 3 (completed 4/5 trials).(**F**) ASMW trials on route 3 (completed 4/5 trials). (**G**) Heatmap showing the minimum of the RIDF when comparing each query image from a route recapitulation (y-axis) with every training route image (x-axis) for the figure-of-eight route shown in C. The black line shows the location of the best match calculated with PM and the blue line as calculated with ASMW. The green lines illustrate the boundaries of the adaptive window used by the ASMW algorithm (**H**) Heatmap for a trajectory from the corridor route from Fig. 3. Conventions as in G.

of generating a new heading compared to PM means that the steering gets corrected more often. In addition, because the view is constrained to match within the window, when the robot does leave the route corridor there is a higher chance of matching with a nearby view and staying on track. Because the PM algorithm can match with anywhere within the route, when the current view is not a great match it can match with images from a long way away in the route. This can be seen in Fig. 2H which shows the matches between test and training for the long corridor route (discussed below). Here, the black line, which shows the image the PM algorithm would match with, which jumps around due to aliasing. Overall, this means that in three figure-of-eight trials with ASMW the robot makes the final corner and completes the route. However, there are still two fails, when the robot gets too far outside the safe route corridor and too close to an obstacle.

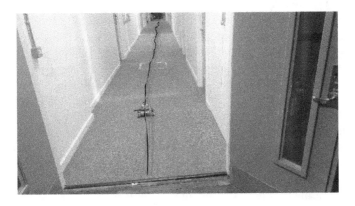

Fig. 3. Example image of the corridor route with overlaid training (thick black line) and ASMW route trajectories (red lines). (Color figure online)

As the figure-of-eight route is artificial and not a path a sensible ant or robot would navigate, we challenged our algorithm with a more complex route (Fig. 2E, F). Both algorithms performed at the same level in this route, with 4/5 successes but ASMW was much quicker due to the reduced computational load. This illustrates that ASMW is also useful where routes pass close to themselves without actually crossing. PM fails at this point (bottom of Fig. 2E) whereas ASMW does not. However, ASMW does fail due to running wide on a convex path and being unable to recover before hitting the wall.

While small in scale, the robot arena is an intentionally challenging environment. As seen in Fig. 1, the artificial plants are complex yet relatively homogeneous shapes, the background walls are white and featureless and we crop the ceiling view. Thus there is no visual information in the distance to stabilise the visual matching algorithms. To assess performance in an environment with more distant features we therefore tested the ASMW algorithm in a corridor. As this route was a lot longer than the previous environments, the PM algorithm was not practical to run live so we just tested the ASMW. However, as mentioned above, an analysis of the images gained from the ASMW algorithm shows that PM would have suffered from aliasing (jumps in black line in Fig. 2H). For the ASMW algorithm, Fig. 3 shows that the repeats of the route are tightly surround the training route. Despite the uniform nature of the corridor, the algorithm follows the small undulations of the training route successfully (compare red test routes to black line in Fig. 3).

4 Conclusion

In this paper, we have shown that a bio-inspired route learning algorithm is capable of robustly navigating an ant-like habitat, containing plastic plants but no distant object information. Further, we showed that the ASMW algorithm – which incorporates sequence information into the navigation process in an adaptive way – was not only more efficient but also more robust than the baseline PM algorithm. We further tested the ASMW algorithm on a longer real-world indoor route and showed that it enabled our robot to navigate reliably. The corridor route is particularly challenging as it traverses a semi-uniform corridor environment while people were walking past the robot during the trials.

However, the ASMW algorithm is, by no means perfect. While it navigated a corridor essentially without errors, the robot did veer off the route on some of the challenging test routes in the robot arena. At the root of many of these errors is the fact that the algorithm has no provisions to converge back to the route if errors accumulate to push it off course. This is compounded by the lack of an obstacle avoidance mechanism, meaning that if the robot gets too close to an obstacle it is doomed to fail. We did not include mechanisms to converge back to the route in our algorithm because we wanted it to reflect ant behaviour; and avoid getting navigation for 'free', particularly in the smaller ant-like robot arena, where obstacle avoidance can easily canalise the robot down a few narrow paths. However, the addition of either mechanism is highly likely to improve performance and is the subject of future work.

Another area for further investigation is the image processing. In this paper, we kept this purposefully simple aside from cropping the image for the ant-like habitat. We did this, because, in indoor environments, the ceiling can provide useful navigational clues. In contrast, outdoors, the ever-changing sky provides a strong cue which can hamper navigational performance. Ants have ultra-violet vision and can identify and remove unreliable signals from the sky [12,21,26,32]. However, our robot does not have this functionality. Thus, we chose to crop out the ceiling in the robot arena (see Fig. 1) to mimic the difficulty of navigating outdoor environments for an ant, but have used the full image for navigating the indoor longer routes as here we were demonstrating our algorithms utility for robotics.

In summary, we have shown that our algorithm for adapting the considered subset of memories based on the agent's confidence in the image match is able to successfully navigate with low computation making it most appropriate for autonomous navigation in applications where computation is severely limited.

Acknowledgement. This work was funded by EPSRC (grants EP/S030964/1 and EP/V052241/1) and BBSRC (BB/X01343X/1). We also thank Dexter Shepherd and Joshua Kybett for their assistance with recording the long-route experiments.

References

1. Amin, A.A., Kagioulis, E., Domcsek, A.D.N., Nowotny, T., Graham, P., Philippides, A.: Robustness of the infomax network for view based navigation of long routes. In: ALIFE 2023: Ghost in the Machine: Proceedings of the 2023 Artificial Life Conference. MIT Press (2023)
2. Baddeley, B., Graham, P., Husbands, P., Philippides, A.: A model of ant route navigation driven by scene familiarity. PLoS Comput. Biol. **8**(1), e1002336 (2012)
3. Baddeley, B., Graham, P., Philippides, A., Husbands, P.: Holistic visual encoding of ant-like routes: navigation without waypoints. Adapt. Behav. **19**(1), 3–15 (2011)
4. Binding, D., Labrosse, F.: Visual local navigation using warped panoramic images. In: Towards Autonomous Robotic Systems 2006, pp. 19–26 (2006)
5. Bradski, G.: The OpenCV Library. Dr. Dobb's J. Software Tools (2000)
6. Dall'Osto, D., Fischer, T., Milford, M.: Fast and robust bio-inspired teach and repeat navigation. In: 2021 IEEE/RSJ International Conference on Intelligent Robots and Systems (IROS), pp. 500–507. IEEE (2021)
7. Fischer, T., Milford, M.: Event-based visual place recognition with ensembles of temporal windows. IEEE Robotics Autom. Lett. **5**(4), 6924–6931 (2020)
8. Furgale, P., Barfoot, T.D.: Visual teach and repeat for long-range rover autonomy. J. Field Robot. **27**(5), 534–560 (2010)
9. Gfrerrer, A.: Geometry and kinematics of the mecanum wheel. Comput. Aided Geometric Des. **25**(9), 784–791 (2008)
10. Godfrey, R.K., Swartzlander, M., Gronenberg, W.: Allometric analysis of brain cell number in hymenoptera suggests ant brains diverge from general trends. Proc. R. Soc. B **288**(1947), 20210199 (2021)
11. Goedemé, T., Tuytelaars, T., Van Gool, L., Vanacker, G., Nuttin, M.: Feature based omnidirectional sparse visual path following. In: 2005 IEEE/RSJ International Conference on Intelligent Robots and Systems, pp. 1806–1811. IEEE (2005)
12. Graham, P., Philippides, A.: Vision for navigation: what can we learn from ants? Arthropod Struct. Dev. **46**(5), 718–722 (2017)
13. Haalck, L., Mangan, M., Wystrach, A., Clement, L., Webb, B., Risse, B.: Cater: combined animal tracking & environment reconstruction. Sci. Adv. **9**(16), eadg2094 (2023)
14. Hausler, S., Jacobson, A., Milford, M.: Multi-process fusion: visual place recognition using multiple image processing methods. IEEE Robot. Autom. Lett. **4**(2), 1924–1931 (2019)
15. Kagioulis, E., Philippides, A., Graham, P., Knight, J.C., Nowotny, T.: Insect inspired view based navigation exploiting temporal information. In: Living Machines 2020. LNCS (LNAI), vol. 12413, pp. 204–216. Springer, Cham (2020). https://doi.org/10.1007/978-3-030-64313-3_20
16. Knaden, M., Graham, P.: The sensory ecology of ant navigation: from natural environments to neural mechanisms. Annu. Rev. Entomol. **61**, 63–76 (2016)
17. Knight, J.C., et al.: Insect-inspired visual navigation on-board an autonomous robot: real-world routes encoded in a single layer network. In: The 2018 Conference on Artificial Life: A Hybrid of the European Conference on Artificial Life (ECAL) and the International Conference on the Synthesis and Simulation of Living Systems (ALIFE), pp. 60–67. MIT Press (2019)

18. Kohler, M., Wehner, R.: Idiosyncratic route-based memories in desert ants, melophorus bagoti: how do they interact with path-integration vectors? Neurobiol. Learn. Mem. **83**(1), 1–12 (2005)
19. Krajník, T., Majer, F., Halodová, L., Vintr, T.: Navigation without localisation: reliable teach and repeat based on the convergence theorem. In: 2018 IEEE/RSJ International Conference on Intelligent Robots and Systems (IROS), pp. 1657–1664. IEEE (2018)
20. Krajník, T., Pedre, S., Přeučil, L.: Monocular navigation for long-term autonomy. In: 2013 16th International Conference on Advanced Robotics (ICAR), pp. 1–6. IEEE (2013)
21. Labhart, T.: The electrophysiology of photoreceptors in different eye regions of the desert ant, cataglyphis bicolor. J. Comp. Physiol. A **158**, 1–7 (1986)
22. Labrosse, F.: Short and long-range visual navigation using warped panoramic images. Robot. Auton. Syst. **55**(9), 675–684 (2007)
23. Lukezic, A., Vojir, T., Cehovin Zajc, L., Matas, J., Kristan, M.: Discriminative correlation filter with channel and spatial reliability. In: Proceedings of the IEEE Conference on Computer Vision and Pattern Recognition (CVPR) (2017)
24. Mangan, M., Webb, B.: Spontaneous formation of multiple routes in individual desert ants (cataglyphis velox). Behav. Ecol. **23**(5), 944–954 (2012)
25. Milford, M.J., Wyeth, G.F.: Seqslam: visual route-based navigation for sunny summer days and stormy winter nights. In: 2012 IEEE International Conference on Robotics and Automation, pp. 1643–1649. IEEE (2012)
26. Mote, M.I., Wehner, R.: Functional characteristics of photoreceptors in the compound eye and ocellus of the desert ant, cataglyphis bicolor. J. Comp. Physiol. **137**, 63–71 (1980)
27. Pepperell, E., Corke, P.I., Milford, M.J.: All-environment visual place recognition with smart. In: 2014 IEEE International Conference on Robotics and Automation (ICRA), pp. 1612–1618. IEEE (2014)
28. Royer, E., Lhuillier, M., Dhome, M., Lavest, J.M.: Monocular vision for mobile robot localization and autonomous navigation. Int. J. Comput. Vision **74**(3), 237–260 (2007)
29. Rozsypálek, Z., Rouček, T., Vintr, T., Krajník, T.: Multidimensional particle filter for long-term visual teach and repeat in changing environments. IEEE Robot. Autom. Lett. **8**(4), 1951–1958 (2023)
30. Schwarz, S., Mangan, M., Webb, B., Wystrach, A.: Route-following ants respond to alterations of the view sequence. J. Exp. Biol. **223**(14), jeb218701 (2020)
31. Smith, L., Philippides, A., Graham, P., Baddeley, B., Husbands, P.: Linked local navigation for visual route guidance. Adapt. Behav. **15**(3), 257–271 (2007)
32. Stone, T., Mangan, M., Ardin, P., Webb, B., et al.: Sky segmentation with ultraviolet images can be used for navigation. In: Robotics: Science and Systems. Robotics: Science and Systems (2014)
33. Vardy, A.: Long-range visual homing. In: 2006 IEEE International Conference on Robotics and Biomimetics, pp. 220–226. IEEE (2006)
34. Vicon: Vicon motion capture system (1979). https://www.vicon.com/applications/
35. Zeil, J., Hofmann, M.I., Chahl, J.S.: Catchment areas of panoramic snapshots in outdoor scenes. JOSA A **20**(3), 450–469 (2003)
36. Zhang, A.M., Kleeman, L.: Robust appearance based visual route following for navigation in large-scale outdoor environments. Int. J. Robot. Res. **28**(3), 331–356 (2009)

Insect-Based Navigation Model Allows Efficient Long-Range Visual Homing in a 3D Simulated Environment

Thomas Misiek[✉][iD], Andrew Philippides[iD], and James Knight[iD]

Sussex AI, School of Engineering and Informatics, University of Sussex, Brighton, UK
{t.misiek,j.knight}@sussex.ac.uk

Abstract. Efficient outdoor navigation remains a challenge for autonomous robots, yet bees excel in robust long-range navigation with minimal computational resources. To do so, they scaffold learning through innate behaviours such as survey flights: loops centred on the nest to explore the environment, which they perform before foraging. While the 2D positions of these flights have been tracked by radar, it has not been tested how well these flights can support subsequent long-range visual homing, nor whether the 3D structure (not captured by the radar) has an effect on homing performance. Using a 6 km^2 3D LIDAR scan of the Rothamsted Research Center – where bumblebee flights were tracked in radar experiments – we recreate the trajectory of bumblebee exploration and foraging flights. We then render panoramic views of the visited coordinates, and use these to test the efficacy of visual homing over large distances, and flying altitudes ranging from 2 to 32 m above the ground. We find that our model can predict the direction of the target from up to 300 m. Additionally, homing improves at higher altitudes, but there is limited transferability of information between flying heights.

Keywords: Spatial navigation · Insect vision · Autonomous systems · Visual homing

1 Introduction

Visual homing – the ability to navigate back to a place of interest using visual information – is a problem of great interest for both roboticists working on autonomous navigation in GPS-denied/unreliable environments, as well as neuroethologists, seeking to understand the neural basis of visual learning in animals [11]. Hymenoptera such as bees and ants are amongst the champion visual navigators in the animal kingdom [4] and can learn in a single trial long visually guided routes across complex landscapes [19]. To accomplish these feats with their small brains (1M neurons) bees and ants have developed specialized behaviors which augment their limited, neural resources. One of the most striking of these behaviours, bumblebee survey flights, has been extensively studied at the Rothamsted Research Center using a harmonic radar to track bees over their whole foraging careers [19,20]. Survey flights occur at the beginning of the foraging career of the bee. They consist in a series of looping flights centered on the nest and are thought to be used to learn the visual information needed to accurately home back to the hive during subsequent exploitation flights. In this paper, we therefore use a visual navigation model to test whether the information from a survey flight is sufficient for subsequent homing. In particular, as bees exploration behaviours have a distinct 3D component that is not captured by the radar, we ask how homing is affected by changes in height. To do this, we combine 2D radar trajectories with a novel 3D reconstruction of the research center [18] to simulate the visual information perceived by bees during exploration and foraging.

View-based homing – where navigation is performed by retinotopic matching between the current view and stored visual memories – can be observed in many species including bees and ants. Strikingly, during exploration bouts, bees and ants can be observed frequently turning toward their hive or nest [5,6], leading to the theory that insects take "snapshots" while facing their goal which could be used to guide their heading during subsequent route traversals by using these views as so-called visual compasses [15,22]. In visual compass style models, the agent uses the stored visual information to recall their movement direction by rotating to align their current view to best match their memories. This allows multiple visual snapshots to be learnt and used for homing, potentially extending the range over which navigation can be performed, replicating visual homing results in ants [12,21] and demonstrating the utility of ant learning walks, [10], the analogue of the bee survey flights.

While variants of these models have been shown to be effective for route navigation in terrestrial robots [13,14] and over large distances [1], most work assessing the information available to flying insects has focused on the information from single images [8,16,22] at heights up to 8 m or route navigation over small scales (2×3 m). While limited in scale, these works showed that snapshot based models can function at different heights and that higher snapshots could be advantageous for increasing the catchment areas of single images [16,22]. In addition, it was shown that information learnt at one altitude could be retrieved when tested at other heights, and that views learned at higher altitudes generalize more to lower heights than the contrary. In this context, [8] hints at the appealing prospect of future robots transferring learning from UAVs to ground-based robots, for search and rescue operations for example.

Here we therefore extend the visual-compass based model of ant learning walks [21] to explore how view-based homing algorithms scale to larger navigation areas and a large wider range of flying heights by modeling the large-scale exploration and foraging flights of bees [18]. Specifically, we investigate the following questions:

– Can our model, when trained solely on initial exploration flights, accurately predict the direction to the hive?
– What is the maximum distance from the hive at which this prediction remains useable for navigation (angular error $<90°$)?

Fig. 1. Aerial view of the 6 km² textured area covered by the 3D model of the Rothamsted Research Center (10 km x 10 km non-textured ring not shown). The position of the hive used by [19] is shown in red. Our 'Test 1' and 'Test 2' navigation experiments were performed within the red rectangle surrounding this location. All rectangles dimensions are 600 m x 600 m. Our additional 'Test 3' experiments were conducted within the areas indicated by the cyan and lime rectangles. (Color figure online)

– Does flying height affect visual homing, and to what extent do navigational snapshots generalize across different altitudes?

In so doing we also introduce the simulated Rothamsted world as a resource to allow the testing of different visual navigation models.

2 Materials and Methods

2.1 The 3D Model

We used a 3D reconstructed model of the Rothamsted Research Center [17], Harpenden, United Kingdom [18] assembled from multiple high-definition aerial pictures taken from a fixed-wing drone by RUAS (https://ruas.co.uk/). The resulting 3D model consists of 55 221 953 vertices textured with two 4096 × 4096 textures. However, to improve rendering performance, we down-sampled this to 2 175 128 vertices, using a voxel-based binning method. Specifically, we aggregated all 3D vertices into 2D bins measuring 2 m × 2 m each. Within each bin, we then only retained the highest vertex and regenerated the mesh using a Delaunay triangulation. The textures remained unchanged. This model spans 3910 m in longitude, 2534 m in latitude and 99 m in altitude, for a total area of about 6 km² (see Fig. 1). However, when generating views from higher altitudes, the boundaries of the simulated environment became visible. The distant horizon would slowly vanish to be replaced with a uniform light gray sky, because the map would cease to extend. This non-naturalistic feature might have been used as a landmark by our model, facilitating navigation and thereby invalidating our results. To mitigate this issue, we expanded the original map with a peripheral lower-resolution 10 km × 10 km map. This texture-less, green extension was designed to mimic the contrast of the horizon actually perceived by bees during the original experiment. It was crafted using the LIDAR Composite Digital Terrain Model (DTM) [7], which is a 1 m resolution elevation map of England. The color was created by blending the full textures of the map into a single hue. To render the 360° panoramic views of the Rothamsted Park from the raw 3D meshes, we used the 'Antworld' rendering pipeline from the Brains on Board robotics library [9] (Fig. 2).

Fig. 2. Two 360° panoramic snapshots rendered at the same longitude and latitude but different altitudes, in front of the Rothamsted Manor, in grey between the trees. It is set near the hive location from [19] (see Fig. 1 in red) (**A**) 32 m above ground.(**B**) 2 m above ground. Notice how different the skylines and ground textures are. (Color figure online)

2.2 Exploratory/Training Data

We extracted the two-dimensional flight coordinates of one particular bumblebee from [19]. We will call her Bee 1 throughout this paper. We selected Bee 1 because she was recorded for the highest number of flights and datapoints. She performed 156 of the 244 flights recorded during the whole experiment and 5947 datapoints. She also performed the highest number of exploitation flights: 142, against 7, 17, and 16 for the three other bees. We split the 156 flights into two categories: exploration and exploitation flights, according to the definition of [19]

> Exploitation flights were defined as consisting of a single loop (i.e. the bee did not return within 15 m of the hive, and then flew out again in a different direction), and including at least one stop in a location the bee had stopped in the past.

Using this definition, 14 flights were categorized as exploratory and 142 as exploitative (three exploratory flights occurred before any exploitation flight was recorded). We decided to train our model using exclusively the data from these three flights, to test if they contained enough information to guide Bee 1 back to its home for her entire subsequent foraging career. We then categorized the 1438 coordinates of exploratory flights as either inward or outward. Inward coordinates were defined as being closer to the hive than the coordinates at the previous time-step, and further from the coordinates at the next time-step. 335 coordinates were labeled as inward. Our reasoning was to keep only coordinates for which we could assume that Bee 1 was heading toward her hive, and thus could take snapshots toward it. From each point on the exploratory route, we determined the heading toward the hive and generated a snapshot in this exact direction. In reality, the ability of bees to determine the heading toward their hive during exploration flights is probably dependant on path-integration, which might be noisy. However, we do not implement any source of noise in the estimation of the heading toward the hive in this study

2.3 Snapshot Based Navigation Model

While numerous extensions have been made to this basic model [2,3,13], this is not the focus of this study so we use the simplest form known as the perfect memory algorithm. In this model, rotated versions of the view experienced by the agent are sequentially compared to all of the stored training snapshots. The best heading is then defined as the one with the lowest image difference, across all training views and rotations. Image difference can be calculated by various functions but here we use the average absolute difference between each of the image pixels to calculate the Image Difference Function (IDF) [22]:

$$IDF(C(\vec{a},\theta), S(\vec{b},\phi)) = \frac{1}{p \times q} \sum_{i=1}^{p} \sum_{j=1}^{q} |C_{i,j} - S_{i,j}| \tag{1}$$

where $C(\vec{a},\theta)$ is a $p \times q$ pixel view captured at location $\vec{a}$ with heading θ, $S(\vec{b},\phi)$ is a $p \times q$ pixel snapshot stored in memory and $C_{i,j}$ and $S_{i,j}$ refers to the intensity of pixels in row i and column j of the captured view and stored snapshot respectively. A Rotational Image Difference Function (RIDF) is then generated by calculating the IDF across a range of θ. Where there is a good match, there will be a minimum in the RIDF which defines the best matching direction.

2.4 Training

Five different agents were simulated, with different flight altitudes during training: 2, 4, 8, 16, and 32 m above the ground at the 335 inward coordinates identified above. To make the agent follow the ground above a set altitude, we added these altitudes to the height of the closest vertex of the 3D reconstructions at each point in the training trajectory. Using the resulting (x, y, z) position and

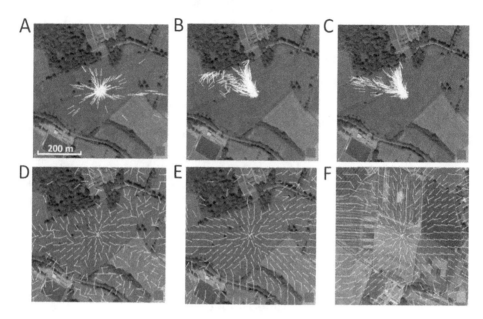

Fig. 3. Overview of the navigational data from training and every test in the paper. Coloured squares in the bottom-right corners refers to the areas highlighted in Fig. 1. **(A)** 335 training coordinates with vectors facing the hive, used to train all agents. The immediate vicinity of the hive is densely sampled, whereas more distant location were sparsely visited during exploration flights. **(B)** Agent trained at 2 m altitude is tested at the same ground height, on 650 actual exploitation coordinates recorded from Bee 1. For her whole foraging career, Bee 1 repeated short flights toward two flower patches close to the hive. The agent fail to predict the position of the hive when flying to the west of the hive. **(C)** Same data, but the Agent was trained and tested at 32 m height above the ground. This agent is not lost when flying away toward the west of the hive, and it correctly predict the position of the hive. (see Fig. 4: A) **(D)** Agent trained at 2 m altitude is tested at 400 coordinates, linearly sampled from a grid of coordinates with 600 m × 600 m size. One sampled point every 30 m. The model only predict the correct direction at close distance from the hive. **(E)** Same as D for agent trained and tested at 32 m altitude. The predictions are accurate even at long distance from the hive. **(F)** Agent trained at 32 m altitude at the lime location (see Fig. 1), at the north of the Rothamsted 3D map. The model is robust and stay accurate in various locations. (Color figure online)

heading toward the hive, we rendered 360° panoramic views at 360 × 100 pixel resolution. All 335 snapshots were then stored in memory and used as $S(\vec{b}, \phi)$ in the perfect memory algorithm.

The set of coordinates used to generate these snapshots was originally acquired by recording Bee 1 flying inside the red area indicated in Fig. 1 (GPS coordinates of the hive's location: 51.803821, −0.368923). We reused the same distribution of points to train the model at different locations of the Rothamsted

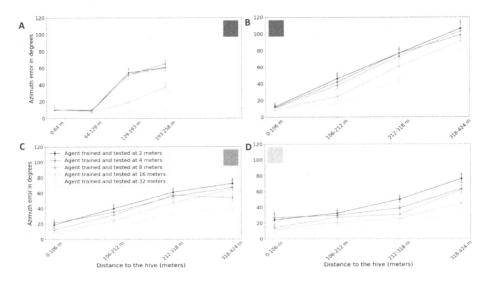

Fig. 4. Average angular error of the agents depending on the distance to the hive and the altitude they were trained at. (**A**) Angular error averaged across 650 coordinates from Bee 1 exploitation flights. The maximum distance between Bee 1 and the hive was 258 m during exploitation flights. The distance bins are sampled linearly. (**B**) Angular error averaged over a 600 m wide, linearly sampled grid. The maximum distance between the agent and the target using this linear grid sampling method is 424 m. The distance bins are sampled linearly. (**C**) Angular error with a 600 m wide, linearly sampled grid at the cyan location, (see Fig. 1). (**D**) Same, but at the lime area (see Fig. 3: F).

3D model: lime (51.812474, −0.375534) and cyan (51.804503, −0.393395) areas (see Fig. 1). The tests protocols are described in the results section.

The code used to modify the 3D Rothamsted model, extract the exploration coordinates of Bee 1 and most importantly perform all the simulations in this paper can be found on GitHub at https://github.com/thomasMisiek/Bumblebee_Rothamsted_SAB2024.

3 Results

3.1 Test 1: Exploitation Flight

We combined the coordinates from all but the last exploitative flights (where Bee 1 supposedly died), resulting in 1556 2D positions. From these, we randomly sampled 650 points and created 20 vectors for each coordinate, facing different directions, each separated by 18°. (see Fig. 1 in red, and Fig. 3:B). At each location and heading, 360° panoramic views were rendered at 2, 4, 8, 16, and 32 m altitude. Each agent was tested on all 65 000 of these images, and the 20 Image-Difference values were calculated for each (x, y, z) coordinate. Because of the low rotational sampling resolution of 18°, the heading error could fall anywhere between 0 and 18° even when facing the correct direction. It is thus not possible to obtain a 0° accuracy in our setup, and a residual error is to be expected. The preferred azimuth of the agent – which is the predicted direction of the hive – is at the angle that is associated with the lowest IDF value. One preferred azimuth was retrieved for each 650 test positions and each of the 5 flying altitudes, for each agent. For each preferred azimuth, we computed the angular error to the actual hive azimuth.

As Fig. 4A show, when the agent is closer to the hive, there is no significant difference in angular error between agents flying at 2 and 32 m. However, beyond 64 m, the agent trained at 32 m altitude has a significantly lower angular error than the one trained at 2 m. Overall, it can be observed that accuracy increases with altitude, but drops with increasing distance to the hive. This can also be observed in Fig. 3B, where the agent trained at 2 m altitude, fails to predict the position of the hive when it is further away. The 32 m altitude agent on the other hand has no problem retrieving the correct heading at any distances from the hive (see Fig. 3:C).

3.2 Test 2: Linearly Sampled Grid

We tested the agent on another set of 400 coordinates, which were linearly sampled from a 600 m wide grid centered on the hive, with a 20 m spacing (see Fig. 1 in red). The purpose of this test was to explore how far the agent was able to predict the correct heading toward the hive, from every direction, without being constrained to following the repetitive trajectories experienced by Bee 1 in exploitation bouts. The terrain was relatively flat within the sampled areas, allowing us to remove test coordinates that were located above trees and buildings, simply by filtering out points where the flying altitude would be higher than a threshold altitude of 140 m above the sea level. As Fig. 4B shows, while the difference in performance between low and high-altitude training is less important than in Test 1, it is still significant. Again, this difference is also clearly visible in Figs. 3: D where the agent trained at a lower altitude clearly fails to predict the position of the hive when it is further away from the hive. In every plots of Fig. 4, a clear gradient can be seen, where performances get better with increasing height.

3.3 Test 3: Linearly Sampled Grid at Other Locations

Finally, we show that these results are generalizable to other locations than the one used in Test 2. We trained and tested 5 agents using the same pipeline as in Test 2, but at two new locations from the Rothamsted 3D world, centered on the lime and cyan areas highlighted in Fig. 1. As Figs. 4C and D illustrate, again, accuracy generally increases with training altitude. Performances in the cyan and lime areas are better than in the red square probably because there are fewer nearby obstacles such as trees.

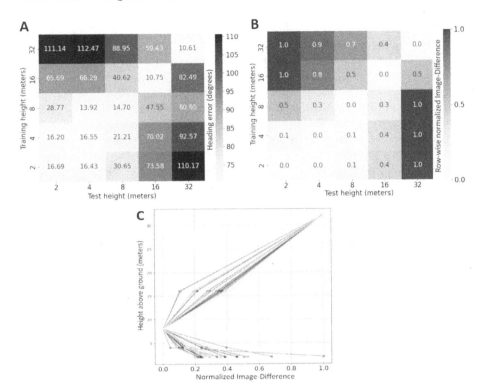

Fig. 5. Additional visual analysis of the navigational data from Test 2 (red location, see Fig. 1), from a distance of 0 to 141 m to the hive (**A**) Generalisation of performances across flying altitude. The diagonal whiter regions represents the average performances of agents trained and tested at the same height. We find no significant difference between the heading error of lower-than-training and higher-than-training tests (**B**) On average, there is a vertical gradient of error to exploit to recover the training altitude: the closer to the training altitude, the lower the image-difference error. (**C**) Height IDFs calculated at 20 (x,y) coordinates, randomly sampled at a distance of 141 m from the hive. Each line shows the normalised height IDF at one location, calculated between a training image at 8 m altitude and test images at 2,4,8,16, and 32 m altitude. In practice, it is possible to navigate back to the training altitude by performing gradient descent on the height IDF. (Color figure online)

4 Conclusions

Generalization is Better at Higher Altitude: Our results clearly indicate that visual homing is facilitated by flying higher. The agent trained at 32 m altitude, consistently demonstrated significantly lower heading errors compared to the agent trained at 2 m altitude. Higher vantage points could aid more accurate navigation in several ways. More comprehensive views of the landscape might be attained, allowing to focus on distant and more stable landmarks. Close obstacles such as trees and buildings might be less likely to obstruct a familiar view, and in particular disrupt the shape of the horizon, which has been proven to be an important landmark.

Poor generalization across altitude: Despite improvements in navigation performance at higher altitudes, as Fig. 5A shows, the model exhibited poor generalization when trained and tested at different altitudes. Thus, to remember a place, the agent needs to revisit it at the correct altitude. This is not necessarily in conflict with results from [8] and [16] as they only explored a smaller range of altitudes between 0 m–8 m, and close to the hive where snapshots are denser. Accordingly, we found some generalization between low flying heights at close distance to the hive. Transfer learning from aerial to ground robots, as discussed in [16] and [8] is thus supported by our results, as long as the aerial robot is not flying too high.

Navigational Strategies for Flying Agents: Our findings suggest that for aerial navigation systems, strategies should include altitude-specific training and possibly the integration of mechanisms to recognize and adapt to altitude changes. In the same way that agents can orient themselves by making turns to descend the gradient of a *rotational* IDF, fig. 5B and C suggests that an agent could also find the correct height by making suitable *translational* movements to descend the gradient of the *height* IDF. The distribution of distances from the hive seen in the exploration flight data shows that bees seem to allocate a large amount of time to explore the vicinity of the hive while only sampling more distant areas in a sparse manner. How do bees cope with this without getting lost? One possibility is that the further they go from the hive, the higher they fly. The increasing generalization at high altitudes would counter the increasing sparsity of memories further from the hive.

Noise: Noise: Robustness to various sources of noise is crucial for navigation systems. In this study, training snapshots were oriented towards the true direction of the hive. In real-life, errors might be introduced in the orientation of these snapshots by noise in the insect's path-integration system. Future research should investigate how variability and errors in the orientation of training snapshots affect navigational performance. Additionally, no noise was added to the visual input in this study. It would be interesting to explore the impact of visual noise on the agents' performance, depending on their flying altitude and distance from the hive. Preliminary results indicate that visual noise impairs performance at higher altitudes and close to the hive. It might be beneficial for robots/bees to fly lower when close to the target and higher when further away. Thus, future work might investigate the impact of visual noise on navigation, and explore the best flight strategy, how to manage flying altitude, to optimize a trade-off between robustness against noise and the ability to generalize.

Future Work: We will soon assess whether a more biologically plausible neuromorphic model of bees memory and visual preprocessing can aid navigation [2], by making the agent focus on the more distant and stable landmarks in the field of view. Further investigations could explore the resilience and robustness of such an algorithm in the real world, rather than in a simulation.

Acknowledgments. JK is funded by EPSRC grant EP/V052241/1. JK and AP are funded by EPSRC grant EP/S030964/1 and BBSRC grant BB/X01343X/1. TM is funded by a School of Engineering and Informatics studentship. All data and code required to reproduce the results in this paper can be accessed through the references in the text, specifically the GitHub repository at https://github.com/thomasMisiek/Bumblebee_Rothamsted_SAB2024.

References

1. Amin, A.A., Kagioulis, E., Domcsek, A.D.N., Nowotny, T., Graham, P., Philippides, A.: Robustness of the infomax network for view based navigation of long routes. In: The 2023 Conference on Artificial Life, MIT Press (2023). https://doi.org/10.1162/isal_a_00645, https://www.mitpressjournals.org/doi/abs/10.1162/isal_a_00645
2. Ardin, P., Peng, F., Mangan, M., Lagogiannis, K., Webb, B.: Using an insect mushroom body circuit to encode route memory in complex natural environments. PLOS Comput. Biol. **12**(2), e1004683 (2016). ISSN 1553-7358,https://doi.org/10.1371/journal.pcbi.1004683
3. Baddeley, B., Graham, P., Husbands, P., Philippides, A.: A model of ant route navigation driven by scene familiarity. PLoS Comput. Biol. **8**(1) (2012). ISSN 1553734X, https://doi.org/10.1371/journal.pcbi.1002336, iSBN: 1553-7358 (Electronic) \n1553-734X (Linking)
4. Collett, M., Chittka, L., Collett, T.S.: Spatial memory in insect navigation. Curr. Biol. **23**(17), R789–R800 (2013)
5. Collett, T.S., Hempel de Ibarra, N.: An 'instinct for learning': the learning flights and walks of bees, wasps and ants from the 1850s to now. J. Exp. Biol. **226**(6) (2023). ISSN 1477-9145, https://doi.org/10.1242/jeb.245278
6. Collett, T.S., Zeil, J.: Places and landmarks: An arthropod perspective (1998)
7. Defra Data Services Platform: Lidar composite digital terrain model (dtm) - 1m. Online (nd). https://environment.data.gov.uk/dataset/13787b9a-26a4-4775-8523-806d13af58fc. Accessed 17 Apr 2024
8. Dewar, A., Graham, P., Nowotny, T., Philippides, A.: Exploring the robustness of insect-inspired visual navigation for flying robots. In: Artificial Life Conference Proceedings (32), pp. 668–677 (2020). https://www.mitpressjournals.org/doi/abs/10.1162/isal_a_00307
9. Dewar, A.D., Knight, J.C.: Brains on board robotics. GitHub (nd). https://github.com/BrainsOnBoard/bob_robotics
10. Dewar, A.D., Philippides, A., Graham, P.: What is the relationship between visual environment and the form of ant learning-walks? Adaptive Behavior **22**(3), 163–179 (2014). ISSN 1741-2633, https://doi.org/10.1177/1059712313516132
11. Graham, P., Philippides, A.: Insect-inspired visual systems and visually guided behavior. Encyclopedia Nanotechnol. 1–9 (2014)
12. Graham, P., Philippides, A., Baddeley, B.: Animal cognition: multi-modal interactions in ant learning. Current Biol. **20**(15), R639–R640 (2010). ISSN 0960-9822, https://doi.org/10.1016/j.cub.2010.06.018
13. Kagioulis, E., Philippides, A., Graham, P., Knight, J.C., Nowotny, T.: Insect inspired view based navigation exploiting temporal information. In: Living Machines 2020. LNCS (LNAI), vol. 12413, pp. 204–216. Springer, Cham (2020). https://doi.org/10.1007/978-3-030-64313-3_20 ISBN 978-3-030-64312-6 978-3-030-64313-3

14. Knight, J.C., et al.: Insect-inspired visual navigation on-board an autonomous robot: real-world routes encoded in a single layer network. In: The 2019 Conference on Artificial Life, pp. 60–67. MIT Press, Cambridge, MA (2019). ISBN 978-0-262-35844-6, https://doi.org/10.1162/isal_a_00141, https://www.mitpressjournals.org/doi/abs/10.1162/isal_a_00141
15. Labrosse, F.: The visual compass: performance and limitations of an appearance-based method. J. Field Robot. **23**(10), 913–941 (2006). ISSN 1556-4967, https://doi.org/10.1002/rob.20159
16. Philippides, A., Steadman, N., Dewar, A., Walker, C., Graham, P.: Insect-inspired visual navigation for flying robots. In: Lepora, N.F.F., Mura, A., Mangan, M., Verschure, P.F.M.J.F.M.J., Desmulliez, M., Prescott, T.J.J. (eds.) Living Machines 2016. LNCS (LNAI), vol. 9793, pp. 263–274. Springer, Cham (2016). https://doi.org/10.1007/978-3-319-42417-0_24 ISBN 9783319424170
17. Rothamsted Research: Advancing sustainable agriculture. Online (nd). https://www.rothamsted.ac.uk/. Accessed 17 Apr 2024
18. Woodgate, J.: 3D model of an agricultural research institute (2019). https://doi.org/10.6084/m9.figshare.7539851.v1, https://figshare.com/articles/online_resource/3D_model_of_an_agricultural_research_institute/7539851
19. Woodgate, J.L., Makinson, J.C., Lim, K.S., Reynolds, A.M., Chittka, L.: Life-long radar tracking of bumblebees. PLoS ONE **11**(8), e0160333 (2016). https://doi.org/10.1371/journal.pone.0160333
20. Woodgate, J.L., Makinson, J.C., Lim, K.S., Reynolds, A.M., Chittka, L.: Continuous radar tracking illustrates the development of multi-destination routes of bumblebees. Scientific Reports **7**(1) (2017). ISSN 2045-2322, https://doi.org/10.1038/s41598-017-17553-1
21. Wystrach, A., Mangan, M., Philippides, A., Graham, P.: Snapshots in ants? new interpretations of paradigmatic experiments. J. Exp. Biol. (2013). ISSN 0022-0949, https://doi.org/10.1242/jeb.082941
22. Zeil, J., Hofmann, M.I., Chahl, J.S.: Catchment areas of panoramic snapshots in outdoor scenes. JOSA A **20**(3), 450–469 (2003)

Vector-Based Navigation Inspired by Directional Place Cells

Harrison Espino[1]( ) and Jeffrey L. Krichmar[1,2]

[1] Department of Computer Science, University of California, Irvine, Irvine, CA, USA
{espinoh,jkrichma}@uci.edu
[2] Department of Computer Science, Department of Cognitive Sciences, University of California, Irvine, Irvine, CA, USA

Abstract. We introduce a navigation algorithm inspired by directional sensitivity observed in CA1 place cells of the rat hippocampus. These cells exhibit directional polarization characterized by vector fields converging to specific locations in the environment, known as ConSinks [8]. By sampling from a population of such cells at varying orientations, an optimal vector of travel towards a goal can be determined. Our proposed algorithm aims to emulate this mechanism for learning goal-directed navigation tasks. We employ a novel learning rule that integrates environmental reward signals with an eligibility trace to determine the update eligibility of a cell's directional sensitivity. Compared to state-of-the-art Reinforcement Learning algorithms, our approach demonstrates superior performance and speed in learning to navigate towards goals in obstacle-filled environments. Additionally, we observe analogous behavior in our algorithm to experimental evidence, where the mean ConSink location dynamically shifts toward a new goal shortly after it is introduced.

Keywords: Navigation · Reinforcement Learning · Hippocampus

1 Introduction

Place cells in the CA1 area of the hippocampus have shown heading based sensitivity in goal-oriented tasks [3,9]. Recent work in rats quantified this sensitivity as a vector field converging to a specific point, known as a ConSink [8]. In an obstacle-free environment, ConSinks of these direction sensitive place cells (which we call "ConSink cells") organized around the goal, and shifted towards a new goal when it was changed. We suggest that this spatial representation may benefit new learning algorithms for navigation and other domains.

Optimal navigation to a goal is also a prominent benchmark in the field of computer science. A common approach to navigating unknown and complex environments is reinforcement learning (RL), such as Deep Q-Networks (DQN) [7] or Proximal Policy Optimization (PPO) [10]. These algorithms aim to learn a policy which determines optimal actions to navigate towards a goal given observations of the agent's surroundings. These algorithms are provided either partial or complete information about the environment. While they have demonstrated efficacy in domains such as robotics [2] and playing games [5], some open issues still persist. Notably, algorithms reliant on training deep neural networks are often computationally intensive, and require substantial training before achieving acceptable performance.

Inspired by the ConSink finding, we propose an algorithm to achieve fast and robust performance on goal-directed navigation tasks in simulated environments. In the Dyna maze environment, our model learns to navigate to the goal from randomly selected start locations in less than 50 trials, which is substantially faster than DQN and PPO. By testing on multiple environments with randomly placed obstacles, we find our that our model performs better than other state-of-the-art (SoTA) RL algorithms. Our model demonstrates the same ability to adapt as alternatives and exhibits similar behavior to real place cells by shifting the population ConSink location towards the new goal.

In summary, our paper makes the following contributions:

1. We introduce a biologically-inspired algorithm for learning goal-oriented navigation tasks that outperforms DQN and PPO, two state-of-the-art RL algorithms. Our model reaches the goal in fewer training epochs, maximizes reward faster, and adapts to changes in goal quickly.
2. We find similar behavior in our model to biological ConSink cells on experiments when the goal changes locations. Individual ConSinks move closer to the new goal, causing the mean ConSink location to shift to the new goal location after training.
3. Based on our model, we predict that ConSink cells in the rat hippocampus may have directional sensitivity away from the goal towards important "subgoals" in environments with obstacles requiring non-goalward movement.

2 Methods

2.1 ConSink Place Cell Model

In [8], rats that were presented two possible choices of travel selected the direction which aligned with the vector from the population of ConSink cells. When there was no available path to the goal, the rat physically surveyed possible directions and were found to choose the direction with the highest population activation, and consequently that nearest towards the goal.

To model this behavior, neurons in the model are characterized by place sensitivity and orientation sensitivity, each of which are calculated individually and multiplied together to a generate the place cell's total activity. In biological neurons, "sensitivity" is characterized by increased firing rate. For our model, it is represented by a scalar value. Place activity at point (x, y) is calculated by (1):

$$v_{place} = \mathcal{N}(\sqrt{(p_x - x)^2 + (p_y - y)^2}; 0, \sigma^2), \tag{1}$$

where the distance between the place cell location (p_x, p_y) and the agent's location (x, y), is passed through a Gaussian activation function. The hyperparameter σ controls how sharply the activity of the place cell decreases as the distance from the cell's preferred location increases. This is dependent on the scale of the environment, and is set to 2.0 in our experiments. Each grid in our environment had a length of 1 unit, so this allowed for sampling in a wide range around the agent's location.

Orientation sensitivity is defined by a response array of 8 values each representing the cardinal and ordinal directions of travel. From these values, a response vector $\hat{n}$ can be generated as:

$$\hat{n} = \frac{\sum_i w_i \hat{u}}{\|\sum_i w_i \hat{u}\|}, \tag{2}$$

where $\hat{u}$ is the unit vector corresponding to one of 8 directions and w_i is the value for this direction. Consistent with the experimental data in [8], values of the response array are initialized to have a preference towards the goal location. This is achieved through:

$$d(p, p_0, d) = \left| \frac{(\mathbf{p} - \mathbf{p_0}) \cdot \mathbf{d}}{\|\mathbf{d}\|} \right| \tag{3}$$

and

$$w_i = \mathcal{N}(d(g_{x,y}, p_{x,y}, \hat{u}_i); 0, \sigma^2) + U(\alpha_{min}, \alpha_{max}). \tag{4}$$

Equation (3) defines a projection function which returns the minimum distance from point p to point p_0 in the direction d and Eq. (4) is the initialization of response vector value w_i as the minimum distance from the place cell's location $p_{x,y}$ to the goal $g_{x,y}$ through the corresponding unit vector $\hat{u}_i$, plus a random amount of noise given by sampling from a random variable U with minimum and maximum values (n_{min}, n_{max}). For our experiments, $(\alpha_{min}, \alpha_{max})$ is set to $(-0.5, 0.5)$ and σ of the activation function is set to 5.0. We found that a shallow-sloped activation function placed sufficient weight on directions that were close but not exactly in the correct direction.

The orientation-based activation for direction θ can then be calculated as in (5), or the cosine similarity between the response vector and the orientation. This value was normalized between 0 and 1 so that we can achieve a total cell activation between 0 and 1 when it is multiplied with the place activation.

$$\left(\frac{\theta \cdot \hat{n}}{\|\theta\| \cdot \|\hat{n}\|} + 1 \right) \cdot 0.5 \tag{5}$$

2.2 Vector Navigation with ConSink Place Cells

To facilitate navigation, a population of directional place cells is initialized, each with place sensitivity for a random location in the environment as defined by Eqs. 1 through 4. Locations of place cells are constrained such that each cell is a minimum distance from its neighbors. This minimum distance is dependent on the size of the environment and the number of place cells, and was necessary to ensure place cells are sufficiently distributed throughout the environment.

At each simulated timestep, an action is chosen by sampling the population of cells at the current location for each possible direction of travel. The values for each direction are passed through a softmax activation function to determine the probability of selecting the corresponding action. The temperature of this softmax function is set to 1.0 in our experiments.

2.3 Eligibility Trace and Reward

Learning is achieved by changing the values of the response array according to the reward signal received by the environment. To do this, each neuron must maintain an eligibility trace e defined by:

$$e = \begin{cases} v_{place} & \text{if } v_{place} > \beta \\ e - \frac{e}{\tau} & \text{otherwise} \end{cases}, \quad (6)$$

where β determines the minimum place activity to set the eligibility trace and τ determines the rate of decay of the eligibility trace. We set β to 0.5 and τ to 100 in our experiments. This β value allowed the eligibility trace of nearby neurons to be set again in the case the agent visited the same location twice in quick succession, and a large τ led to a slower decay and greater updates to neurons involved in earlier decisions.

The eligibility trace of each neuron is updated after an action is taken. The purpose of this eligibility trace is to determine the contribution of each cell to the previous action, as well as solve the credit assignment problem when a reward signal is used to update the modeled ConSink neurons. When the neuron's eligibility trace is updated to v_{place}, a unit vector $\hat{l}$ between the neuron's preferred location and the agent's location is saved so that it can be used to update the directional sensitivity.

Upon receiving a positive or negative reward, the values of each neuron's response arrays are calculated by:

$$w_i = w_i + \frac{\hat{u}_i \cdot \hat{l}}{\|\hat{u}_i\| \cdot \|\hat{l}\|} * r * e_i * \alpha, \quad (7)$$

where r is a negative or positive reward from the environment and α is a learning rate (set to 0.001 in our experiments). The cosine similarity between the saved direction of travel and the response array value's unit vector determine its contribution to the agent's action. The eligibility trace additionally scales the learning by how active the neuron was during traversal.

2.4 Environment

We use the OpenAI Gym framework to build environments capable of training our models and Reinforcement Learning models for comparison [1]. Agents traverse a grid-world environment and are allowed to move to neighboring squares

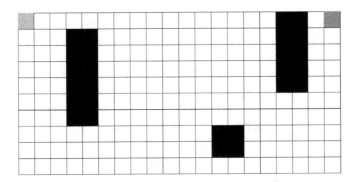

Fig. 1. The Dyna maze environment. Black squares represent intraversable obstacles. Red square is the goal. The agent starts randomly along the leftmost column (the green square represents one starting location). (Color figure online)

in 8 directions if that square is not occupied by an obstacle or out of bounds. We test our algorithm in three different environments.

The first environment is the Dyna maze, which was introduced to test the Dyna Reinforcement Learning algorithm and has been used in subsequent navigation papers [6,12]. Our implementation can be seen in Fig. 1. The agent (green) started randomly at any space in the leftmost column of the environment, and the goal (red) was always in the top-right space.

To further assess the robustness of our algorithm, we train multiple instances of the model on environments with randomly placed obstacles. Obstacles are placed randomly according to a density percentage (10% in our experiments) and then iteratively moved to ensure a valid path from the start to end exists. In this case, the agent's start position is always on the top-left space, and the goal on the bottom-right space. A sample environment of this kind can be seen in Fig. 2.

Lastly, we train models in a completely open environment similar to that used in the rat experiments [8], which was a 10 X 10 grid. The environment and models are initialized with the goal in the center of the environment (5, 5). The goal is moved to a different location (2, 3) after 100 epochs of training. At each episode, the agent started randomly at the edge of the environment.

Agents received an observation in the form of its current grid location (x, y) and 8 values corresponding to the distance to the closest boundary or wall in the directions of movement. The reward signal for the environments is as follows. Reaching the goal results in a reward of 1.0. Attempting to move into a space occupied by an obstacle results in a reward of -0.8, and -0.75 for attempting to move out of bounds. If the agent moves to a previously visited space, it receives a reward of -0.25. The agent receives a reward of -0.04 in all other cases. An epoch in this environment lasted until the goal is reached, or the total reward went below -100. This reward structure was chosen to encourage fast planning towards the goal while discouraging wasteful actions and encountering obstacles.

Individual values were fine-tuned for convergence on our model's comparisons. To ensure fairness, all models had the same observation space and rewards.

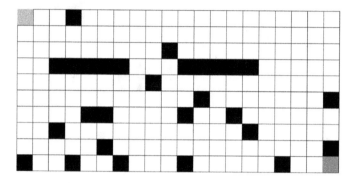

Fig. 2. A sample environment with randomly placed obstacles at 10% obstacle density. The agent's start location is always on the top left, and the goal is always on the bottom right.

2.5 Comparisons

We assess the performance of our model by comparing two other SoTA RL algorithms on the same environment.

Deep Q-Networks: An RL algorithm combining the Q-learning technique and deep learning [7]. A neural network is trained to approximate the Q-function, which estimates the expected reward for taking an action at a given state. Past experiences are replayed to stabilize training and improve data efficiency. We implemented DQN with a 2 hidden layer network, with 256 neurons at each layer.

Proximal Policy Optimization: An RL algorithm with the objective of approximating an optimal policy which maps specific states to actions [10]. A neural network is trained to learn a probability distribution of each action given the agent's current state. PPO has found great success in robotics and game playing for it's simplicity and efficiency. We implemented PPO with a 2 hidden layer network, with 64 neurons at each layer.

3 Results

3.1 Maze Learning

The result of 500 epochs of training on the Dyna maze are shown in Fig. 3a. Solid lines and shaded areas represent the mean and standard error of 10 agents. Since the reward signal is overwhelmingly negative, it is expected that an agent's

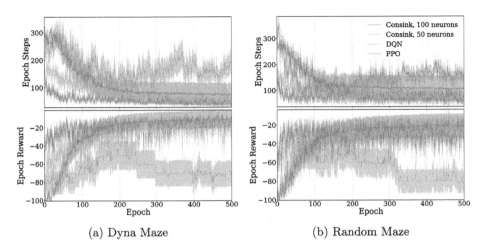

Fig. 3. DQN, PPO, and Consink (ours) agents trained on the (a) Dyna maze environment and (b) random mazes for 500 epochs. Top graphs are the total number of steps before reaching the goal or failing, bottom graphs are the total reward at each epoch. The line and shaded area are the mean and standard error of 10 different agents each.

total reward approaches zero as it learns to avoid obstacles and take minimal steps to reach the goal. Agents using our ConSink model learned faster than both PPO and DQN agents, as evidenced by larger reward epochs in a shorter time. This can also be seen when measuring the epoch step count, where our models consistently learned shorter paths. One reason for this is that ConSink place cells have preferred directions initially, as observed by [8].

We see similar results on randomly generated mazes in Fig. 3b. Our ConSink models with both 100 or 50 neurons outpaced DQN and PPO on both total reward per epoch and steps per epoch. Despite random initialization for the environment, the location of cells, and directional sensitivity of each cell, we find striking consistency over multiple runs.

Figure 4 depicts the difference in the neurons' directional sensitivities from initialization to the final training epoch. Place cells are drawn at their specified location, with lines indicating the direction of the response vector. Neurons in the model initially showed scattered directional sensitivity due to the noise from their initial preference towards the goal as described in Sect. 2.1. After training, populations of nearby neurons show similar directional sensitivity to each other. Near obstacles such as the bottom-most square, the neurons' directional sensitivities appear to curve around it, suggesting the model learns the optimal action is to avoid the obstacle while moving towards the goal.

Tables 1 and 2 report the average number of epochs before each model reaches the goal for the first time, as well as the percentage of time the goal is reached after 10, 50, 100, and all epochs. Each epoch in the environment lasts until the agent reaches the goal, or the total reward goes below -100. The results are the mean of 10 different agents.

ConSink models are first to reach the goal from initialization, and reach the goal a larger percentage of the time throughout training. The speed in which our models successfully reach the goal can be attributed to a number of factors. Our models do not rely on training densely connected neural networks like DQN and PPO. Our learning rule facilitates adjustments to the model completely online, rather than during specific training times between epochs. Additionally, initializing each neuron's directional sensitivity to be towards the goal means the model will often start by choosing actions in a straight line towards the goal, which may be correct in many cases.

Table 1. Success reaching goal on Dyna maze. Mean±standard deviation of 10 runs reported. Top row is the epochs to reach the goal. Other rows are the percentage of time the goal is reached after 10, 50, 100, and all epochs.

Model	ConSink-100	ConSink-50	DQN	PPO
First Goal Epoch	**0.7 ± 1.0**	2.11 ± 2.85	29.8 ± 55.4	24.2 ± 28.0
First 10 Goal %	**80.0 ± 21.9%**	35.6 ± 29.5%	22.0 ± 36.3%	2.0 ± 6.0%
First 50 Goal %	**91.2 ± 5.8%**	70.4 ± 24.0%	23.8 ± 37.7%	29.8 ± 23.1%
First 100 Goal %	**93.0 ± 4.5%**	76.1 ± 21.8%	26.9 ± 37.3%	52.8 ± 28.5%
Total Goal %	**96.4 ± 3.1%**	91.1 ± 6.5%	36.0 ± 26.1%	82.1 ± 27.8%

Table 2. Success reaching goal on randomly generated mazes over 10 runs (mean ± stdev). Top row is the number of epochs to reach the goal. Other rows are the percentage of time the goal is reached after 10, 50, 100, and all epochs.

Model	ConSink-100	ConSink-50	DQN	PPO
First Goal Epoch	2.3 ± 3.78	**1.7 ± 2.33**	10.5 ± 8.16	19.3 ± 33.47
First 10 Goal %	**62.0 ± 42.4%**	58.0 ± 36.3%	10.0 ± 16.1%	12.0 ± 18.3%
First 50 Goal %	**83.2 ± 15.8%**	75.6 ± 23.0%	29.4 ± 24.7%	34.2 ± 29.3%
First 100 Goal %	**85.6 ± 15.4%**	77.5 ± 25.1%	40.0 ± 33.6%	51.5 ± 32.4%
Total Goal %	**86.5 ± 15.2%**	79.7 ± 32.6%	36.4 ± 32.1%	74.3 ± 37.0%

3.2 Goal-Switching and Adaptability

Our experiments evaluating adaptability in an open environment are shown in Fig. 5. The agents are trained for 100 epochs with the goal at (5, 5), after which the goal is moved to (2, 3). The point at which the goal is moved is indicated by the dotted red line. As with the previous experiments, the line and shaded areas represent the mean and standard error of 10 agents. Because the environment

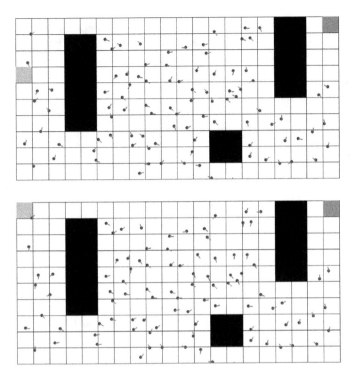

Fig. 4. Top image shows the initial directional sensitivity of place cells. Bottom image shows directional sensitivity after 500 epochs of training. Place cells are drawn in purple, with a line indicating the direction of highest activity.

contains no obstacles and the agent can reach the goal in minimal steps, our models and DQN quickly found success navigating. After the goal switch, we see a dramatic decrease in epoch reward and an increase in epoch steps as our models initially fail to reach the new goal. However, after less than 20 epochs, they return to performing better on reward and steps taken than DQN and PPO.

The high epoch step count and low reward from our models immediately following the goal shift can be attributed to a lack of exploratory behavior from the trained model. This could potentially be alleviated by implementing a variable softmax temperature for action selection based on the total accumulated reward in an epoch. This way, the model exploits its learned action policy unless it has accumulated negative rewards, in which case it will switch to a more explorative strategy.

Figure 6a shows the change in distance of the mean ConSink location to the goal before and after the goal switch. A neuron's ConSink location is calculated by multiplying the neuron's response vector by its distance to the goal. ConSinks are initially tuned to the location of the original goal, and over the first 100 epochs the mean ConSink location continues to shift closer to the center of the goal. After the goal switch we see the mean ConSink location shift in

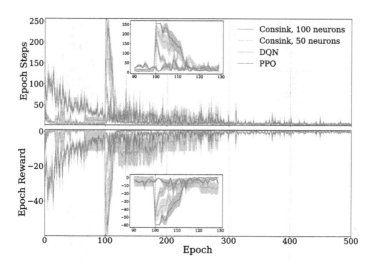

Fig. 5. DQN, PPO, and Consink (ours) agents trained on an obstacle-free open environment for 500 epochs. The goal is moved to an alternate location after 100 epochs, shown by the dotted line. Top graph is the total number of steps before reaching the goal or failing, bottom graph is the total reward at each epoch. The line and shaded area are the mean and standard error of 10 different agents each. Inset graph shows training epochs from 90 to 130.

the direction of the new goal with training, as the place neurons' directional sensitivities begin to change due to learning.

Similar to Fig. 2e from the rodent ConSink experiments [8], we plot the shift of individual ConSinks before and after training in Fig. 6b. The gray square is the location of the initial goal. Filled gray circles are the locations of ConSinks for 10 candidate neurons, and the white-filled gray circle is the mean ConSink location for the entire population. The same is true for the red square and red circles, which represent the new goal and ConSinks after the goal is changed. Gray lines connect the ConSinks of the same neuron. We find that individual ConSinks move closer to the new goal during training, leading the mean ConSink location to shift as well.

4 Discussion

We introduce a navigation algorithm that is capable of rapidly and accurately reaching a goal in obstacle-filled environments. Our model outperforms SoTA reinforcement learning models in maximizing accumulated reward, minimizing steps taken, consistently reaching the goal throughout training. When the goal is moved after training in a completely open environment, our model is able to adapt to this change in few epochs of exploration. Cells in our model also exhibit behavior reminiscent of biological data, as the mean ConSink location moves towards the new goal during training.

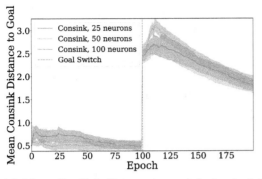 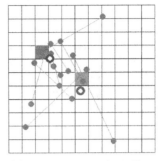

(a) Mean ConSink distance to goal during training (b) Shift in individual ConSinks

Fig. 6. (a) Distance of the mean ConSink location after the goal is changed in an open environment. Consink agents with 100, 50, and 25 neurons are trained for 100 epochs before the goal is changed. The line and shaded area are the mean and standard error of 10 different agents each. (b) Visualization of Consinks before and after the goal is changed. Gray circles represent the initial location of Consinks and the gray square is the location of the first goal. Red circles represent the Consink location of the same place cell after training on the new goal. (Color figure online)

Our findings mirror biological evidence showing adaptation in the ConSink cells recorded in rats. Our novel learning rule achieves this using a combination of eligibility trace, saved trajectories, and reward signal. This suggests that ConSink place cells in the hippocampus may be employing similar reward-based learning to adjust their directional sensitivity.

Our results also raise interesting questions about the role of directional sensitivity in place cells for complex environments. Previous work investigating place cell directionality in rats were in environments in which a direct route to the goal was always possible [3]. Ormond and O'Keefe used a unique "honeycomb maze" in which the rat was required to make a sequence of binary choices on raised platforms, sometimes requiring the rat to move away from a goal [8].

In both cases, the environment did not have obstacles and was fully visible to the rat. The honeycomb maze removed the ability for the rat to move directly to the goal, however the correct binary choice was always more in the direction of the goal than the incorrect one. Our model predicts that, in environments filled with obstacles, populations of ConSinks near obstacles blocking direct routes to the goal may need to point towards sub-goals in order to facilitate navigation. Behavioral studies have shown that mice utilize sub-goal strategies when learning new environments [11]. Future work might investigate if sub-goal memorization is encoded in place cell directionality.

Our model is similar to traditional Q-learning in that each place neuron is learning an optimal action for the location it is sensitive to, much like a Q-table learns optimal actions for each state. However, our model uniquely takes advantage of the fact that environment states in a navigational problem can be related over two axes of position. This allows for the generalization of unlearned areas by sampling the learned actions of place cells representing nearby areas. Future work may aim to apply our model to non-navigational problems which can be represented in this way. The model may also employ heterogeneous learning rates and eligibility trace functions across neurons to represent dynamic environments.

Deep Q-learning uses a form of experience replay in which previously saved state-action-reward sequences are used to train the model multiple times. PPO employs a similar technique, maintaining a growing replay buffer which is used to train the model every set number of epochs. While such a replay system is not yet implemented in our model, recordings of the rat hippocampus have shown sequential reactivation of place cells resembling past trajectories during sleep or rest states [4]. By saving past experiences, replay can potentially be used to further improve the efficiency of our models. Additionally, the model can serve as a generator for novel trajectories by continually sampling and simulating actions offline.

Acknowledgement. This work was supported by NIH award R01 NS135850-02.

Disclosure of Interests. The authors have no competing interests.

References

1. Brockman, G., et al.: Openai gym. CoRR abs/1606.01540 (2016)
2. Gu, S., Lillicrap, T., Sutskever, I., Levine, S.: Continuous deep q-learning with model-based acceleration. In: Balcan, M.F., Weinberger, K.Q. (eds.) Proceedings of The 33rd International Conference on Machine Learning. Proceedings of Machine Learning Research, vol. 48, pp. 2829–2838. PMLR (2016)
3. Jercog, P.E., Ahmadian, Y., Woodruff, C., Deb-Sen, R., Abbott, L.F., Kandel, E.R.: Heading direction with respect to a reference point modulates place-cell activity. Nat. Commun. **10**(1), 2333 (2019)
4. Lee, A.K., Wilson, M.A.: Memory of sequential experience in the hippocampus during slow wave sleep. Neuron **36**(6), 1183–1194 (2002)
5. Lin, C.J., Jhang, J.Y., Lin, H.Y., Lee, C.L., Young, K.Y.: Using a reinforcement q-learning-based deep neural network for playing video games. Electronics **8**(10) (2019)
6. Mattar, M.G., Daw, N.D.: Prioritized memory access explains planning and hippocampal replay. Nat. Neurosci. **21**(11), 1609–1617 (2018)
7. Mnih, V., et al.: Playing ATARI with deep reinforcement learning. CoRR abs/1312.5602 (2013)
8. Ormond, J., O'Keefe, J.: Hippocampal place cells have goal-oriented vector fields during navigation. Nature **607**(7920), 741–746 (2022)
9. Sarel, A., Finkelstein, A., Las, L., Ulanovsky, N.: Vectorial representation of spatial goals in the hippocampus of bats. Science **355**(6321), 176–180 (2017)
10. Schulman, J., Wolski, F., Dhariwal, P., Radford, A., Klimov, O.: Proximal policy optimization algorithms. CoRR abs/1707.06347 (2017)
11. Shamash, P., Olesen, S.F., Iordanidou, P., Campagner, D., Banerjee, N., Branco, T.: Mice learn multi-step routes by memorizing subgoal locations. Nat. Neurosci. **24**(9), 1270–1279 (2021)
12. Sutton, R.S.: Dyna, an integrated architecture for learning, planning, and reacting. SIGART Bull. **2**(4), 160–163 (1991)

The Cost of Behavioral Flexibility: Reversal Learning Driven by a Spiking Neural Network

Behnam Ghazinouri and Sen Cheng

Faculty of Computer Science, Ruhr University Bochum, Bochum, Germany
{behnam.ghazinouri,sen.cheng}@rub.de
http://www.rub.de/cns

Abstract. To survive in a changing world, animals often need to suppress an obsolete behavior and acquire a new one. This process is known as reversal learning. The neural mechanisms underlying RL in spatial navigation have received limited attention and it remains unclear what neural mechanisms maintain behavioral flexibility. Here, we extend a closed-loop simulator of spatial navigation and learning, based on spiking neural networks [7]. In this model, activity of place cells and boundary cells are fed as inputs to action selection neurons, which drive the movement of the agent. Upon reaching the goal, behavior is reinforced with spike-timing-dependent plasticity (STDP) coupled with an eligibility trace which marks synaptic connections for future reward-based updates. We model a task with an ABA design, where the goal is switched between two locations A and B after 10 trials. Agents using symmetric STDP excel initially on finding target A, but fail to find target B after the goal switch, persevering on target A. Either using asymmetric STDP, using many small place fields, or injecting short noise pulses to action selection neurons are effective in driving spatial exploration, which ultimately leads to finding target B. However, this flexibility comes at the price of slower learning and lower performance. Our work shows three examples of neural mechanisms that achieve flexibility at the behavioral level, each with different characteristic costs.

Keywords: Spatial navigation · Reversal learning · Behavioral flexibility · closed-loop simulation

1 Introduction

Both spatial navigation and reversal learning are fundamental skills for mobile animals. Spatial navigation involves identifying and maintaining a path between two locations [16]. Research over the past 50 years has identified key brain regions for spatial navigation in the medial temporal lobe and has discovered spatially selective cells like place cell (PC) [15], boundary cells (BC) [9], and head direction cells [20]. Recent studies also point to the hippocampus' role in reversal learning in spatial navigation tasks [22], where an obsolete behavior has to be replaced by a new one. While many computational models have been developed for spatial navigation, they focus on static environments [3, 4, 7, 12] and the dynamic aspects of reversal learning have received limited attention.

Here, we focus on an ABA reversal learning task where an agent has to learn to navigate to a fixed target (A), then to a different target (B), and finally to target A again. This task touches on two trade-offs: stability vs. plasticity, well known in the neural network literature [21], and exploitation vs. exploration, which is central in the reinforcement learning literature [18]. Behavioral flexibility is essential in the reversal learning task and requires exploration of new locations. For new behaviors to be learned, plasticity in the neural network is required. The cost of this flexibility is a decrease in exploitation and stability. The challenge is understanding how a biologically plausible spiking neural network maintains flexibility and its associated costs.

Using a spiking neural network, we find that a combination of symmetric spike-timing dependent plasticity (STDP) and optimized place field parameters performs well on the first target, but lacks flexibility for the second. In three other cases, the agent remains flexible, but incurs different costs. Asymmetric STDP [14, 19] results in highly variable behavior. Using many small place fields leads to low overall performance. Providing an external supervisory signal (injecting noise when unrewarded for too long) results in slow reversal learning and variable performance on the second target, but better performance on the first. Our results suggest multiple neural mechanisms can lead to behavioral flexibility, each with a distinct profile, allowing biological agent to evolve or develop solutions that fit its needs.

2 Methods

To model reversal learning in a biologically plausible network, we adopted the CoBeL-spike (Closed-Loop Simulator of **Co**mplex **Be**havior and **L**earning Based on Spiking Neural Networks) framework[1] [7].

Behavioral Task. The spatial navigation task was similar to the Morris water maze task [13]. Simulations were conducted in a 2.4 m × 2.4 m open field. Each simulation included 30 trials, in which the agent started at the center and freely navigated to find a hidden target, a circular area of diameter 0.6 m. A trial ended when the agent reached the target or after 5 s (timeout). Upon reaching the target, the agent received a reward. In the first phase (trials 1–10), the target was centered at [0.5, 0.5] (target A). In the second phase (trials 11–20), the target moved to [−0.5, −0.5] (target B), requiring the agent to unlearn the previous location and learn the new one. In the third phase (trials 21–30), the target returned to location A, and the agent had to remember the original location or re-learn it.

---

[1] https://github.com/sencheng/CoBeL-spike.

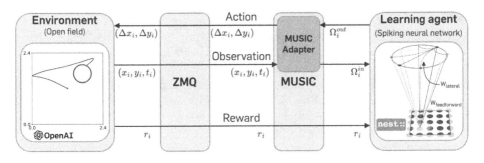

Fig. 1. Toolchain components. The CoBeL-spike toolchain consists of two components: a virtual environment (based on OpenAI Gym) and a learning agent (two-layered spiking neural network). ZMQ and MUSIC provide interfaces between the two model components. The diagram in the right box illustrates the network architecture.

Overview of the Computational Model. The CoBeL-spike toolchain comprises two components: a virtual environment and a learning agent (Fig. 1). The virtual environment, an open field Gym environment powered by OpenAI [2], monitors the agent's location and ensures that actions adhere to physical limitations. At each time step (0.01 s), the environment transmits the current trial time, Cartesian coordinates, and reward signal ($[x_i, y_i, t_i], [r_i]$) to the agent via ZMQ (**Ze**ro **M**essage **Q**ueuing) [10] to MUSIC (**MU**lti **SI**mulator **C**oordinator) [5]. MUSIC integrates the Gym environment with the learning agent, enabling real-time data transmission.

The learning agent is a two-layer spiking neural network (Fig. 1) implemented in NEST (**NE**ural **S**imulation **T**ool) [6]. The network's first layer comprises PC, BC, and noise generators (NG). MUSIC Adapters calculate the firing rates of these cells: $\Omega_{PC}, \Omega_{BC}, \Omega_{NG}$. The output layer contains 40 action selection neurons, whose activity is mapped to movement in Cartesian space by MUSIC Adapters via ZMQ to the environment. The action selection neurons drive movement and the function is similar to neurons in striatum [8]. The environment updates the agent's location based on the action, provided the new location is valid; otherwise, the agent remains in place. In the example trial shown in Fig. 1, the agent started at the center (green marker), moved through the environment (blue curve), and reached the target (purple circle).

Simulation of Neural Activity. Each unit in the first layer is represented by a Poisson generator and a parrot neuron in NEST [6]. The number of PC (N_{PC}) varies across simulations. The firing fields of evenly distributed PC $\Omega_{PC}^i(\mathbf{x}(t))$ are Gaussian 2D distributions with width σ_{PC} [7]. The maximum firing rate of all cells have the same and unchanged value within a simulation ($\eta_{PC}^i = \eta_{PC}$) and is such that the summed firing rate at the center of the environment is 3500 Hz [7]. Each PC has an excitatory connection to all actions selection neurons in the second layer. These feedforward weights are plastic (see below, with initial weights $W \sim \mathcal{N}(30, 5)$.

To prevent the agent from getting stuck at the edges and corners of the environment, the first layer has eight BC [9] that project to the second layer and that drive the agent towards the environment's interior. Each BC is active ($\Omega^i_{BC} = 500$) when the agent is inside its field [7].

To examine the effect of noise on the agent's performance, in some simulations, noise generators ($N_{NG} = 40$) were added to the network with static one-to-one excitatory connections to the second-layer neurons. In most successful trials, the agent was able to find the target within the first 2 s of the trial. Therefore, if the trial duration exceeded 2.5 s, it was likely that the agent engaged in repetitive unsuccessful behavior. To drive more exploration, noise was injected by activating a random set of five neighboring noise cells at time steps 2.5 s, 3.5 s, and 4.5 s, each noise injection lasting 0.4 s.

The second layer consists of 40 action selection neurons modeled as Leaky Integrate-and-Fire (LIF). Each neuron represents a movement direction evenly distributed across 360°. These neurons form a continuous ring attractor network [3,7], with local excitation and global inhibition [7].

Synaptic Plasticity and Learning. Only the weights between place cells and action selection neurons are plastic according to an STDP rule with an eligibility trace. The weight of the feedforward synapse between the i-th place cell and the j-th action selection neuron changed according to the following rules [6,7,11,17]:

$$\dot{w}^{ij} = c\,(n - b) \tag{1}$$

$$\dot{c} = -\frac{c}{\tau_c} + \Psi(\Delta t)\delta\left(t - s_{pre/post}\right) \tag{2}$$

$$\dot{n} = -\frac{n}{\tau_n} + \frac{1}{\tau_n}\delta\left(t - s_n\right) \tag{3}$$

c is an eligibility trace that ensures that connections of place cells that were activated closer to reaching the target have a larger weight update [3]. n is the dopamine concentration, and b is the dopamine baseline concentration. τ_c and τ_n are the time constants of the eligibility trace and the dopamine concentration, respectively. $\delta(t)$ is the Dirac delta function, $s_{pre/post}$ the time of a pre- or post-synaptic spike, s_n the time of a dopamine spike. The STDP rule $\Psi(\Delta t)$ depends on the temporal difference between a post- and a pre-synaptic spike, $\Delta t = t_{post} - t_{pre}$, according to:

$$\Psi(\Delta t) = \begin{cases} A_+ e^{-|\Delta t|/\tau} & \text{if } \Delta t > 0 \\ -A_- e^{-|\Delta t|/\tau} & \text{if } \Delta t \leq 0 \end{cases} \tag{4}$$

where A_+ and A_- are the amplitudes of the weight change, and τ is a time constant. In all simulations $A_+ = 0.1$ and $A_- = 0.1$ for asymmetric STDP and $A_- = -0.1$ for symmetric STDP. The reward signal is sent to the dopaminergic neurons through MUSIC.

Measuring the Agent's Performance. The agent's navigation performance was measured using three parameters: First, the time from the start of a trial to when the agent finds the target, i.e., escape latency. This was our default measure due to its simplicity and common use in animal experiments [7,13]. Second, to capture cases of partial learning (the agent approaches, but does not reach, the target), we determined the minimum distance between the agent's trajectory and the target. However, there are differences between the trajectory of the agent and the ideal path that neither latency nor minimum distance can capture. Third, we therefore quantified the similarity between the agent's trajectory and the ideal path by using dynamic time warping (DTW),

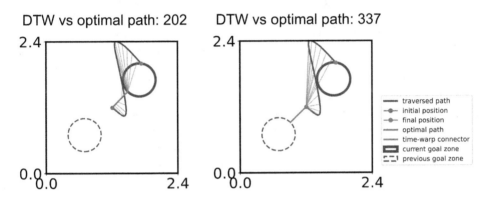

Fig. 2. Illustration of dynamic time warping (DTW). DTW quantifies the similarity between the agent's trajectory in a trial (blue curve) and the optimal path (red line), the shortest path from start to target. To study which target the agent is most likely aimed for, we compare DTW for navigation towards the current target (left) and the previous target (right). Gray lines represent the time-warp relationship. A path closer to the ideal has a lower DTW value. (Color figure online)

DTW is a technique for measuring the similarity between two sequences, which may differ in length, speed and/or timing [1]. Given two sequences $X = \{x_1, x_2, \ldots, x_n\}$ and $Y = \{y_1, y_2, \ldots, y_m\}$, DTW constructs an $n \times m$ distance matrix D, where each element $D[i, j]$ represents the distance between x_i and y_j, typically calculated using the Euclidean norm. The accumulated cost matrix C is computed as follows:

$$C[i, j] = D[i, j] + \min(C[i - 1, j], C[i, j - 1], C[i - 1, j - 1]) \quad (5)$$

This recurrence relation determines the minimal cumulative distance to align the subsequences $X[1 : i]$ and $Y[1 : j]$. The optimal warping path is the monotonic path from $[1, 1]$ to $[n, m]$, which always moves from one element of $C[i, j]$ to the neighboring one with the lowest value. The DTW distance between the two sequences is given by $C[n, m]$. Figure 2 demonstrates DTW between a sample trajectory and two optimal paths to targets A and B. The grey lines connect the points along the optimal warping path.

3 Results

3.1 Studying Behavioral Flexibility in a Learning Agent Based on Spiking Neural Networks

To investigate behavioral flexibility in reversal learning in spatial navigation tasks, we utilized the CoBeL-spike toolchain. As the reference agent, we chose parameters (symmetric STDP, that were most effective in the task with a single target $\sigma_{PC} = 0.2$ m and $N_{PC} = 21^2$) [7]. The agent successfully learned to efficiently navigate to target A after 10 trials (Fig. 3A). However, in the second phase of the ABA reversal learning task, the agent consistently returned to target A and failed to find target B. Performance improved again in the third phase.

The learning curves based on the average escape latency revealed that these features of spatial learning are consistent across simulations (Fig. 4A, top). The agent learned to navigate to target A within the first 10 trials, but the escape latency in the second phase increased to nearly 5 s, the time-out value. Good performance returned in the third phase, when the goal was target A again. To confirm that the agent persevered on target A in the second phase, we analyzed the agent's trajectories (Fig. 4A, middle and bottom). Proximity and DTW metrics with respect to the current target B are represented by solid green lines and to the former target A by dashed green lines. Indeed, the agent continues to seek target A in the second phase.

We hypothesized that the reference agent lacks behavioral flexibility for two main reasons. 1. Its behavior is focused on exploitation and neglects exploration. 2. Symmetric STDP can only potentiate synapses and a new behavior could only emerge, if the new association was stronger than the previous ones. However, the established associations prevent new behaviors from emerging (stability).

3.2 Network Features that Introduce Behavioral Flexibility

To test our hypotheses, we introduced an asymmetric STDP learning rule while the other parameters remained identical to the reference agent. This agent was indeed more flexible, learning target B (Fig. 3B) and performing much better than the reference agent in the second phase (Fig. 4B). However, this flexibility comes at a cost—during the first phase, the agent's performance is reduced compared to the reference agent (Fig. 4A,B).

We confirmed this pattern across varying place cell parameters for symmetric (Fig. 5A) and asymmetric (Fig. 5B) STDP rules. We scaled the field size, the PC number, or both simultaneously. In all scenarios, agents with symmetric STDP performed better on target A, while those with asymmetric STDP reached target B mostly independently of place cell parameters. The lower performance is the result of more variable behavior, as is evident in the example trajectory (Fig. 3B), the higher average latency and higher DTW, but the minimum distance is not worse (Fig. 4B). Hence, the behavioral flexibility afforded by the asymmetric STDP comes at the cost of more variable behavior.

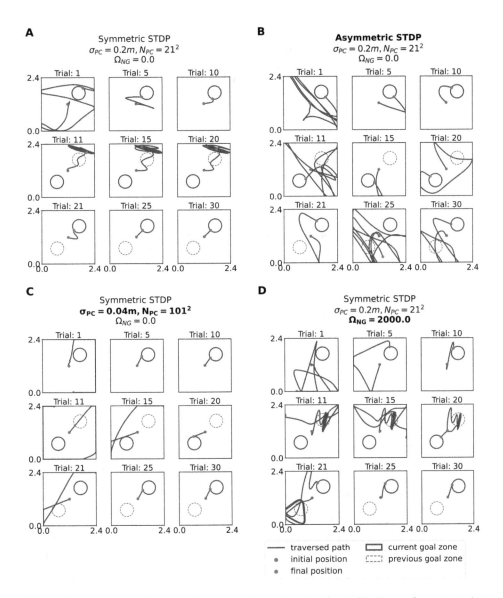

Fig. 3. Behavior in simulated reversal learning task. A–D: Example trajectories from four simulations, each comprising 30 trials split into three phases (rows). Within each row, plots illustrate trials 1, 5, and 10. In each trial, the agent starts at the center of the environment, seeks, and might find the target (solid purple circle). The target is located in the top-right corner (target A) in the first and third phases and in the bottom-left corner (target B) in the second phase. A dashed circle marks a former target location. **A:** Symmetric STDP. **B:** Asymmetric STDP. **C:** Symmetric STDP with many smaller fields. **D:** Symmetric STDP with noise pulses injected to the action neurons. (Color figure online)

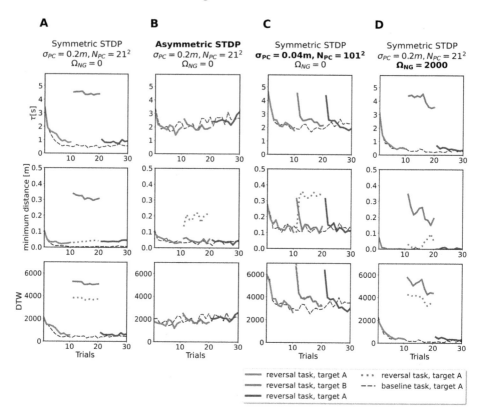

Fig. 4. Learning curves during reversal learning. A–D: Each panel illustrates different measures of learning for three phases of a reversal learning task (represented by red, green, and blue curves, respectively) compared to a baseline task with single target (dashed black line). Results were averaged over 50 repetitions. Top: Escape latency. Middle: Minimum distance to the current target (solid lines) or former target (dashed green line). Bottom: Dynamic-Time-Warping (DTW). Additionally, DTW measure is presented for target A (dashed green line) in the second phase, when the goal was target B (solid green line). **A:** Symmetric STDP. The agent learns target A quickly during the first phase, but fails to explore and learn target B after the target switch. **B:** Asymmetric STDP. The agent learns the location of the target A in the first phase and target B in the second. **C:** Symmetric STDP with many smaller fields. In each phase, the agent adapts to a new target and unlearns the previous one. **D:** Symmetric STDP combined with noise injections. The agent learns target B in the second phase. (Color figure online)

Close inspection revealed that using symmetric STDP, the agent performed well on target B, particularly when there are many and small place fields (e.g. $N_{PC} = 101^2$ and $\sigma_{PC} = 0.04\,m$, Fig. 5A, bottom). We hypothesized that in this case, only a subset of place cells are involved in the learning of target A, hence leaving some place cells with unpotentiated synapses to learn target B (Fig. 3C). However, the agent performed more poorly at target A compared to the baseline

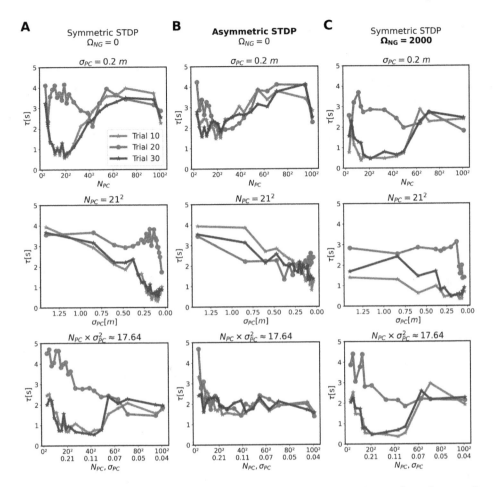

Fig. 5. Dependence of the reversal learning on plasticity rule, place cell parameters and external noise: Each panel shows the average escape latency across 30 simulations on the last trial in each phase (red: first phase, green: second phase, blue: third phase) as a function of some place cell parameters for **A**: symmetric STDP, **B**: asymmetric STDP, and **C**: symmetric STDP with noise. Each row represents different scaling of place cell parameters. Top: Scaling place cell number N_{PC}, while keeping field size fixed $\sigma_{PC} = 0.2$ m. Middle: Scaling σ_{PC}, while keeping N_{PC} fixed. Bottom: Simultaneous scaling of both variables, such that $N_{PC} \times \sigma_{PC}^2 = 17.6$. (Color figure online)

agent on all three learning measures (Fig. 4C). Even more dramatically, learning target B in the second phase interfered with previously learned behavior and the agent had to relearn target A in the third phase as if it was naïve. This indicates that the behavioral flexibility, in this case, comes at the cost of destabilizing the already learned information.

3.3 Driving Behavioral Flexibility Through an External Signal

Another alternative is using an external supervisor to detect when the agent has not been rewarded for too long and inject noise activity into the network ($\Omega_{NG} = 2000$, see Methods) to disrupt repetitive loops and promote exploration (Fig. 3D, middle row). In the third phase, the agent initially gets stuck at target B, but escapes this loop to find target A again. Comparing learning curves with and without noise injection shows that noise enhanced performance in the second phase, however, reversal learning (target B) is not as consistent as the initial learning (target A) (Fig. 4A and D). Notably, noise injection improved performance in the first and third phases, suggesting that noise helps the agent find more efficient solutions, similar to the ε-greedy in reinforcement learning.

4 Discussion

We modeled reversal learning in spatial learning using spiking neural networks in a closed-loop simulation with reward-driven STDP. An agent with symmetric STDP and optimized place cell properties excels at learning a specific target, but struggles to explore, unlearn the first target, and learn new targets. The agent's behavior is dominated by exploitation and stability at the expense of exploration and plasticity. We studied three potential solutions to encourage exploration and reduce behavioral stability, but each comes with its own costs.

There is an analogy to attractor dynamics that can help interpret our results. The reference agent persevering on the first target it found is like an attractor capturing the agent's behavior. Destabilizing the attractor with asymmetric STDP or making it shallower with many small place fields can encourage exploration. Alternatively, injecting noise can help the agent escape the attractor. Each mechanism has drawbacks: destabilized attractors lead to low performance, shallower attractors result in slower convergence and faster forgetting, and injecting noise leads to a variable success rate during reversal learning. Since none of the methods are optimal, we hypothesize that the brain uses one or a combination of these mechanisms depending on the current condition or tasks, and there is no universal mechanism to maintain behavioral flexibility.

The closest simulation frameworks to ours are those our model is based on [3]. While that paper touched on goal relocation, it primarily focused on sequential neuromodulation of synaptic plasticity rules in learning navigation tasks, not on reversal learning. Another framework simulates closed-loop behavior using reinforcement learning but emphasizes the role of context in ABA extinction learning [23] and is less biologically plausible compared to ours.

In conclusion, our modeling suggests three methods that may contribute to behavioral flexibility. This provides experimental neuroscientists with clues for designing experiments to study behavioral flexibility and for interpreting results relating reversal learning behavior to neural recordings.

Acknowledgement. This work was supported by the Deutsche Forschungsgemeinschaft (DFG, German Research Foundation) – project number 316803389 – through

SFB 1280, projects A14. We thank Amany Omar, Paul Adler and Raymond Black for help with the figures and the CoBeL-spike software framework.

Code Availability. The CoBeL-spike framework is freely available at https://github.com/sencheng/CoBeL-spike.

References

1. Al-Naymat, G., Chawla, S., Taheri, J.: SparseDTW: A Novel Approach to Speed up Dynamic Time Warping (2012). https://doi.org/10.48550/arXiv.1201.2969
2. Brockman, G., et al.: OpenAI Gym. arXiv arXiv:1606.01540 (2016). https://doi.org/10.48550/arXiv.1606.01540
3. Brzosko, Z., Zannone, S., Schultz, W., Clopath, C., Paulsen, O.: Sequential neuromodulation of Hebbian plasticity offers mechanism for effective reward-based navigation. eLife **6**, e27756 (2017). https://doi.org/10.7554/eLife.27756
4. Diekmann, N., Vijayabaskaran, S., Zeng, X., Kappel, D., Menezes, M.C., Cheng, S.: CoBeL-RL: a neuroscience-oriented simulation framework for complex behavior and learning. Front. Neuroinform. **17**, 1134405 (2023). https://doi.org/10.3389/fninf.2023.1134405
5. Djurfeldt, M., et al.: Run-time interoperability between neuronal network simulators based on the MUSIC framework. Neuroinformatics **8**(1), 43–60 (2010). https://doi.org/10.1007/s12021-010-9064-z
6. Gewaltig, M.O., Diesmann, M.: NEST (NEural Simulation Tool). Scholarpedia **2**(4), 1430 (2007)
7. Ghazinouri, B., Nejad, M.M., Cheng, S.: Navigation and the efficiency of spatial coding: insights from closed-loop simulations. Brain Struct. Funct. **229**(3), 577–592 (2024). https://doi.org/10.1007/s00429-023-02637-8
8. Goodroe, S.C., Starnes, J., Brown, T.I.: The complex nature of hippocampal-striatal interactions in spatial navigation. Front. Hum. Neurosci. **12** (2018). https://doi.org/10.3389/fnhum.2018.00250
9. Hartley, T., Burgess, N., Lever, C., Cacucci, F., O'Keefe, J.: Modeling place fields in terms of the cortical inputs to the hippocampus. Hippocampus **10**(4), 369–379 (2000). https://doi.org/10.1002/1098-1063(2000)10:4<369::AID-HIPO3>3.0.CO;2-0
10. Hintjens, P.: ZeroMQ: Messaging for Many Applications. O'Reilly Media Inc., Sebastopol (2013)
11. Izhikevich, E.M.: Solving the distal reward problem through linkage of STDP and dopamine signaling. Cereb. Cortex **17**(10), 2443–2452 (2007). https://doi.org/10.1093/cercor/bhl152
12. Jordan, J., Weidel, P., Morrison, A.: A closed-loop toolchain for neural network simulations of learning autonomous agents. Front. Comput. Neurosci. **13**, 46 (2019). https://doi.org/10.3389/fncom.2019.00046
13. Morris, R.G.M.: Spatial localization does not require the presence of local cues. Learn. Motiv. **12**(2), 239–260 (1981). https://doi.org/10.1016/0023-9690(81)90020-5
14. Morrison, A., Diesmann, M., Gerstner, W.: Phenomenological models of synaptic plasticity based on spike timing. Biol. Cybern. **98**(6), 459–78 (2008). https://doi.org/10.1007/s00422-008-0233-1

15. O'Keefe, J., Dostrovsky, J.: The hippocampus as a spatial map. Preliminary evidence from unit activity in the freely-moving rat. Brain Res. **34**(1), 171–175 (1971). https://doi.org/10.1016/0006-8993(71)90358-1
16. Parra-Barrero, E., Vijayabaskaran, S., Seabrook, E., Wiskott, L., Cheng, S.: A map of spatial navigation for neuroscience. Neurosci. Biobehav. Rev. **152**, 105200 (2023). https://doi.org/10.1016/j.neubiorev.2023.105200
17. Potjans, W., Morrison, A., Diesmann, M.: Enabling functional neural circuit simulations with distributed computing of neuromodulated plasticity. Front. Comput. Neurosci. **4** (2010)
18. Rhee, M., Kim, T.: Exploration and exploitation. In: Augier, M., Teece, D.J. (eds.) The Palgrave Encyclopedia of Strategic Management, pp. 543–546. Palgrave Macmillan UK, London (2018). https://doi.org/10.1057/978-1-137-00772-8_388
19. Roberts, P.D., Bell, C.C.: Spike timing dependent synaptic plasticity in biological systems. Biol. Cybern. **87**(5), 392–403 (2002). https://doi.org/10.1007/s00422-002-0361-y
20. Taube, J.S., Muller, R.U., Ranck, J.B.: Head-direction cells recorded from the postsubiculum in freely moving rats. I. Description and quantitative analysis. J. Neurosci. **10**(2), 420–435 (1990)
21. Verbeke, P., Verguts, T.: Learning to synchronize: how biological agents can couple neural task modules for dealing with the stability-plasticity dilemma. PLoS Comput. Biol. **15**(8), e1006604 (2019). https://doi.org/10.1371/journal.pcbi.1006604
22. Walsh, C.M., Booth, V., Poe, G.R.: Spatial and reversal learning in the Morris water maze are largely resistant to six hours of REM sleep deprivation following training. Learn. Mem. **18**(7), 422–434 (2011). https://doi.org/10.1101/lm.2099011
23. Walther, T., et al.: Context-dependent extinction learning emerging from raw sensory inputs: a reinforcement learning approach. Sci. Rep. **11**(1), 2713 (2021). https://doi.org/10.1038/s41598-021-81157-z

A Behavior-Based Model of Foraging Nectarivorous Echolocating Bats

Thinh H. Nguyen[✉] and Dieter Vanderelst

University of Cincinnati, Cincinnati, USA
{nguye2t7,vanderdt}@mail.uc.edu

Abstract. We propose a simulation-based model of flower finding in echolocating nectarivorous bats. In particular, we propose a behavior-based model that uses two sensorimotor loops to dock with flowers. The EchoVr, as we have termed our echo simulator, uses a bank of echoes collected by ensonifying real objects with a physical (bat-like) sonar device. Using the EchoVr, we built a 2D environment consisting of simulated objects. We trained a neural network to activate the correct sensorimotor loop based on the echoes received by the simulated bat. The model guides the simulated bat to dock successfully with the flower opening (95% success rate) by computing control commands solely from echoic inputs.

Keywords: Nectarivorous bats · Echolocation · Sonar · Behavior-based

1 Introduction

More than 500 species in 67 plant families depend on bats as their primary pollinators [4]. Certain flower species have been shown experimentally to be approached by bats using sonar [5,7,8,18], and many others are believed to be located using echolocation. Several studies have investigated the acoustic cues bats use to approach flowers [8,16,18], but no sensorimotor model has been proposed or tested to explain how bats exploit these cues.

Previously, we argued [13,14,22] that behavior-based control architectures [2,11] offer promising models to understand bat sonar-based behavior. These architectures assume minimal reliance on internal representations, which are challenging to infer from echo data. Using sensorimotor loops that directly connect sensory input to actions should allow bats to react effectively in real-time situations [1], regardless of environmental complexity.

In this paper, we set out to build a model of the sensorimotor behavior that guides echolocating bats to the opening of flowers while avoiding obstacles. In particular, we present a behavior-based model that uses separate sensorimotor loops to complete the task. Our model uses two sensorimotor loops: an approach loop and an avoid loop. The approach loop uses acoustic cues from echoes to guide the bat toward a target, while the avoid loop steers the bat away from potential obstacles. The simulated bat will learn to activate an appropriate loop on the currently received echoes.

© The Author(s), under exclusive license to Springer Nature Switzerland AG 2025
O. Brock and J. Krichmar (Eds.): SAB 2024, LNAI 14993, pp. 51–62, 2025.
https://doi.org/10.1007/978-3-031-71533-4_4

2 Methods

2.1 Targets and Arena

We set up a foraging task in a virtual arena for a simulated echolocating bat. The goal is for the bat to navigate to flower openings while avoiding collisions with flower sides and arena walls. In the current work, we reuse our previously developed method to simulate the echoes received by an echolocating bat, termed EchoVR, [14]. This simulator uses a bank of echoes collected by ensonifying real objects with a physical (bat-like) sonar device. The sonar device comprises a single ultrasonic emitter and two microphones embedded in 3D-printed bat pinnae.

The EchoVR allows one to build a 2D environment consisting of the ensonified objects and calculates the binaural echoes for arbitrary positions and orientations of the simulated bat in the environment. Note that by feeding the EchoVR a large set of echoes collected from real objects, the environment also incorporates sources of noise, i.e., the noise inherent in the physical devices used to collect the echo data. For example, the EchoVR models variations in the echo that occur even if the same object is ensonified by the bat from the same position and orientation. As the signal-to-noise ratio of the physical ensonfication devices is lower than that of the bat's sonar system, the noise produced by the EchoVR can be seen as a worst-case scenario when modeling bat echolocation [3].

The EchoVR takes into account the head-related transfer function of the bat, as its echo database was collected using microphones embedded in 3D printed pinnae of the bat *Micronycteris microtis* [20]). It also accounts for the acoustic phenomena of spreading and atmospherical attenuation that weaken the echoes as distance increases and interference between echoes. The EchoVr also includes a model of the auditory periphery of bats [24]. This model converts simulated echoes to a representation that is a proxy of the bat's cochlear activation pattern. This model returns a low-passed, logarithmically compressed envelope of the received echoes (an example is shown in Fig. 1). Therefore, the cochlear model reduces the temporal resolution of the echoes but accentuates weaker echoes. More details on the EchoVR can be found in [14].

For the present work, we ensonified two objects: a plant and a cardboard pole. These two objects vary greatly in geometric complexity. The plant consists of many stochastically oriented surfaces that result in complex echoes, typical of vegetation, e.g., [23,27]. In contrast, the cardboard pole has a basic geometry, resulting in a simple echo. The echoes of the two objects are depicted in detail in [14].

The bat is modeled as emitting narrowband calls with a frequency of 42 kHz and a duration of about 2 ms. These parameters are determined by the ensonification device used to collect the physical echoes from the pole and the plant, as detailed in [14]. Furthermore, we model the bat as emitting 40 calls per second. This emission rate is biologically plausible and corresponds to the approximate call rate used by Nectarivorous bats approaching targets [5,6].

Without access to the echoes of real flowers visited by bats, we used the two objects to construct target flowers for the simulated bat. In doing so, we attempted to create a target to test the hypothesis that a behavior-based model can be trained to approach a complex target (returning complex echoes) from a range of appropriate angles. We constructed a proxy for a flower composed of one plant and two poles. The poles are spaced at a distance of 30 cm from the center of the plant, creating an opening with an angular span of 60°. The opening of the flower defines the successful docking zone. This arrangement recreates the typical bilateral symmetry of flowers. The diameter of 30 cm of the proxy flower is larger than for the real flowers visited by nectarivorous bats [16]. However, by making the flower larger, we compensate for our simulated bat's (current) lack of spectral information (see Discussion for details).

The arena walls, which the bat should avoid, are made of plants arranged in a rectangle. Figure 1 shows the arrangement of the flower and the arena. Note that this arrangement of the flower results in considerably complex echoes (see Fig. 1c for an example). The plant is a complex reflector in itself. The two poles add two more strong reflectors to this. Furthermore, the relative strength and time of arrival of the echoes of the plants and poles are determined by the position and orientation of the bat in the arena.

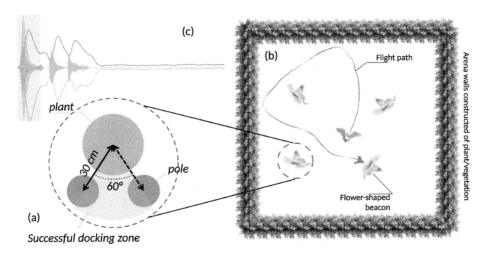

Fig. 1. (a) The arrangement of the proxy for a flower used in this paper. The arrangement consists of a central plant and two symmetrically placed cardboard poles. (b) The arena (16 × 16 m²) in which we test our simulated bat. The walls of the arena consist of plants. (c) An example of echoes received by the bat from the flower in the left (orange) and right (blue) ears. The envelopes extracted using the cochlear model are also depicted (the envelope for the left ear is plotted inverted for clarity). The highlighted part of the waveforms is the bat's emission. These are followed by a complex echo (consisting of multiple copies of the emitted pulse). (Color figure online)

2.2 Sensorimotor Loops

In this paper, we propose that bats can find flowers using only two sensorimotor loops: an approach loop and an avoid sensorimotor loop. By activating the correct loop at each point in time, the bat can approach targets (i.e., the flower's docking zone) or avoid objects (i.e., the walls and parts of the flower outside the allowed docking zone). In the following, we introduce these control loops. Next, we explain how we trained the model to activate the correct loop based on the current echoes.

Approach Loop. The approach sensorimotor loop uses two acoustic cues extracted from the echoes: the delay in the onset of the echo t_o and the Interaural Level Difference (ILD) ΔI. In the following, we explain how these cues are used to set the linear velocity ν and the angular velocity ω when the approach sensorimotor loop is active.

Following each call generated by the bat, the echo onset delay t_o is extracted from the simulated echoes by determining the time at which the response of the cochlear model crosses a threshold. Therefore, the delay in the echo t_o gives the delay between the generation of the call and the arrival of the first (sufficiently strong) echo. The echo onset delay t_o is related to the distance d_o of the object that returns the echo as $t_o = 2d_o/c$ with c the speed of sound in air (here set to 343 m/s).

The ILD ΔI is obtained by integrating the response of the cochlear model for the left and right ears, starting at t_o for the duration of the echo (i.e., until the response drops below a threshold). This results in the echo intensity I_L and I_R for the left and right ears, respectively. This assumes that the auditory system of a bat functions approximately as an energy integrator [25]. Next, the ILD ΔI is calculated as follows,

$$\Delta I = 10 \times \log_{10} \frac{I_L}{I_R} \quad (1)$$

A positive (negative) ILD indicates that the echo reaching the left (right) ear is louder. Indeed, the ILD inherently contains azimuthal position information about the object from which the echo is returned, e.g., [21].

After extracting the ILD ΔI and the onset delay t_o, we use these parameters to calculate the control commands, that is, the linear velocity ν and the angular velocity ω. The equation (derived from tau-based control principles [10]) for the linear velocity ν is,

$$\nu = (\nu_{\max} - \nu_{\min})[1 - (1 - K \times A \times (d_o))^{\frac{1}{K}-1}] \quad (2)$$

with K set to 0.1 and A representing the braking coefficient, determining the rate at which the slowing occurs. We set $\nu_{\min} = 1$ m/s, $\nu_{\max} = 4$ m/s, and $A = 1$ m/s^2 as conservative flight speeds and acceleration for nectarivorous bats [26]. Equation 2 results in a non-linear deacceleration of the bat as it approaches the target. After calculating the linear velocity, we calculate the bat's angular velocity as follows,

$$\omega = \nu/R_{\text{trn}} \quad (3)$$

with R_{trn}, the bat's turn radius, given as,

$$R_{trn} = -\text{sign}(\Delta I)\left(\log(|\Delta I|) - R_{min}\right) \quad (4)$$

Eqs. 3 and 4 allow the bat to steer towards a target. The bat must minimize the difference in loudness between the left and right ears to approach the source of the returning echo. In other words, ΔI should be regulated to zero. When $\Delta I > 0$ ($\Delta I < 0$), the echo source is left (right) of the bat and the turning radius R_{trn} should be set to a positive (negative) value to turn left (right). Moreover, when the magnitude of the ILD ($|\Delta I|$) is large, the source of the echo is further away from the central azimuth, and the bat must make a sharper turn. In contrast, when the magnitude of the ILD ($|\Delta I|$) is smaller, the bat will make a smaller correction turn to gradually guide itself towards the source of the returning echo. When $|\Delta I| = 0$, the turning radius R_{trn} is infinite and the bat flies straight. In Eq. 4, R_{min} is the minimum allowed turning radius, which is determined based on the bat's current velocity ν,

$$R_{min} = \nu^2/a \quad (5)$$

to allow us to restrict the angular velocity of the bat to a biologically realistic range. Holderied [9] reported that the flight speed determines the smallest turning radius of different species of bats. In particular, he suggested that bats' turning radii are limited such that the g-force they experience is (most often) below about 3g. Hence, in Eq. 5, we set $a = 3$.

In summary, the approach sensorimotor loop controls the angular and linear velocity of the simulated bat. It turns the bat towards the source of the echo and reduces its speed as it approaches the target. The angular and linear velocities and the acceleration are limited to biologically plausible values, resulting in realistic dynamics.

Avoid Loop. When activated, the avoid sensorimotor loop steers the bat away from obstacles. In this section, we describe how this loop is implemented. Similarly to the approach sensorimotor loop, the avoid sensorimotor loop uses the onset distance d_o and ILD ΔI to set the linear velocity ν and the angular velocity ω when activated. When using the avoid sensorimotor loop, the linear velocity ν is calculated in the same way as when using the approach loop.

The main difference between the approach and avoid loops lies in how the angular velocity is set. Equation 6 sets the turning radius R_{trn} when the avoid loop is activated. The turning radius depends on the distance to the object that returns the echo (and not on the magnitude of the ILD as in the approach loop). If the distance d_o is smaller, the bat is more likely to collide with the object and should perform a sharper turn (that is, employ a smaller turning radius). Equation 6 implements this logic.

$$R_{trn} = \text{sign}(\Delta I)\left[\log\left(1 - \frac{d_o}{d_{0,max}}\right) - R_{min}\right] \quad (6)$$

In Eq. 6, $d_{0,\max}$ is the maximium detection delay (distance), here set to 3 m. The turn radius is converted to angular velocity using Eq. 3. Moreover, the minimum turning radius $R_{\min}$ is, as for the approach loop, set using Eq. 5.

In summary, the avoid loop sets the bat's linear velocity in the same way as the approach loop. Nearer to obstacles, the linear velocity is reduced but kept within biologically realistic bounds. The angular velocity is set so that closer to obstacles the bat takes sharper turns.

2.3 Training the Bat

Having specified the sensorimotor loops, we now explain how we train a neural network to select the correct loop to activate based on current echoes. Since the network takes the current binaural echoes (after cochlear processing) as input and returns a selected loop to be activated, the network converts the echoes into suitable motor commands by activating the right loop.

To train the neural network, we need to collect echoes from the arena and label each echo with the suitable sensorimotor loop to be activated. We design a *teacher model* to allow us to automate the echo labeling task. Although training can be set up in a reinforcement learning framework (see [12, 14] for our previous work using reinforcement learning to train artificial echolocators), the current approach allows us to avoid the complexity of using reinforcement learning by leveraging heuristics about the problem. In the following, we explain the functioning of the *teacher model* before describing how the *teacher model* was used to train the neural network.

Teacher Model. The *teacher model* takes the bat's current pose and the spatial arrangement of the walls and flowers to determine the appropriate sensorimotor loop for this spatial arrangement. This sets the *teacher model* apart from the final neural network that controls the bat. The bat's control neural network will only receive the echoes as input and is not provided with any information about the bat's current pose and spatial arrangement of objects.

The *teacher model* operates as follows. First, it assesses the proximity of the bat to the arena walls. If the bat is within 1.5 m of a wall and is oriented toward it, the *teacher model* returns the avoid loop as the appropriate loop to activate. If the bat is not within 1.5 m of a wall or is not oriented toward it, the *teacher model* lists all flowers in the arena in order of their distance from the bat. Next, the *teacher model* assesses the relative position between the bat and the closest flower. It feeds this relative pose to a Support Vector Machine classifier (SVM). This classifier tries to predict the outcome of activating the approach loop for a given relative pose between a bat and a flower. More details on training the SVM classifier will be discussed in the following section.

If the SVM classifier suggests that the approach loop would lead to a *dock* (i.e., a successful outcome), the *teacher model* returns the approach sensorimotor loop as appropriate to activate. If the SVM predicts that the approach loop would result in a *hit* with the non-opening side of the flower, the avoid sensorimotor loop is selected. When activating the approach loop is not predicted to result in contact with the flower (*dock* or *hit*), the SVM classifies the outcome at that pose to be *miss*. If the classifier returns a *miss* output, the *teacher model* proceeds to evaluate the next flower, farther away. The avoid loop is returned if no *dock* is predicted after the *teacher model* has evaluated the four closest flowers.

Support Vector Machine Classifier. As mentioned in the previous section, the *teacher model* used to scaffold the training of the neural network controlling the bat contains an SVM that suggests which sensorimotor loop to use based on the pose of the bat and the configuration of the arena. In this section, we explain how this SVM was trained.

To gather training data, we performed a simulation in which a single flower is placed in the center of the arena, and a bat is spawned in a random pose within a range of 10 m around the flower. The bat uses the approach sensorimotor loop until one of three outcomes occurs: (i) the bat successfully arrives at the proxy flower opening, (ii) the bat collides with the flower outside the docking zone, or (iii) the bat moves beyond the 10-m radius around the circle. We refer to these outcomes as *dock*, *hit*, and *miss*, respectively.

Running 200,000 trials starting from randomized poses, we observed that 5,659 poses result in *docking*, 29,421 poses result in *hitting*, and the remaining 164,920 poses result in *miss*. The poses leading to the three outcomes are shown in Fig. 2. This figure shows that most poses pointing toward the flower opening guide the bat to the flower opening (panel a).

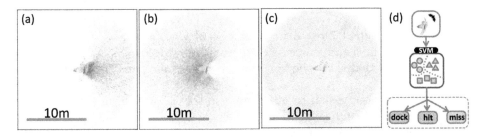

Fig. 2. Results of 200,000 trials in which the bat ran the approach sensorimotor loop starting from random poses with respect to the flower. Each data point in panels a-c represents a different pose of the bat (location and orientation). (a) Poses that lead to *docking*. (b) Poses that lead to *hits*. (c) Poses that lead to *misses*. The poses in panels a-c are used to train an SVM classifier (d) to predict the outcome if the approach loop is used at a given pose.

Using the data generated using the 200,000 runs, we fitted an SVM capable of classifying the current pose as one of the three outcomes based on the echoes received at that pose. Due to a disproportionally large number of *miss* poses, we impose a limit of 100,000 *miss* poses within our data set. This results to a data

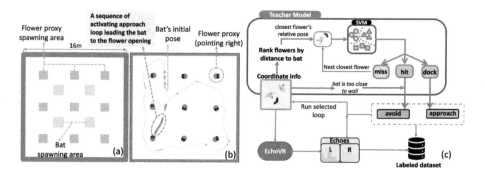

Fig. 3. (a) schematic layout of the environment used to train the neural network. The arena measured 16 × 16 m and contained 9 zones in which a flower was placed (the flower's center was jittered) with a random orientation. The bat spawned in one of four areas with a random orientation. (a) One example run used to collect the training data for the neural network. (c) *teacher model* architecture and how it is used to collect a data set of echoes labeled with sensorimotor loops.

set consisting of 135,080 poses with 4% *dock* poses, 22% *hit* poses, and 74% *miss* poses.

The SVM (Figs. 2d and 3c) was trained on these data using a regularization parameter $C = 0.1$ and a kernel of the radial basis function (RBF), using a kernel coefficient $\gamma = 10$. We adopted an 80/20 train-test split for evaluating the model's performance. After training, the SVM classifier could correctly classify 91% of poses as leading to *dock*, *hit*, or *miss* when using the approach loop.

Training the Neural Network. In this section, we explain how the *teacher model* presented in the previous sections was used to train the bat's controller: a neural network that takes the binaural echoes and activates one of two sensorimotor loops.

We created a setup featuring nine flowers in a 16 × 16 m arena surrounded by walls consisting of plants (Fig. 3). The flowers are placed in nine square regions. The bat spawns in one of four regions. At each step of the simulation, the *teacher model* is used to select the approach or the avoid sensorimotor loop. Figure 3b shows one run that ends with the bat docking with a flower. Using the *teacher model*, we collected 5000 runs ending in the bat docking with the flower. For each step of each run, we recorded the binaural echoes as received by the bat and the sensorimotor loop suggested by the *teacher model*. Therefore, we constructed a large data set of echoes labeled with the sensorimotor actions suggested by the *teacher model*. The data set consisted of 2,156,782 binaural echoes. Approximately 22% of these were associated with the approach loop. The remaining 78% were associated with the avoid loop. We partitioned the data set into an 80:20 train-test split. This data set was used to train a neural network that directly converted the echoes to a selected sensorimotor loop.

We used a fully connected neural network model that features six hidden layers with sizes of 128, 256, 256, 128, 64, and 16, each using ReLU activation. The output layer has a softmax function applied over one-hot vector the size of 2 corresponding to the two sensorimotor loops. Training is done using the Adam optimizer with a learning rate of 5×10^{-5} and weight decay of 5×10^{-5}. The network input layer receives the echo envelopes. The output layer consists of two units that assess the probability associated with each sensorimotor loop for a given echoic input. We trained the model for 200 epochs using the Adam optimizer and a mini-batch size of 2048. The model converges to 84% accuracy.

3 Results

We evaluated the bat's performance controlled by the neural network for 1000 episodes, during which the simulated bat autonomously determined the sensorimotor loop to execute at each step solely based on the echoic input from the scene. It is important to note that while the *teacher model* had access to the bat's relative position to the flowers and the walls, the neural network only used the echoes as input. An evaluation episode is run for 2000 steps (50 s at 40 calls per second) (*miss* outcome), or until the bat arrives at a flower opening (*dock* outcome) or collides with a wall or the side of a flower (*dock* outcome). We compared our 'trained bat' with a 'baseline bat' that randomly selected a new sensorimotor loop at each simulation step.

The trained bat using the neural network for sensorimotor loop selection outperforms the 'baseline bat' selecting sensorimotor loops randomly (Fig. 4g). The trained model yields a *dock* rate of 93%. On average, the trained model takes around 600 steps (or 15 s at 40 calls per second) to collide with an object (either *dock* or *miss*), which is approximately 2.5 times longer than the duration that the random model takes (see Fig. 4h). We show examples of each outcome in Fig. 4 a-f.

Dock outcomes occur when the bat recognizes it is in a favorable position (based on the echoes) to approach the flower opening and engage the approach loop. In both Figs. 4a and 4b, the bat turns on the approach loop when facing the opening of flower *C* to *dock*. In contrast, the bat tends to turn away if it comes close to a non-opening side of the flower (as depicted with flower *B* in Fig. 4a). When the bat is not facing any particular object, it consistently opts for the avoid loop, as observed at the beginning of the episode in Fig. 4b.

The most common scenario resulting in a *hit* outcome is when the bat becomes trapped in the corner of the arena (see Fig. 4c). Other occurrences of the *hit* outcome, happening at a rate of 0.5%, are when the bat collides with a non-opening part of a flower (see Fig. 4d). As the bat attempts to navigate the arena while looking for the front of a flower, there are instances where it *bounces around* until the time limit expires. One of these *miss* episodes is illustrated in Figs. 4e.

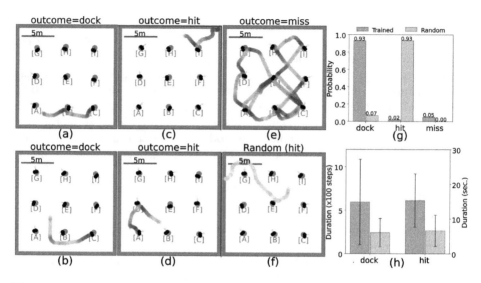

Fig. 4. Examples of runs where the bat is guided by the trained neural network, resulting in outcomes categorized as *dock* (a-b), *hit* (c-d), and *miss* (e). The network chooses between the avoid and approach sensorimotor loop at each step based on the echoes. Avoidance steps are shown in shades of blue, while approach steps are shown in shades of red, with darker shades indicating higher confidence. An example where the bat is guided by a random policy is shown in (f). (g) The likelihood of each outcome for the trained neural network and the random policy. (h) The duration of each outcome for both the trained and random models.

4 Discussion

We developed a behavior-based reactive navigation model trained to guide a bat in foraging flowers using narrowband echolocation calls as input. The model demonstrates high success rates in this task, approaching flowers closely and veering away if the bat nears the non-opening side of a flower while continuing to approach if facing the flower opening. This strategy effectively prevents collisions with walls or the non-opening sides of flowers. However, a notable inefficiency arises: when the bat veers away from the non-opening side of a flower, it misses the opportunity to investigate the other side, resulting in the *miss* outcome. In the future, we will explore an additional sensorimotor loop that allows the bat to focus on a single flower target to improve efficiency.

Our model introduces several simplifications. The flowers were modeled as composite objects, and the bat operated only in 2D. Furthermore, the environment was simpler than the environment typically faced by nectarivorous bats. Nevertheless, the model demonstrates the feasibility of modeling the complex task of approaching flowers at the correct angle using sensorimotor loops activated under the correct conditions and justifies further expanding our work. We are working on a more elaborate model that includes more realistic flower echoes and a more sophisticated arbitration approach to select the appropriate sensorimotor loop. Future iterations of the model could also include more specific behavior observed in bats. For example, bats are capable of selecting flowers to approach based on their size [17] or morphology [7].

Previous explanations of foraging in nectarivorous bats suggested that bats use the spectral content of the echoes to find and approach flowers (see [7, 16–18] for example). Our current model does not use spectral cues, nor does it have access to them, as the echoes were narrowband signals. Instead, the current model solves the problem in the time domain using the time-intensity profiles of the echoes. To allow the use of temporal cues, our proxy flower had a diameter larger than is typical for flowers (here: 30 cm; see [19] for examples of real flowers). Hence, in our model, we trade spectral information for temporal information (see, for example, [15] for a discussion about the relationship between spectral and temporal information in echolocating bats). In the future, collecting broadband echoic data to populate the EchoVR would enable us to create a model that exploits the spectral cues presumably used by bats.

Acknowledgments. This work was funded by NSF grant 2034885, Principles of intelligent sensorimotor behavior under informational constraints.

References

1. Arkin, R.C.: Behavior-Based Robotics. Intelligent Robotics and Autonomous Agents, 3rd ed, 2000 edn.. The MIT Press (1998). http://gen.lib.rus.ec/book/index.php?md5=bdd0766feef0ca8774ea8cf6646add72
2. Brooks, R.A.: Cambrian Intelligence: The Early History of the New AI. MIT Press (1999)
3. Chitradurga Achutha, A., Peremans, H., Firzlaff, U., Vanderelst, D.: Efficient encoding of spectrotemporal information for bat echolocation. PLoS Comput. Biol. **17**(6), e1009052 (2021)
4. Fleming, T.H., Geiselman, C., Kress, W.J.: The evolution of bat pollination: a phylogenetic perspective. Ann. Bot. **104**(6), 1017–1043 (2009)
5. Gonzalez-Terrazas, T.P., et al.: How nectar-feeding bats localize their food: echolocation behavior of Leptonycteris yerbabuenae approaching cactus flowers. PLoS ONE **11**(9), 1–18 (2016). https://doi.org/10.1371/journal.pone.0163492
6. Gonzalez-Terrazas, T.P., Martel, C., Milet-Pinheiro, P., Ayasse, M., Kalko, E.K.V., Tschapka, M.: Finding flowers in the dark: nectar-feeding bats integrate olfaction and echolocation while foraging for nectar. Roy. Soc. Open Sci. **3**(8) (2016). https://doi.org/10.1098/rsos.160199
7. Helversen, D.V., Helversen, O.V.: Object recognition by echolocation: a nectarfeeding bat exploiting the flowers of a rain forest vine. J. Comparat. Physiol. Neuroethol. Sens. Neural Beh. Physiol. **189**(5), 327–36 (2003)
8. von Helversen, D., von Helversen, O.: Acoustic guide in bat-pollinated flower. Nature **398**(6730), 759–760 (1999). https://doi.org/10.1038/19648
9. Holderied, M.W.: Akustische Flugbahnverfolgung von Fledermäusen: Artvergleich des Verhaltens beim Suchflug und Richtcharakteristik der Schallabstrahlung. Ph.D. thesis, Friedrich-Alexander-Universität Erlangen-Nürnberg (2001)
10. Lee, D.N., Reddish, P.E., Rand, D.T.: Aerial docking by hummingbirds. Naturwissenschaften **78**(11), 526–527 (1991). https://doi.org/10.1007/BF01131406

11. Mataric, M., Michaud, F.: Behavior-based systems. In: Siciliano, B., Khatib, O. (eds.) Springer Handbook of Robotics, pp. 891–909. Springer, Heidelberg (2008). https://doi.org/10.1007/978-3-540-30301-5_39
12. Mohan, A.V., Vanderelst, D.: Training a simulated bat: Modeling sonar-based obstacle avoidance using deep-reinforcement learning. In: 2020 IEEE Symposium Series on Computational Intelligence (SSCI), pp. 2241–2249 (2020). https://doi.org/10.1109/SSCI47803.2020.9308555
13. Nguyen, T., Vanderelst, D., Peremans, H.: Sensorimotor behavior under informational constraints: a robotic model of prey localization in the bat micronycteris microtis. In: ALIFE 2021: The 2021 Conference on Artificial Life. MIT Press (2021)
14. Nguyen, T.H., Vanderelst, D.: Toward behavior-based models of bat echolocation. In: 2022 IEEE Symposium Series on Computational Intelligence (SSCI), pp. 1529–1536 (2022). https://doi.org/10.1109/SSCI51031.2022.10022100
15. Schmidt, S.: Perception of structured phantom targets in the echolocating bat, Megaderma lyra. J. Acoust. Soc. Am. **91**(4), 2203–2223 (1992). https://doi.org/10.1121/1.403654
16. Simon, R., et al.: Acoustic traits of bat-pollinated flowers compared to flowers of other pollination syndromes and their echo-based classification using convolutional neural networks. PLoS Comput. Biol. **17**(12), e1009706 (2021)
17. Simon, R., Holderied, M.W., von Helversen, O.: Size discrimination of hollow hemispheres by echolocation in a nectar feeding bat. J. Exp. Biol. **209**(18), 3599–3609 (2006)
18. Simon, R., Holderied, M.W., Koch, C.U., von Helversen, O.: Floral acoustics: conspicuous echoes of a dish-shaped leaf attract bat pollinators. Science **333**(6042), 631–633 (2011)
19. Simon, R., Rupitsch, S., Baumann, M., Wu, H., Peremans, H., Steckel, J.: Bioinspired sonar reflectors as guiding beacons for autonomous navigation. Proc. Nat. Acad. Sci. **117**(3), 1367–1374 (2020). https://doi.org/10.1073/pnas.1909890117
20. Vanderelst, D., De Mey, F., Peremans, H., Geipel, I., Kalko, E., Firzlaff, U.: What noseleaves do for FM bats depends on their degree of sensorial specialization. PLoS ONE **5**(8), e11893 (2010)
21. Vanderelst, D., Holderied, M.W., Peremans, H.: Sensorimotor model of obstacle avoidance in echolocating bats. PLoS Comput. Biol. **11**(10), e1004484 (2015)
22. Vanderelst, D., Peremans, H.: A computational model of mapping in echolocating bats. Anim. Behav. **131**, 73–88 (2017)
23. Vanderelst, D., Steckel, J., Boen, A., Peremans, H., Holderied, M.W.: Place recognition using batlike sonar. Elife **5**, e14188 (2016)
24. Wiegrebe, L.: An autocorrelation model of bat sonar. Biol. Cybern. **98**(6), 587–595 (2008)
25. Wiegrebe, L., Schmidt, S.: Temporal integration in the echolocating bat, megaderma lyra. Hear. Res. **102**(1), 35–42 (1996)
26. Winter, Y.: Flight speed and body mass of nectar-feeding bats (glossophaginae) during foraging. J. Exp. Biol. **202**(14), 1917–1930 (1999)
27. Yovel, Y., Franz, M.O., Stilz, P., Schnitzler, H.U.: Plant classification from bat-like echolocation signals. PLoS Comput. Biol. **4**(3) (2008). https://doi.org/10.1371/journal.pcbi.1000032

Benefit of Varying Navigation Strategies in Robot Teams

Seyed A. Mohaddesi[1(✉)], Mary Hegarty[2], Elizabeth R. Chrastil[1], and Jeffrey L. Krichmar[1]

[1] University of California Irvine, Irvine, CA 92697, USA
{smohadde,chrastil,jkrichma}@uci.edu
[2] University of California Santa Barbara, Santa Barbara, CA 93106, USA
hegarty@ucsb.edu

Abstract. Inspired by recent human studies, this paper investigates the benefits of employing varying navigation strategies in robot teams. We explore how mixed navigation strategies impact task completion time, environment exploration, and overall system effectiveness in multi-robot systems. Experiments were conducted in a simulated rectangular environment using Clearpath PR2 robots and evaluated different navigation strategies observed in humans: 1) Route (RT) knowledge where agents follow a predefined path, 2) Survey (SW) knowledge where agents take the shortest path while avoiding obstacles, 3) Mixed strategies with varying proportions, such as 40% RT and 60% SW (0.4RT 0.6SW) and 60% RT and 40% SW (0.6RT 0.4SW), and 4) An additional strategy where agents switch from RT to SW 10% of the time (0.9RT 0.1SW). While SW strategy is the most time-efficient, RT strategy covers more of the environment. Mixed strategies offer a balanced trade-off. These findings highlight the advantages of variability in navigation strategies, suggesting benefits in both biological and robotic populations. Additionally, we have observed that human participants in a similar study would start on a route, and then 10% of the time switch to survey. Therefore, we investigate a 90% Route 10% Survey (0.9RT 0.1SW) strategy for individual team members. While a pure Survey strategy is the most efficient regarding time taken and a pure Route strategy covers more of the environment, a mixture of strategies appears to be a beneficial tradeoff between time taken to complete a mission and area coverage. These results highlight the advantages of population variability, suggesting potential benefits in both biological and robotic populations.

Keywords: Navigation · Teams · Route · Survey · Multi-robot systems

---

This work was supported by the Air Force Office of Scientific Research (AFOSR) Contract No. FA9550-19-1- 0306, and by the National Science Foundation (NSF-FO award ID IIS-2024633.

© The Author(s), under exclusive license to Springer Nature Switzerland AG 2025
O. Brock and J. Krichmar (Eds.): SAB 2024, LNAI 14993, pp. 63–77, 2025.
https://doi.org/10.1007/978-3-031-71533-4_5

1 Introduction

Multi-robot navigation plays a vital role in advancing robotic technology, providing a dynamic solution to complex tasks that exceed the capabilities of individual robots. Its significance lies in enhancing efficiency, collaboration, and adaptability across various domains, including manufacturing, search and rescue operations, and autonomous vehicles. Coordinated movement among multiple robots enables them to collectively navigate intricate environments, share information, and optimize routes, addressing challenges that a single robot might find overwhelming.

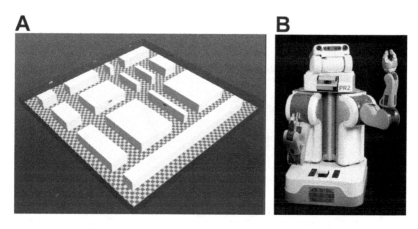

Fig. 1. (**A**) Overall view of our maze in the Webots environment (**B**) Clearpath's PR2 robot

Research suggests that people employ different types of knowledge to navigate [2]. Survey knowledge contains metric information that includes distances and directions between locations. This knowledge enables flexible path planning resulting in shortcuts or planned trajectories over never-experienced paths. In contrast, route knowledge consists of sequences of actions associated with places or decision points. Typically, the routes are fixed paths and inflexible.

Boone and colleagues [1] investigated the knowledge people use during navigation in the Dual Solutions Paradigm. In the DSP, participants follow a fixed loop around a virtual environment that has several landmarks along the way. After several laps, they are tested by placing participants at a landmark and telling them to go to another landmark. If they take the fixed loop, they are applying *route* knowledge. If they take a novel shortcut, they are applying *survey* knowledge. When told to take the shortest path, the proportion of participants applying survey knowledge increased. This suggests that many participants had survey knowledge, but they might find it easier to take a learned route.

In a metadata analysis, Krichmar and He showed that the variation observed in human navigation is both between and within subjects. This variability might be explained by taking the cost of traversing an environment into consideration [8]. They found that when told to find a goal, roughly 60% of participants used a route strategy and 40% used a survey strategy. However when told to take the shortest path to a goal, those proportions were reversed (40% route, 60% survey). Furthermore, they found that subjects starting on a route switched to a survey 10% of the time.

In the present paper, we simulate teams of robots to test whether there are advantages to using human-inspired strategies for navigating (Fig. 1). Our present work makes the following contributions:

- Simulating human variation has advantages for robot navigation, and possibly for planning algorithms in self-driving vehicles and robotic swarms.
- Incorporating a simple navigation strategy inspired by human subjects, rather than finding an optimal solution, makes multi-agent systems easier to scale.
- A mix of route and survey strategies leads to a tradeoff between time to find goals and more exploration of an environment that may be advantageous for biological and robotic populations.
- In human studies, it is technically challenging to monitor multiple participants navigating at the same time. Simulating a human-sized environment with navigating robots can overcome this limitation.

To better assess how these findings transfer to the real-world, we used a physical robot simulation with human sized robots and local sensing. In the following sections, background and methods are described in further detail.

2 Related Work

Several studies have explored the benefits of varying navigation strategies in teams of robots, addressing various aspects of multi-agent path planning and cooperative behavior. Most of the related work falls into two categories: 1) Multi-Agent Path Finding (MAPF), and 2) Heterogeneous swarm navigation.

2.1 Multi-Agent Path Finding (MAPF)

In MAPF, the goal is to plan collision-free paths for many agents to reduce mission duration and maximize team productivity. An example is the Conflict Based Search (CBS) algorithm which addresses the challenge of finding optimal paths in multi-agent scenarios by considering conflicts among agents and employing efficient exploration techniques [19]. Another example is the branch-and-cut-and-price (BCP) algorithm that incorporates a shortest path pricing problem for finding paths for every agent independently and constraints for resolving conflicts [9,10]. Both BCP and CBS are optimal but because of computational complexity, they don't scale well to a large number of agents [12]. Sub 1.5 MAPF algorithm on grids, optimizing time with a 1.5x longest-to-shortest path ratio constraint. This is a sub-optimal solution but it scales better [5]. Other algorithms scale well, by applying simple movement rules but are far from optimal [17,21]. MAPF-LNS is a hybrid approach that first creates a fast planning solution and then optimizes with a heuristic called large neighborhood search [13]. In the above cases, the solution is applied once for all agents. But a more realistic situation is a warehouse in which robots need to plan efficient, collision-free paths for long duration. The Rolling-Horizon Collision Resolution (RHCR) approach addressed the issue of lifelong multi-agent pathfinding in large-scale warehouses by applying their MAPF algorithm over different window periods [13].

This paper acknowledges the inherent limitations often found in human decision-making processes, in contrast to an emphasis on optimal solutions. While conflict resolution among robots remains a necessary aspect, it does not stand as the central objective within this paper's scope. However, future versions could take advantage of conflict resolution policies in the work discussed above. Unlike most MAPF algorithms, a centralized planner is absent in this work. Instead, each robot independently charts its path, driven by individual objectives, and limited knowledge of the other robot's state and intention.

2.2 Heterogeneous Swarm Navigation

Swarm robotics typically assumes that agents can communicate or interact with others in the vicinity. Heterogeneous teams of agents have been used to solve a wide range of problems [4,6,20]. In a heterogeneous robotic swarm, certain tasks can be solved efficiently through cooperation and functional specialization [3]. The agents or robots have different shapes or capabilities (e.g., different sensors or different locomotion). Unlike MAPF, the planning is decentralized. In a typical navigating swarm, there might be leaders and followers and the task is to find a goal while circumventing obstacles and preventing collisions [14,18]. Because of the locality requirement, the swarm often resembles a flock of birds.

Although there are similarities to our approach, these swarms by definition stay together. Furthermore, the majority of the prior work in swarm navigation assumes a 'bird's eye' view, rather than local sensing that would be required in many field operations. We are interested in heterogeneous foraging, where the robots are independent, decentralized and use different strategies.

3 Methods

3.1 Path Planning Algorithms

We simulated the Dual Solutions Paradigm (DSP) developed to study human behavior with teams of 1, 3, and 5 robots navigating the environment given in Fig. 2 which had 14 landmarks. To simulate survey knowledge, any path planner that finds the shortest path between a start and destination would be sufficient [11]. The present paper uses a spiking wavefront propagation algorithm, which has been described in [7], to calculate the survey paths. The spikewave algorithm takes into consideration the cost of traversal, which is encoded as a

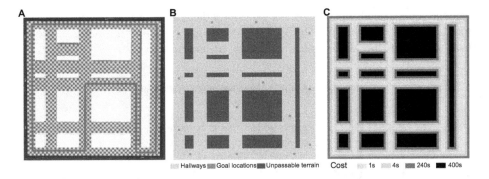

Fig. 2. (**A**) A bird's eye view of the map, The blue line indicates the specified route (**B**) Grid map of the environment. Purple cells show impassable terrain locations, goal locations are indicated in red, and beige cells indicate the floor of the aisles. (**C**) Heatmap depicting traversal delay costs measured in seconds for each cell on the map. The delay increases as cells approach the walls. This map serves as a universal guide utilized by all robots.

propagation delay. Difficult to traverse regions have long delays and impassable areas have very long delays (Fig. 2(c)). It is important to note that every path generated by spikewave is a sequence of straight line segments, meaning that there are no curves in generated paths. The yaw needed for the robot to face towards and move forward along the line segment is then determined. In order to simulate route knowledge, the robots followed a fixed path that took them past all the goal locations (Fig. 2(a)). Specifically, route knowledge was simulated with spikewave by giving the robot closely spaced waypoints to ensure it stayed on the specified route. As described below, the robots could either use route knowledge, survey knowledge or a mixture of these to find the goal locations.

3.2 Simulation Environment

In this experiment, a 3D simulated environment was created using Webots [15], an open-source robot simulator. The environment contains a square-shaped maze, which is made of 4096(64*64) cells each representing a square meter, as shown in Fig. 1(a). Up to 5 simulated PR2 robots, provided in the Webots environment, were used to explore the maze (Fig. 1(b)). The PR2 robot is a mobile robot with advanced sensors and software that can perform various tasks such as navigation and perception with high accuracy and precision. Its unique design features a mobile base with two wheels and a caster and a robotic arm with seven degrees of freedom. The maze size was scaled such that the robot fit within a cell.

3.3 Experimental Control and Design

The experimental design followed the DSP [1], but with multiple agents simultaneously navigating the environment. Based on the team's size, each robot is assigned a task involving a subset of randomly designated goals chosen randomly from the 14 landmarks (Fig. 2(b)). These goals are distributed among the robot team to ensure that each goal is visited by at least one robot.

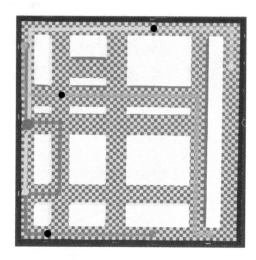

Fig. 3. Example of 3 Robots navigating through the environment, Black spots indicate the current location of each robot. Bold colored lines show the path each robot has taken till now, and pale colored lines indicate the path planned. The blue circle is the conflict spot between the yellow and red robots (Color figure online)

To conduct the simulation, several challenges need to be addressed. The first was to develop a subroutine that could handle movement between targets. This controller had to be capable of calculating a route between two targets that were free of obstructions and could be traversed by the robots. For the route strategy, an unobstructed path was provided to the robot as in the Dual Solution Paradigm used in human studies [1]. For the survey strategy, the spikewave algorithm uses the map with traversal delays to find the optimal path (Fig. 2(c)) as in [7]. An obstacle or environment boundary would have a long delay that would slow the wave propagation, and a traversable path would have a short delay that results in the wave finding a short, obstacle-free path.

Due to the requirement of simulating physical robots in the present work, several challenges needed to be addressed. One challenge is that sometimes agents come so close to colliding with a wall. The PR2's base Lidar, with its scanning range of 270°C, was used to measure the distance to the wall. If this length is less than 60cm, a subroutine is triggered to take the robot back to its original path. The spikewave algorithm then computes a new path to the goal. Another challenge was that a rule had to be devised and implemented to handle bottleneck conflicts and determine which robot has the right of way. Figure 3 is an example of the paths of 3 robots. For the purposes of this experiment, an ID was assigned to each robot (varying from 1 to 5). Two robots are in conflict when the distance between them is less than 6 m. When conflicts arise (Fig. 4(a)), the robot with the higher ID pauses, allowing the other robot to move a distance of at least 12 m away. This distance was empirically determined to effectively reduce the potential for subsequent collisions involving the PR2 robots. In cases where our defined route strategy is applicable, prohibiting rerouting, conflict resolution favors the robot that holds a positional lead on the route. Finally, since the problem of navigation strategies is similar to real-life scenarios, a single management unit could not be used to handle all conflicts between the robots. Instead, all decisions to control the disputes had to be either through predetermined rules or message passing among the agents. When a collision occurs, a signal is sent to all other robots within 25 m of the collision point. The robots that receive this signal increase the delay of the cells around the collision point on the delay matrix. The delay is increased by 500 s for a 3×3 square block around the collision point. Each agent then runs the spikewave algorithm to reroute to the temporary goal using the new matrix. This helps robots avoid the collision point and prevents a third robot from interfering with two conflicting robots (Fig. 4(b)).

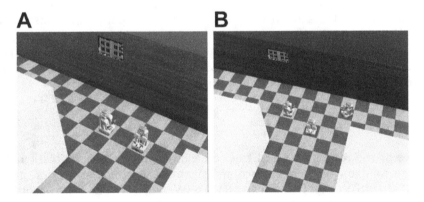

Fig. 4. Types of robot collisions (**A**) Collision of two robots. (**B**) Collision of three robots.

In order to automate these routines we designed a standalone controller with real-time path planning and online conflict resolution for all the robots. The pseudo-code that demonstrates how this controller works is provided (see Supplementary Algorithm 1). All agents used the same C++ implementation of this pseudo-code in their controllers.

To evaluate and compare different navigation strategies, we used these metrics:

- Time taken in seconds (s) till all goals have been visited.
- Total number of collisions between robots.
- Overall area occupied by all robots as a percentage of the entire environment (%).

– Overall intersection of the occupied area between robots as a percentage of the entire environment (%).

We ran 5 trials with 1, 3 or 5 robots in the environment using these strategies:

– Route (RT): Familiar route strategy. All the paths to the targets have to be aligned to a pre-specified route.
– Survey (SW): The shortest, obstacle-free path to the target is calculated using the spikewave (SW) algorithm.
– Mixed (0.9RT 0.1SW): Agents start on a route and 10% of the time the paths they switch to the spikewave algorithm at the halfway point.
– Mixed (0.6RT 0.4SW): 40% of the time the paths to the targets are calculated using the spikewave algorithm and 40% of the time it has to go along the route.
– Mixed (0.4RT 0.6SW): 60% of the time (i.e., 3 of the 5 trials) the paths to the targets are calculated using spikewave algorithm and 40% of the time (i.e., 2 out of the 5 trials) it has to go along the route.

The 3 mixed strategies represent variability observed within and between individuals. The percentages for the mixed strategies were derived from a DSP meta-analysis [8] that showed when asked to go to a goal participants roughly used SW 40% and RT 60% and when told to take the shortest path participants roughly used SW 60% and RT 40%. The (0.9RT 0.1SW) strategy was derived from the observation that when starting on a route, 10% of the time participants switch to a survey strategy halfway towards their goal.

4 Results

We present the results obtained from a comprehensive set of simulations that aimed to compare the performance of 5 distinct navigation strategies, namely RT, 0.9RT 0.1SW, 0.6RT 0.4SW, 0.4RT 0.6SW, and SW. The experiments were conducted using different team sizes of 1, 3, and 5 robots, allowing us to investigate the impact of team size and strategies on navigation performance. To ensure the reliability of our findings, each combination of strategy and robot number was subjected to five independent trials. Various metrics were measured throughout the trials, encompassing the time taken, the number of collisions, the area covered (occupancy), and team intersection. The metrics obtained from the trials were then averaged across the five trials for each experimental condition.

Time taken, defined as the duration in seconds from the start of the simulation until the visit of the last remaining goal, served as a crucial performance indicator. The SW strategy had a significantly shorter time to complete the task across all tested robot numbers (Fig. 5(A)). Conversely, the RT strategy exhibited the longest average time. This suggests that robots employing the SW strategy were able to devise innovative shortcuts, enabling them to complete the task more swiftly. For the complete statistical comparisons, see Supplementary Table 1. Additionally, we observed that increasing the number of robots led to

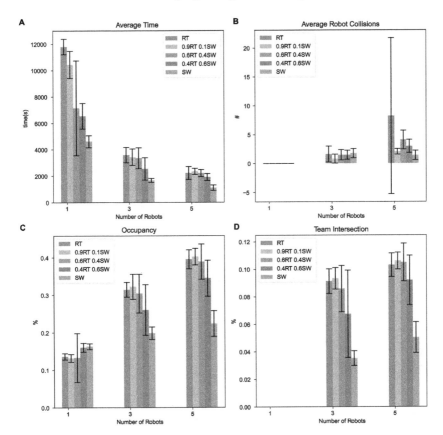

Fig. 5. Results of experiments for 1, 3, and 5 robots employing 5 different strategies, Route (RT), Survey (SW), and the mixed strategies (0.9RT 0.1SW, 0.6RT 0.4SW, 0.4RT 0.6SW). These values were measured and averaged over 5 trials. (**A**) Time to completion. Cumulative time taken to accomplish all goals (**B**) Robot collisions (**C**) Occupancy, which is the proportion of the environment covered by the robot teams (**D**) Intersection, which quantifies the percentage of cells visited by two or more robots

a significant reduction in the average time taken (see Supplementary Table 8 for statistical comparisons). This outcome can be attributed to the distribution of the goal locations among multiple robots, thereby enhancing efficiency.

We examined the metric of robot collisions, which denotes the total number of encounters between two robots. Notably, for a single robot scenario, this value is zero. The RT strategy exhibited the highest average number of robot collisions when tested with five robots (Fig. 5(B)). Although this may be due to an RT trial with 5 robots in which many collisions occurred, Increasing the number of robots tended to increase the number of collisions. It appeared that mixed strategies and the SW strategy had fewer collisions. However, it is hard to compare due to high variability from trial to trial (see Supplementary Tables 3 and 4 for statistical comparisons).

We analyzed the metric of occupancy, which reflects the extent of the map area covered by the robots. As anticipated, an increase in the number of robots resulted in significantly more area coverage (Fig. 5(C)) for all strategies except for SW when comparing teams of 3 to teams of 5 (see Supplementary Table 6). The RT strategy and mixed strategies covered more area than SW (Supplementary Table 5).

Finally, we evaluated the metric of team intersection, which quantifies the extent of intersected areas between robots on the map. Note that there is no intersection with only one robot. As the navigation strategies went from pure RT to mixed strategies to pure SW, the intersected areas between robots diminished significantly (Fig. 5(D)). Conversely, an increase in the number of robots resulted in a significant rise in this metric (see Supplementary Tables 7 and 8 for statistical comparisons).

Taken together, these results suggest that there may be a sweet spot where varying navigation strategies, like 0.9RT 0.1SW, 0.6RT 0.4SW and 0.4RT 0.6SW, can lead to shorter search times than a pure RT strategy and more coverage of the environment than a pure SW strategy. Moreover, increased intersections, as seen with RT and mixed strategies, may facilitate team communication, which is especially important in situations where lines of communication are down (e.g., disaster zones).

5 Discussion

The main finding of the paper was that although survey knowledge was the most time-efficient strategy and route-based strategies explored most regions of the environment, mixtures that varied strategies struck a balance between being efficient and covering more of the environment. As the number of robots increases tasks are completed faster, and the area covered expanded. The mixed strategies were a sweet spot of efficient navigation, area coverage, and collisions as the team size increased. Robot teams that had varied strategies completed the task of finding all the goal locations in less time than a pure route strategy while visiting more of the environment than a pure survey strategy. This heterogeneous strategy increases efficiency while gaining more knowledge compared to a homogeneous team strategy. In human populations, and potentially in robot teams, the choice of strategy depends on the context. In conditions where efficiency is critical, shortcuts that take more cognitive processing should be favored [8]. In contrast, a familiar route alleviates cognitive load and leads to more interactions with others.

Our results have significant implications for the design of heterogeneous swarms. Rather than varying the agent's sensing or locomotion capabilities [14,18], the agents used a mixture of strategies, which demonstrated clear population benefits. Unlike many multi-agent studies, the present study required to incorporate the physics of the environment with an accurate simulation of commercially available robots [15].These computationally intensive simulations limited team size. In the future, it would be of interest to test the benefits of varying strategies with larger, physical robot teams, which might be possible in an environment like the Robotarium [16].

In summary, the present results explain why human populations vary their navigation strategies and demonstrate that this variation is beneficial. By employing a physical robot simulation that used realistic local sensing and spacing of physical robots, we can better assess how the these results could transfer to the real world.

6 Supplementary Materials

6.1 Algorithms

Algorithm 1. Agent's controller pseudo-code

1: **procedure** CONTROLLER(self, robots, source, destination, strategy, costMap):
2:   **while** self → timeStep() **do**
3:     d[self] ← ∞
4:     newCM ← costMap
5:     **for** agent: all agents except self **do:**
6:       d[agent] ← agent's distance from self
7:     **end for**
8:     [r,i] ← [min(d), argmin(d)]
9:     checkForObstructionSignal(&newCM)   ▷ *checks if other robots have sent any obstruction signal, newCm was passed as reference to apply changes accordingly*
10:    **if** r ≤ 6 **then**
11:      collision[self]++
12:      sendObstructionSignalTo(robots, 25)   ▷ *Sends a signal to other robots within 25 meters of the conflict point to update their costMap subsequently*
13:      **if** self → id ≤ i **then**
14:        newCM ← newCM + block(3, robots[i])   ▷ *block(size,robot) makes an all-zero 64*64 matrix except for a size * size square block centering at the location of robot with infinity value, used size=3 as an arbitrary value to avoid further collisions between these two robots*
15:      **else** self→ waitFor(robot[i],12)   ▷ *waitFor would stop this agent until its distance to robot[i] is at least 12 and then restores the original costMap for both robots*
16:      **end if**
17:    **end if**
18:    path ← spikeWave (source, destination, strategy, newCM)   ▷ *Shortest path from source to destination obeying the strategy*
19:    navigate(path)   ▷ *Navigate functions as a helper method that ensures traversal along a given path*
20: **end while**
21: **end procedure**

6.2 Statistical Comparisons

All statistical comparisons and p-values were generated using the two-sample Kolmogorov-Smirnov goodness-of-fit hypothesis test. Bonferroni corrections based on the number of comparisons were used to test for significance level (Table 2).

Table 1. Time to Completion. Comparing effects of strategy within same robot team. ** denotes $p < 0.01$ and * $p < 0.05$ after Bonferroni correction.

Strategy	1 Robot	3 Robots	5 Robots
RT vs. RT60%,SW40%	0.0038*	0.8899	0.9999
RT vs. RT40%,SW60%	0.0038*	0.0515	0.237
RT vs. SW	0.0038*	0.00001**	0.00001**
RT60%, SW40% vs. RT40%, SW60%	0.209	0.1359	0.237
RT60%, SW40% vs. SW	0.0361	0.00001**	0.00001**
RT40%, SW60% vs SW	0.0348	0.00001**	0.00001**

Table 2. Time to Completion. Comparing effects of strategy between different robot team sizes. ** denotes $p < 0.01$ and * $p < 0.05$ after Bonferroni correction.

Strategy	1 vs. Robots	1 vs. 5 Robots	3 vs. 5 Robots
RT	0.0003**	0.0001**	0.0006**
RT60%, SW40%	0.0066*	0.0037*	0.0098*
RT40%, SW60%	0.0003**	0.0001**	0.0006**
SW	0.0003**	0.0001**	0.00001**

Table 3. Collisions. Comparing effects of strategy within same robot team. ** denotes $p < 0.01$ and * $p < 0.05$ after Bonferroni correction.

Strategy	3 Robots	5 Robots
RT vs. RT60%, SW40%	0.9983	0.237
RT vs. RT40%, SW60%	0.8899	0.4141
RT vs. SW	0.9983	0.237
RT60%, SW40% vs. RT40%, SW60%	0.9999	0.4141
RT60%, SW40% vs. SW	0.9983	0.0038*
RT40%, SW60% vs SW	0.9983	0.0038*

Table 4. Collisions. Comparing effects of strategy between different robot team sizes. ** denotes $p < 0.01$ and * $p < 0.05$ after Bonferroni correction.

Strategy	3 vs. 5 Robots
RT	0.5898
RT60%, SW40%	0.0904
RT40%, SW60%	0.0363
SW	0.8634

Table 5. Occupancy. Comparing effects of strategy within same robot team. ** denotes $p < 0.01$ and * $p < 0.05$ after Bonferroni correction.

Strategy	1 Robot	3 Robots	5 Robots
RT vs. RT60%, SW40%	0.209	0.6974	0.9996
RT vs. RT40%, SW60%	0.0361	0.209	0.209
RT vs. SW	0.0038*	0.0038*	0.0038*
RT60%, SW40% vs. RT40%, SW60%	0.9996	0.6974	0.209
RT60%, SW40% vs. SW	0.6974	0.0038*	0.0038*
RT40%, SW60% vs SW	0.6974	0.209	0.0361

Table 6. Occupancy. Comparing effects of strategy between different robot team sizes. ** denotes $p < 0.01$ and * $p < 0.05$ after Bonferroni correction.

Strategy	1 vs. 3 Robots	1 vs. 5 Robots	3 vs. 5 Robots
RT	0.0038*	0.0038*	0.0038*
RT60%, SW40%	0.0038*	0.0038*	0.0361
RT40%, SW60%	0.0361	0.0038*	0.0361
SW	0.0038*	0.0361	0.209

Table 7. Intersection. Comparing effects of strategy within same robot team. ** denotes $p < 0.01$ and * $p < 0.05$ after Bonferroni correction.

Strategy	3 Robots	5 Robots
RT vs. RT60%, SW40%	0.1359	0.0258
RT vs. RT40%, SW60%	0.1359	0.0001**
RT vs. SW	0.00001**	0.00001**
RT60%, SW40% vs. RT40%, SW60%	0.1359	0.0001**
RT60%, SW40% vs. SW	0.00001**	0.00001**
RT40%, SW60% vs SW	0.0047*	0.00001**

Table 8. Intersection. Comparing effects of strategy between different robot team sizes. ** denotes $p < 0.01$ and * $p < 0.05$ after Bonferroni correction.

Strategy	3 vs. 5 Robots
RT	0.0012**
RT60%, SW40%	0.0012**
RT40%, SW60%	0.00001**
SW	0.00001**

References

1. Boone, A.P., Maghen, B., Hegarty, M.: Instructions matter: individual differences in navigation strategy and ability. Mem. Cognit. **47**(7), 1401–1414 (2019). https://doi.org/10.3758/s13421-019-00941-5
2. Chrastil, E.R., Warren, W.H.: Active and passive spatial learning in human navigation: acquisition of graph knowledge. J. Exp. Psychol. Learn. Mem. Cogn. **41**(4), 1162–1178 (2015). https://doi.org/10.1037/xlm0000082
3. Dorigo, M., et al.: Swarmanoid: a novel concept for the study of heterogeneous robotic swarms. IEEE Robot. Automation Magaz. **20**(4), 60–71 (2013). https://doi.org/10.1109/MRA.2013.2252996
4. Engelbrecht, A.P.: Heterogeneous particle swarm optimization. In: Dorigo, M., Birattari, M., Di Caro, G.A., Doursat, R., Engelbrecht, A.P., Floreano, D., Gambardella, L.M., Groß, R., Şahin, E., Sayama, H., Stützle, T. (eds.) ANTS 2010. LNCS, vol. 6234, pp. 191–202. Springer, Heidelberg (2010). https://doi.org/10.1007/978-3-642-15461-4_17
5. Guo, T., Yu, J.: Sub-1.5 time-optimal multi-robot path planning on grids in polynomial time. CoRR abs/2201.08976 (2022)
6. Hara, A., Shiraga, K., Takahama, T.: Heterogeneous particle swarm optimization including predator-prey relationship. In: The 6th International Conference on Soft Computing and Intelligent Systems, and The 13th International Symposium on Advanced Intelligence Systems, pp. 1368–1373 (2012). https://doi.org/10.1109/SCIS-ISIS.2012.6505194
7. Hwu, T., Wang, A.Y., Oros, N., Krichmar, J.L.: Adaptive robot path planning using a spiking neuron algorithm with axonal delays. IEEE Trans. Cognitive Dev. Syst. **10**(2), 126–137 (2018). https://doi.org/10.1109/Tcds.2017.2655539
8. Krichmar, J.L., He, C.: Importance of path planning variability: a simulation study. Top. Cogn. Sci. **15**(1), 139–162 (2023). https://doi.org/10.1111/tops.12568
9. Lam, E., Le Bodic, P.: New valid inequalities in branch-and-cut-and-price for multi-agent path finding. In: Proceedings of the International Conference on Automated Planning and Scheduling 30(1), pp. 184–192 (2020). https://doi.org/10.1609/icaps.v30i1.6660
10. Lam, E., Le Bodic, P., Harabor, D., Stuckey, P.J.: Branch-and-cut-and-price for multi-agent path finding. Comput. Oper. Res. **144**, 105809 (2022). https://doi.org/10.1016/j.cor.2022.105809
11. LaValle, S.M.: Motion planning part i: the essentials. IEEE Robot. Autom. Mag. **18**(1), 79–89 (2011). https://doi.org/10.1109/Mra.2011.940276

12. Li, J., Chen, Z., Harabor, D., Stuckey, P.J., Koenig, S.: Anytime multi-agent path finding via large neighborhood search. In: Zhou, Z.H. (ed.) Proceedings of the Thirtieth International Joint Conference on Artificial Intelligence, IJCAI-21, pp. 4127–4135. International Joint Conferences on Artificial Intelligence Organization (8 2021). https://doi.org/10.24963/ijcai.2021/568, main Track
13. Li, J., Tinka, A., Kiesel, S., Durham, J.W., Kumar, T.K.S., Koenig, S.: Lifelong multi-agent path finding in large-scale warehouses. CoRR abs/2005.07371 (2020)
14. Maeda, R., Endo, T., Matsuno, F.: Decentralized navigation for heterogeneous swarm robots with limited field of view. IEEE Robot. Automation Lett. **2**(2), 904–911 (2017). https://doi.org/10.1109/LRA.2017.2654549
15. Michel, O.: Webots: professional mobile robot simulation. J. Adv. Robot. Syst. **1**(1), 39–42 (2004)
16. Pickem, D., et al.: The robotarium: a remotely accessible swarm robotics research testbed. In: 2017 IEEE International Conference on Robotics and Automation (ICRA), pp. 1699–1706 (2017). https://doi.org/10.1109/ICRA.2017.7989200
17. Sajid, Q., Luna, R., Bekris, K.E.: Multi-agent pathfinding with simultaneous execution of single-agent primitives. In: Fifth Annual Symposium on Combinatorial Search (2012)
18. Sano, Y., Endo, T., Shibuya, T., Matsuno, F.: Decentralized navigation and collision avoidance for robotic swarm with heterogeneous abilities. Adv. Robot. **37**(1–2), 25–36 (2023). https://doi.org/10.1080/01691864.2022.2117996
19. Sharon, G., Stern, R., Felner, A., Sturtevant, N.R.: Conflict-based search for optimal multi-agent pathfinding. Artif. Intell. **219**, 40–66 (2015). https://doi.org/10.1016/j.artint.2014.11.006
20. Valle, Y.d., Venayagamoorthy, G.K., Mohagheghi, S., Hernandez, J., Harley, R.G.: Particle swarm optimization: basic concepts, variants and applications in power systems. IEEE Trans. Evol. Comput. **12**(2), 171–195 (2008). https://doi.org/10.1109/TEVC.2007.896686
21. Wang, K.H.C., Botea, A.: Mapp: a scalable multi-agent path planning algorithm with tractability and completeness guarantees. J. Artif. Int. Res. **42**(1), 55–90 (2011)

Biomimetic Robots

No-brainer: Morphological Computation Driven Adaptive Behavior in Soft Robots

Alican Mertan[✉][iD] and Nick Cheney[iD]

University of Vermont, Burlington, VT 05401, USA
{alican.mertan,ncheney}@uvm.edu

Abstract. It is prevalent in contemporary AI and robotics to separately postulate a brain modeled by neural networks and employ it to learn intelligent and adaptive behavior. While this method has worked very well for many types of tasks, it isn't the only type of intelligence that exists in nature. In this work, we study the ways in which intelligent behavior can be created without a separate and explicit brain for robot control, but rather solely as a result of the computation occurring within the physical body of a robot. Specifically, we show that adaptive and complex behavior can be created in voxel-based virtual soft robots by using simple reactive materials that actively change the shape of the robot, and thus its behavior, under different environmental cues. We demonstrate a proof of concept for the idea of closed-loop morphological computation, and show that in our implementation, it enables behavior mimicking logic gates, enabling us to demonstrate how such behaviors may be combined to build up more complex collective behaviors.

Keywords: Soft robotics · Adaptive behavior

1 Introduction and Background

Recent advances in artificial intelligence and machine learning have benefited greatly from the rise of modern deep learning systems, ultimately aimed at artificial general intelligence [22]. The coming-of-age of these artificial neural network systems includes a long history of bio-inspiration, dating back to Mcculloch and Pitts [26]. Yet the processes behind biological intelligence reach far beyond systems and processes confined to the brain of living organisms.

Our bias toward attributing intelligent behavior to the mind is far from new. Descartes' mind-body-dualism dates back to the 1600 s [13]. However, Bongard warns that "thinking about thinking is misleading" [4]. Moravec's paradox, the observation that reasoning takes less computational resources than sensory-motor skills (contrary to the expectations of experts), is an example of how intuition could fail in introspection on thinking [30]. Examples of complex, perhaps intelligent-seeming, behavior stemming from non-neural processes are abound in both engineered and natural systems. Perhaps the most notable example of a system built to mimic a complex behavior in simple and purely mechanical form is the passive walking robot, which gracefully walks down an incline

© The Author(s), under exclusive license to Springer Nature Switzerland AG 2025
O. Brock and J. Krichmar (Eds.): SAB 2024, LNAI 14993, pp. 81–92, 2025.
https://doi.org/10.1007/978-3-031-71533-4_6

plane simply due to the well-tuned structure of its joints and limbs interacting with the physics of its environment [27]. In natural systems, the growth and shape change towards stimuli in plants serve as an obvious example of non-neural behavior [20,38].

More generally, these phenomena – of the body's complex and "intelligent" behavior, with or without a brain – are widely studied within the fields of embodied cognition [39] and morphological computation [34]. The literature argues that the brain is not the only source of adaptive and complex behavior, and that the body "shapes the way we think" [33]. Even for humans, who naturally consider their brains as the source of their intelligent behavior, there are strong indications that the body plays an important role in intelligence, for example, natural language, something we consider one of the fundamental products of our intelligence, is deeply connected to our bodies [21,35].

Prior works have demonstrated the concept of morphological computation in physical bodies, designing or evolving a body plan which executes a given behavior with as little neural or control intelligence as possible [1,9,11,12,16,18,32], but these body plans tend to focus on finding an effective morphology for a single desired behavior. Conversely, very simple adaptive behaviors in neural circuits demonstrate minimally cognitive systems that adapt the behavior of a fixed body plan to environmental stimuli via extraordinary simple neural circuitry [2,7]. Bridging the gap between the two, prior work on brain-body co-optimization attempt to simultaneously evolve simple controllers and evolving robot bodies [5,17,19,23,36] or develop approaches to rapidly adapt a controller to an ever-evolving body plan [10,28,29]. But, to the best of our knowledge, prior work thus far has failed to demonstrate an example of adaptive real-time behavior stemming entirely from morphological computation (i.e. with no controller or neural circuitry).

Inspired by the diverse forms of complex behaviors in nature, here we study ways in which adaptive behavior can be created in an artificial agent without a separately postulated brain. Particularly, we evolve voxel-based soft robots [3] for the task of locomotion to study how morphology could be a source of adaptive behavior. Four stimuli-responsive materials change shape in response to environmental cues, resulting in a gross morphological change to the robot. Robots are optimized via an evolutionary algorithm to produce body-plans that adaptively change direction on-the-fly in response to combinations of these environmentally-driven morphology changes (Fig. 1), resulting in closed-loop robot behaviors with no controller or neural circuitry (Figs. 2, 3 and 4), that can be combined to create complex morphological logic gates (Fig. 5).

2 Methodology

Simulation: We run our experiments on the EvoGym simulator [3]. It is a 2D soft robot simulator where robots are represented by voxels that can actuate by changing their areas ($a \in [0.6 \times r, 1.6 \times r]$ where r is the resting area). Designing robots' bodies means placing voxels with varying materials in a grid layout and

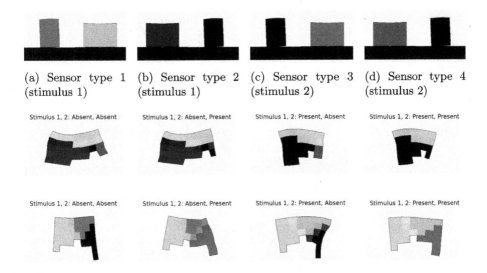

(a) Sensor type 1 (stimulus 1) (b) Sensor type 2 (stimulus 1) (c) Sensor type 3 (stimulus 2) (d) Sensor type 4 (stimulus 2)

(e) Example morphologies under different stimuli patterns. The behavior of sensory voxels molds the morphology of the robot, potentially changing its behavior.

Fig. 1. Images from the simulator environment [3] that shows the behavior of sensory voxels. Voxels in (a) and (b) respond to stimulus 1 and voxels in (c) and (d) respond to stimulus 2. Voxels in (a) and (c) shrink to their minimum width in response to the absence of their respective stimulus (left), and expand to their maximum width in response to the presence of their respective stimulus (right). Voxels in (b) and (d) respond the opposite way. These materials enable robots (e) that actively undergo shape change to produce new behaviors according to cues from some sensory stimuli.

specifying connections with their neighbors. In our experiments, we assume all neighboring voxels are connected to simplify the design space.

Environment: We use EvoGym's Python API to define a custom environment with two stimuli. These stimuli work in a binary fashion – they are either present or absent in the environment. For our robots to sense these stimuli and adapt their behavior, we hand-design "sensory voxels". We consider two environmental stimuli, each with one voxel type that expands in response to that stimuli and one that contracts in the presence of that stimuli – for a total of 4 sensory voxel types (Fig. 1; green, red, magenta, blue). We also hand-designed two active voxels (orange and teal) that horizontally expand and contract to provide energy for locomotion following a sinusoidal signal with a 180° phase offset, similar to the designs in [11], but do not provide any neural or informational connectivity between them to organize a coordinated controller by means other than physical forces traveling through the robot's body. Lastly, we provide evolution with two passive voxels, soft (gray) and rigid (black).

Evolutionary algorithms [14] are the most common approach to optimize robot body plans in the above-cited prior work, as they work by evaluating the empiric and holistic phenotype/behavioral effect of random mutations of a robot's artificial genome, selecting the most fit robots to survive and reproduce – and thus do not require a differentiable model of the robot's morphology for credit assignment of the robot's fitness to individual voxels. Here we employ the Map-Elites algorithm [31] to evolve an archive of diverse robots. Specifically, the archive consists of seven feature dimensions. Three of these dimensions are based on morphological features – the number of active voxels, sensory voxels, and total number of voxels. Following the common practice [11], we limit the morphology space with a $H \times W$ bounding box. Limiting the total number of voxels in this way allows us to discretize the three morphological dimensions of the archive into four bins each. The next four dimensions are based on the behavior of the robot, the direction of locomotion in the x-axis, under different stimuli patterns. The two binary stimuli result in four patterns, and a robot can move in either the positive x or the negative x direction, resulting in 16 possible behaviors in total. By utilizing QD algorithms and defining the dimensions of the archive in the described way, we allow evolution to search for all possible behaviors with varying morphological structures in a single run, and keep the best-performing robot for each unique combination of morphological-behavioral traits, via the MAP-Elites algorithm.

Evaluation and Selection: Individuals are evaluated under 5 simulated conditions. First, they are evaluated for 40 actuation cycles (i.e. 40 periods of the sinusoidal actuation signal) where at each 10th actuation cycle the environmental stimuli change as follows: (Stimulus 1, Stimulus 2) = {(Absent, Absent), (Absent, Present), (Present, Absent), (Present, Present)}. In the next 4 simulations, individuals are simulated under each of these 4 fixed stimuli patterns for 10 actuation cycles each. We use multi-objective selection to select robots that maximize total displacement in the 40-cycle variable stimuli simulation and maximize the minimum displacement across the 4 single stimuli 10-cycle simulations. Inspired from [25], an offspring replaces an individual from the archive, only if it dominates the individual, i.e. is better than the individual in both fitness values. In order to also select for robots with dynamically stable gaits that are not overly dependent on initial conditions, we disregard all robots that move in different directions for the same stimuli pattern in combined vs. single stimuli simulations.

Genome: Similar to [11,12], robot morphologies are represented by compositional pattern-producing networks (CPPNs) [37]. CPPN is a directed acyclic graph, where each node represents a math function. Here we use the sine, absolute value, square, and square-root as activation functions within each node. The CPPN takes the normalized coordinates of a voxel as well as its distance to the center of the bounding box as an input and outputs a scalar value for each possible voxel type, including the empty voxel. The type indicated by the max output value is assigned to the queried location. During evolution, offspring are created

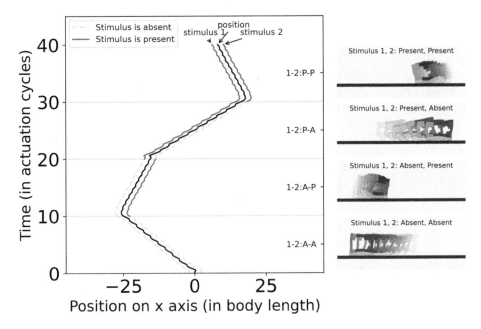

Fig. 2. The spacetime diagram of a run champion (left) and its gait under different stimuli patterns as snapshots from the simulation (right). In the spacetime diagram, the solid black line in the middle keeps track of the robot's center of mass over time. The two accompanying lines show the current stimuli pattern over time. The robot moves toward the positive x-direction when one stimulus is present, otherwise, it moves in the opposite direction. The different gait behaviors are achieved by the morphological changes caused by sensory voxels – an example of morphological computation and how the body could be the source of adaptive behavior. See more robots in action at our repository.

through mutation by the addition/removal of nodes/links, change of activation functions at nodes, and change of weights in edges.

Parameters: We experiment with three different bounding boxes, $(H, W) \in \{(5, 5), (7, 7), (10, 10)\}$, and repeat each experiment 10 times. Evolution was run for 3000 generations where each generation 16 individuals were chosen as parents from the archive uniformly randomly to create 16 offspring through mutation. Code to repeat our experiments is available at github.com/mertan-a/no-brainer.

3 Responding to Binary Stimuli

Firstly we investigate whether purely morphological changes can result in behavior that adapts to environmental stimuli. Over the 4 stimuli patterns presented to the robot, our MAP-Elites archive collects different combinations of behavioral patterns that different morphologies produce (labeled for whether a robot moves (L)eft or (R)ight for each of the 4 stimuli). Figure 2 visualizes the behavior of a robot exhibiting LRRL behavior as a spacetime diagram to investigate how

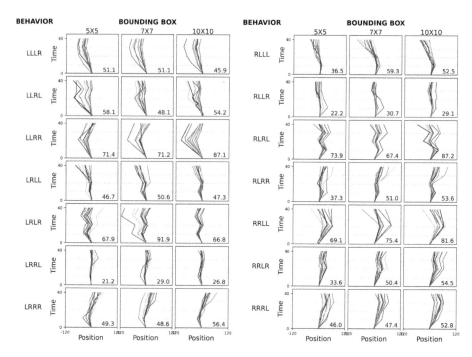

Fig. 3. The spacetime diagrams of all run champions from independent runs, plotted together. We omit the behaviors where the robot doesn't respond to any stimulus (LLLL and RRRR behaviors). Time is normalized to reflect actuation cycles and position is normalized to reflect the robot's body length. The numbers in the bottom right corner show the average distance traveled by run champions.

the morphology gives rise to adaptive behavior. The solid line tracks the position of the robot's center of mass during simulation. The two lines accompanying the solid line display the stimuli pattern at each time step. The dotted horizontal line shows the moments when the stimuli pattern changes. The robot's morphology and its gait under different stimuli can be seen on the right. The robot moves to the right only if one of the two stimuli is present (the middle two settings). In the case where both stimuli are present or absent (the first and last setting), the robot moves to the left. This behavior is achieved with the help of sensory voxels – the change in the sensory voxels alters the robot's shape and thus changes its gait. The robot exhibits an adaptive behavior without the help of a dedicated brain. Reactive materials are enough to unlock morphological computation.

Figure 3 displays spacetime diagrams of all run champions successfully exhibiting different adaptive behaviors. Each run champion is drawn in a different color and the average normalized space traversed by run champions in each plot can be seen in the bottom right corner. We see that given adequate environmentally-sensitive building blocks to work with, evolution can find morphologies that exhibit various adaptive behaviors – responding to stimuli in different ways via shape change alone (without a learned and/or stimuli-aware controller).

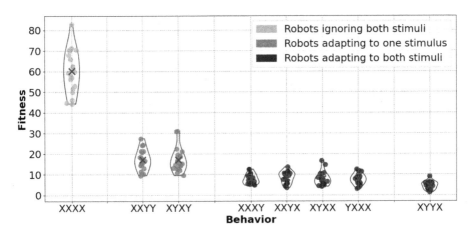

Fig. 4. Performances of run champions from 10 independent runs with a 10×10 bounding box. Symmetric robot behaviors are grouped by the pattern of their movement direction across stimuli (e.g. LRLR and RLRL are both represented as XYXY). Colors represent the number of stimuli a robot's morphology adapts to. Each data point is plotted and mean values are marked with an 'x'. Behaviors are separated with empty columns based on statistically significant performance differences (at $P < 0.001$) – each distribution in a group is statistically significantly better compared to each other distribution belonging to any group to their right. Bounding boxes of size 5×5 and 7×7 are qualitatively consistent.

To analyze the effect of different behavior patterns, we group the run champions from independent runs based on the directions of their movement trajectories across the 4 stimuli. For this analysis, we group behaviors that are symmetric in the x-axis (e.g. LLLL and RRRR are grouped as XXXX, LLRR and RRLL are grouped as XXYY, etc.) as the challenge of achieving symmetric behaviors is equal. As can be seen in Fig. 4, robots ignoring both stimuli and moving in the same direction throughout (behavior XXXX; mustard color) are able to locomote further compared to robots that exhibit a sensory-dependent behavior. Similarly, robots that respond only to a single stimulus (behavior patterns XXYY and XYXY; light blue color) perform better compared to robots adapting their behavior following both stimuli (green). As expected, there is no statistically significant difference between following the first stimulus (XXYY) or the second (XYXY). Adapting behavior only when one of the two stimuli is present but not both (behavior XYYX; right-most green) is harder compared to all other behaviors that use both stimuli (other green). Interestingly, this behavior (XYYX) is analogous to XOR and XNOR binary boolean functions, which are the only ones that are not linearly separable. All comparisons are statistically significant at the level of $P < 0.001$ and hold for all experimented bounding boxes.

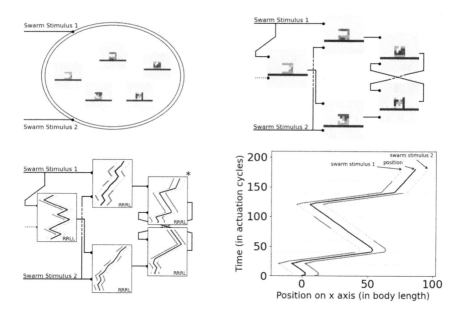

Fig. 5. Top row High-level schematic of the swarm (left) and its inner wiring (right). Lines connecting to robots from the left represent stimuli and the ones on the right represent outputs. We assume that the behavior of one robot could determine the stimulus of another. **Bottom row** The behavior of each robot with respect to their immediate stimuli (left) and the behavior of the starred robot with respect to swarm stimuli (right). While each robot successfully performs the behavior that it evolved for with respect to its immediate stimuli, examination of the selected robot with respect to swarm stimuli shows complex behavior that would otherwise require memory to exhibit.

4 Making a Swarm for More Complex Behavior

The above section demonstrated the evolution of robots exhibiting adaptive behavior in response to the two stimuli. To the best of our knowledge, this is the first example of a closed-loop fully-morphological behavior in evolved robots. The simplicity of the chosen behavior (more right or left) allows us to clearly see how shape change due to the stimuli-sensing materials affects the robots' behavior. But this also invites a thought experiment: how could this proof of concept scale to create complex "intelligent" behaviors from this simple framework?

In this section, we study how these robots can be grouped to create a swarm exhibiting more complex behavior. As a proof of concept, we use a subset of previously evolved robots and hand-design the swarm to show a certain behavior.

We start with an observation. Given the binary nature of the stimuli and the possible behaviors of the robot (moving in a positive or negative direction on the x-axis), the robots could be considered implementing logic gates from the perspective of an outside observer. Above, we evolved robots implementing all possible boolean functions by evolving an archive of robots exhibiting all possible behaviors (Fig. 3). For instance, if we assume that a robot moving in the positive x direction on the x-axis is outputting a 1, robots exhibiting the RRRL behavior can be considered as implementing the NOT-AND boolean function.

We can organize robots, by assuming that the behavior of one robot could determine the stimulus of another robot. While this is arbitrary given our manually and abstractly designed stimuli, it's not a great stretch of the imagination to envision some analogous physical or chemical sensory apparatus that would allow voxels to sense the relative change in distance of a neighboring robot similar to how animals do (e.g. it became "louder" or its "smell" became stronger).

To demonstrate a more complex function derived from these morphological logic gates, we manually design a swarm implementing a D-type latch[1]. Figure 5 top left, shows the schematic of the swarm from an outside observer's perspective. The swarm consists of a single robot with behavior RRLL (leftmost robot, implements AND gate), and four robots with behavior RRRL (implement NAND gate). The swarm has two swarm stimuli, observed by three of the robots. The rest of the stimuli that robots observe are determined by others' behavior. The inner wiring of the swarm can be seen in Fig. 5 top right, where wires on the left side of the robots represent stimuli and the wires on the right side of the robots represent the output of the robot (which determines the stimulus of another robot). If a robot's center of mass has moved in the positive x direction in the last actuation cycle it is considered to be outputting 1 and turns the stimulus on for another robot and vice versa.

We plot the behavior of the swarm as a spacetime diagram in Fig. 5 bottom row. The behavior of each individual and their immediate stimuli can be seen on the left. As expected, robots successfully perform the behaviors that are evolved for, seamlessly responding to changing stimuli patterns. However, if we observe the behavior of the starred robot (top-right robot) with respect to swarm stimuli (as opposed to the robot's immediate stimuli) in Fig. 5 bottom right, we see that the robot is moving following the swarm stimulus 1, only when the swarm stimulus 2 is present, otherwise the robot keeps moving in the direction of last observed swarm stimulus 1. Given the right frame of reference (observing the robot's behavior with respect to the swarm stimuli) and certain assumptions (behavior of one robot determining the stimulus of another), we show that robots exhibit complex adaptive behavior, without any separately postulated brain, that would typically require memory to achieve. This achievement is, certainly, not a surprise, given the equivalence of robots' behavior with boolean functions. Once we have robots that can implement the NOT-AND function, we could build a swarm that would perform any computable behavior. We leave the design of such swarms as an exercise for the reader.

---

[1] While we hand-design the swarm as a proof of concept, one could evolve the design of the swarm towards achieving any target behavior as well.

5 Discussion

The above results demonstrate "intelligent" morphological computation that adapts the robot's behavior via stimuli-educed shape change rather than by a dedicated control or brain-like mechanism. While we feel that this proof of concept is an important, and hopefully entertaining, demonstration, we fully acknowledge that the most complex and adaptive behaviors in natural systems are not fully morphological, but are an interplay of embodied cognition connecting the body and the brain.

Recent literature had demonstrated the challenges of brain-body co optimization and noted fragile co-adaptation to be a major challenge of evolving these systems [8,29], further suggesting that increasing the adaptability and robustness of the brain and body to modification in the other could result in less fragile and thus more effective brain-body systems [10,28]. We hope that by creating more intelligent and adaptive bodies, this work helps to open doors to transfer additional computation to the morphology and enable less fragile co-adaptation.

Lastly, we want to connect this research to recent ideas in artificial and natural evolution. Specifically, it has been argued that a key feature of natural evolution is that it works with material that is active. Cells that already have an array of functionalities come together to create multi-cellular structures, tissues, organs, and systems, creating a "multiscale competence architecture" where each scale exploits and regulates the already-existing functionalities of the scale below to achieve its goals [24]. This is considered to be an important feature of natural evolution and understanding/exploiting this in artificial evolution could have substantial effects [6,15]. Inspired by these ideas, here we show a first pass at demonstrating that simple materials each with straightforward behaviors can be optimized in their emergent design to result in a complex and adaptive system. In future work, we hope to examine how we can consider such a system as a substrate for further evolution (potentially in combination with separately postulated brain models).

6 Conclusion

We investigate ways in which complex, adaptive behavior can be created in virtual voxel-based soft robots without the use of a separately postulated brain. We hand-design simple muscle (that continuously expands and contracts) and sensory (that expands/shrinks in the presence or absence of a binary stimulus) voxel materials and show that evolution can create designs that exhibit closed-loop adaptive behavior, responding to two binary stimuli. Moreover, a swarm of such robots can exhibit even more complex behavior. In future work, we hope to study more ways of creating adaptive and complex behavior.

Acknowledgements. This material is based upon work supported by the National Science Foundation under Grants No. 2008413 and 2239691. Computations were performed on the Vermont Advanced Computing Core supported in part by NSF Award No. 1827314.

References

1. Auerbach, J.E., Bongard, J.C.: Evolving cppns to grow three-dimensional physical structures. In: Proceedings of the 12th Annual Conference on Genetic and Evolutionary Computation, pp. 627–634 (2010)
2. Beer, R.D.: Toward the evolution of dynamical neural networks for minimally cognitive behavior. In: From Animals to Animats 4: Proceedings of the Fourth International Conference on Simulation of Adaptive Behavior, vol. 4, p. 421. MIT Press (1996)
3. Bhatia, J., Jackson, H., Tian, Y., Xu, J., Matusik, W.: Evolution gym: a large-scale benchmark for evolving soft robots. Adv. Neural. Inf. Process. Syst. **34**, 2201–2214 (2021)
4. Bongard, J.: Evolutionary robotics lectures. https://www.youtube.com/playlist?list=PLAuiGdPEdw0iyApypcLk_xBKjLchQM-Mg
5. Bongard, J.C., Pfeifer, R.: Evolving complete agents using artificial ontogeny. In: Morpho-Functional Machines: The New Species: Designing Embodied Intelligence, pp. 237–258. Springer (2003)
6. Bongard, J., Levin, M.: There's plenty of room right here: biological systems as evolved, overloaded. Multi-Scale Machines. Biomimetics **8**(1), 110 (2023). https://doi.org/10.3390/biomimetics8010110
7. Braitenberg, V.: Vehicles: Experiments in synthetic psychology. MIT press (1986)
8. Cheney, N., Bongard, J., Sunspiral, V., Lipson, H.: On the difficulty of co-optimizing morphology and control in evolved virtual creatures. In: Proceedings of the Artificial Life Conference 2016, pp. 226–233. MIT Press, Cancun, Mexico (2016). https://doi.org/10.7551/978-0-262-33936-0-ch042
9. Cheney, N., Bongard, J., Lipson, H.: Evolving soft robots in tight spaces. In: Proceedings of the 2015 Annual Conference on Genetic and Evolutionary Computation, pp. 935–942 (2015)
10. Cheney, N., Bongard, J., SunSpiral, V., Lipson, H.: Scalable co-optimization of morphology and control in embodied machines. J. R. Soc. Interface **15**(143), 20170937 (2018). https://doi.org/10.1098/rsif.2017.0937
11. Cheney, N., MacCurdy, R., Clune, J., Lipson, H.: Unshackling evolution: evolving soft robots with multiple materials and a powerful generative encoding. In: Proceedings of the 15th Annual Conference on Genetic and Evolutionary Computation, pp. 167–174 (2013)
12. Corucci, F., Cheney, N., Lipson, H., Laschi, C., Bongard, J.: Material properties affect evolutions ability to exploit morphological computation in growing soft-bodied creatures. In: Proceedings of the Artificial Life Conference 2016, pp. 234–241. MIT Press, Cancun, Mexico (2016). https://doi.org/10.7551/978-0-262-33936-0-ch043
13. Descartes, 1596-1650, R.: Discourse on method ; and, Meditations on first philosophy. Third edition. Indianapolis : Hackett Pub. Co., [1993] 1993 (1993) https://search.library.wisc.edu/catalog/999718190702121
14. Eiben, A.E., Smith, J.E.: Introduction to evolutionary computing. Springer (2015)
15. Hartl, B., Risi, S.L., Levin, M.: Evolutionary Implications of Multi-Scale Intelligence, April 2024. https://doi.org/10.31219/osf.io/sp9kf
16. Hauser, H., Ijspeert, A.J., Füchslin, R.M., Pfeifer, R., Maass, W.: Towards a theoretical foundation for morphological computation with compliant bodies. Biol. Cybern. **105**, 355–370 (2011)

17. Hornby, G.S., Pollack, J.B.: Body-brain co-evolution using l-systems as a generative encoding. In: Proceedings of the 3rd Annual Conference on Genetic and Evolutionary Computation, pp. 868–875 (2001)
18. Joachimczak, M., Suzuki, R., Arita, T.: Artificial metamorphosis: evolutionary design of transforming, soft-bodied robots. Artif. Life **22**(3), 271–298 (2016)
19. Joachimczak, M., Wróbel, B.: Co-evolution of morphology and control of soft-bodied multicellular animats. In: Proceedings of the 14th Annual Conference on Genetic and Evolutionary Computation, pp. 561–568 (2012)
20. Karban, R.: Plant behaviour and communication. Ecol. Lett. **11**(7), 727–739 (2008)
21. Lakoff, G., Johnson, M.: Metaphors we live by. University of Chicago press (2008)
22. LeCun, Y., Bengio, Y., Hinton, G.: Deep learning. Nature **521**(7553), 436–444 (2015)
23. Lehman, J., Stanley, K.O.: Evolving a diversity of virtual creatures through novelty search and local competition. In: Proceedings of the 13th Annual Conference on Genetic and Evolutionary Computation, pp. 211–218 (2011)
24. Levin, M.: Darwin's agential materials: evolutionary implications of multiscale competency in developmental biology. Cell. Mol. Life Sci. **80**(6), 142 (2023)
25. Macé, V., Boige, R., Chalumeau, F., Pierrot, T., Richard, G., Perrin-Gilbert, N.: The quality-diversity transformer: generating behavior-conditioned trajectories with decision transformers. In: Proceedings of the Genetic and Evolutionary Computation Conference, pp. 1221–1229 (2023)
26. McCulloch, W.S., Pitts, W.: A logical calculus of the ideas immanent in nervous activity. Bull. Math. Biophys. **5**, 115–133 (1943)
27. McGeer, T.: Passive dynamic walking. Int. J. Robot. Res. **9**(2), 62–82 (1990)
28. Mertan, A., Cheney, N.: Modular controllers facilitate the co-optimization of morphology and control in soft robots. In: Proceedings of the Genetic and Evolutionary Computation Conference, pp. 174–183 (2023)
29. Mertan, A., Cheney, N.: Investigating premature convergence in co-optimization of morphology and control in evolved virtual soft robots. In: European Conference on Genetic Programming (Part of EvoStar), pp. 38–55. Springer (2024)
30. Moravec, H.: Mind children: The future of robot and human intelligence. Harvard University Press (1988)
31. Mouret, J.B., Clune, J.: Illuminating search spaces by mapping elites. arXiv:1504.04909 [cs, q-bio], April 2015. http://arxiv.org/abs/1504.04909
32. Paul, C.: Morphological computation: a basis for the analysis of morphology and control requirements. Robot. Auton. Syst. **54**(8), 619–630 (2006)
33. Pfeifer, R., Bongard, J.: How the body shapes the way we think: a new view of intelligence (2006)
34. Pfeifer, R., Gómez, G.: Morphological computation–connecting brain, body, and environment. Creating brain-like intelligence: From basic principles to complex intelligent systems, pp. 66–83 (2009)
35. Pulvermüller, F., Fadiga, L.: Active perception: sensorimotor circuits as a cortical basis for language. Nat. Rev. Neurosci. **11**(5), 351–360 (2010)
36. Sims, K.: Evolving virtual creatures. In: Seminal Graphics Papers: Pushing the Boundaries, vol. 2, pp. 699–706 (2023)
37. Stanley, K.O.: Compositional pattern producing networks: a novel abstraction of development. Genet. Program Evolvable Mach. **8**(2), 131–162 (2007). https://doi.org/10.1007/s10710-007-9028-8
38. Trewavas, A.: Aspects of plant intelligence. Ann. Bot. **92**(1), 1–20 (2003)
39. Wilson, M.: Six views of embodied cognition. Psychonomic Bull. Rev. **9**, 625–636 (2002)

CuttleBot: Emulating Cuttlefish Behavior and Intelligence in a Novel Robot Design

Michael A. Pfeiffer[1,2](✉), Sriskandha Kandimalla[2], Jiahe Liu[3], Katherine Hsu[1], Eleanore J. Kirshner[4], Alina Yuan[5], Hin Wai Lui[3], and Jeffrey L. Krichmar[2,3]

[1] Department of Electrical Engineering and Computer Science,
University of California, Irvine, CA 92697, USA
kaishih@uci.edu
[2] Department of Cognitive Sciences, University of California, Irvine, CA 92697, USA
{mapfeiff,skandima,jkrichma}@uci.edu
[3] Department of Computer Science, University of California, Irvine, CA 92697, USA
{jiahel6,hwlui}@uci.edu
[4] Department of Mechanical and Aerospace Engineering, University of California,
Irvine, CA 92697, USA
ekirshne@uci.edu
[5] Department of Art, University of California, Irvine, CA 92697, USA
yuanax@uci.edu

Abstract. The CuttleBot project aspires to encapsulate the sophisticated behavior of cuttlefish in a neurorobot. The long-term goal is to construct a machine that mirrors the unique intelligent behavior demonstrated by this invertebrate. The current CuttleBot prototype represents an early step towards realizing a robotic system capable of advanced environmental interaction and decision-making. Its custom-made shell demonstrates the camouflaging and signaling observed in cephalopods in response to environmental stimuli. Similar to cuttlefish, the CuttleBot hunts for prey and responds to predators with defensive behaviors. Cuttlefish are impressive learners. Therefore, reinforcement learning was implemented to learn the appropriate behavioral responses to predators (e.g., camouflage or hide) and prey (e.g., confuse and attack). By creating cognitive systems with insights from the natural world, the CuttleBot project lays the groundwork for an era of robotics that comprehends and interacts with the environment in ways that are as dynamic and complex as the biological entities that inspire it.

Keywords: Behavioral Signaling · Camouflage · Cephalopods · Reinforcement Learning · Robotics

1 Introduction

The cuttlefish is a marine mollusc under the class Cephalopoda with extraordinary cognitive capabilities and versatile behavior [1,14]. For example, experi-

M. A. Pfeiffer, S. Kandimalla and J. Liu—Contributed equally to the paper.

ments typically tested on vertebrates have displayed cuttlefishes' ability to exert self-control in delayed gratification tasks [15] and solve spatial tasks [6]. Cuttlefish, like other cephalopods, evolved to shed their protective shell. Therefore they developed an ability to change the color and texture of their skin [17]. This incredible skill is integral to evade predators, capture prey, and communicate [4,22].

Our end goal is to build a biomimetic robot that emulates the sophisticated behaviors and cognitive processes of cuttlefish. In the present work, we focus on the ability to change skin coloring in response to environmental cues by subjecting the CuttleBot to a reinforcement learning paradigm. A high level view of the robot design is given in Fig. 1. A video of the CuttleBot can be viewed at: https://www.youtube.com/watch?v=UYVzGVnPNL0 The CuttleBot, with its embodiment, learning capability, and behavioral trade-offs, follows the design principles of neurorobotics [5]. In the present work, we introduce a neurorobot that follows the cuttlefish by:

- Learning the appropriate responses to predators and prey.
- Camouflaging in the presence of a distant predator.
- Blanching and hiding when too close to a predator.
- Using dynamic skin patterns to confuse prey before attacking and grabbing.

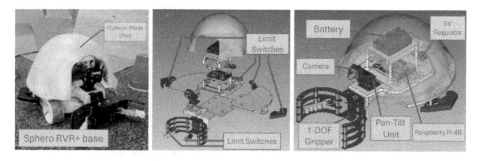

Fig. 1. Left panel. The CuttleBot has a custom-made shell, a camera with pan-tilt, limit switches, and a 1°C of freedom gripper all on top of a Sphero RVR+ base. Right panels. CAD drawing showing parts and electrical components.

2 Related Work

Research in animal cognition has used machine learning simulations and robotics to understand the interplay between neural networks and natural behavior. Most of this work has focused on mammals as the model organism. One group however focused on the economic decisions and foraging behavior of the predatory sea slug Pleurobranchaea using an approach versus avoidance system based on an appetite incentive and separately integrated sensory stimuli [2]. This illuminated the neural mechanisms determining choices while foraging, enhancing the understanding of decision-making processes in biological organisms. Shih and their team were inspired by the distinct neurophysiology of octopuses and proposed a hierarchical control framework for coordinating multiple soft arms during foraging tasks [16]. Separating control into distinct levels while getting sensory feedback created a more realistic replication of observed behavior in nature. Our work takes a similar approach in a foraging predator-prey scenario with the cephalopod cuttlefish as a model organism. There have been attempts to create octopus robots, but most of these efforts have focused on the unique features of octopus arms in reaching and grasping tasks. [9,10,18].

An important cephalopod feature is its ability to change the color and shape of its skin [22]. Attempts have been made to mimic the color-changing chromatophores in special materials [12,13,24]. These synthetic systems face challenges in mimicking the color-changing abilities of a mobile agent or robot. One team employed simple microfluidic networks to change the surface color and shape of a soft robot [11]. These materials are exotic and difficult to program and have not been realized in a behaving robot. However, there have been attempts to demonstrate camouflage in a biomimetic chameleon robot [7], and a remote control cuttlefish robot that could change colors to communicate was created for the BBC documentary *Spy In the Ocean*. We take a practical approach to changing skin colors by embedding strips of programmable LEDs into a custom-made, translucent shell that fits on top of the CuttleBot. This allows for rapid, dynamic displays in response to environmental stimuli as observed in cuttlefish in their natural habitat.

3 Methods

3.1 CuttleBot Design

The robot was built on a Sphero RVR+[1] having tank treads and a downward-facing RGB (Red, Green, and Blue) color sensor (Fig. 1). This base interfaced with a Raspberry Pi 4B[2] through a four-pin UART (Universal Asynchronous Receiver / Transmitter) communication protocol. The RGB color sensor was included to detect different colors on the ground, which provides a basic parallel to the cuttlefish's ability to perceive and respond to patterns and contrasts in their environment. The robot utilized a wide-angle camera to capture visual data of its surroundings, interfacing with the Raspberry Pi through the CSI (Camera Serial Interface) port. With its 160-degree FOV (Field of View), the camera enabled the robot to detect objects far off to the side, ensuring efficient navigation. This wide FOV mimics the visual capabilities of real cuttlefish, which have large, well-developed eyes that allow them to see a wide area around them [20]. The camera was mounted on a pan-tilt unit driven by a

---

[1] https://sphero.com/products/rvr.
[2] https://www.raspberrypi.com/products/raspberry-pi-4-model-b/.

micro servo, granting it two degrees of freedom. This configuration allowed the robot to survey its environment without necessitating substantial movement of its base, similar to how cuttlefish can independently move their eyes to scan their surroundings [3]. However, during testing, the pan-tilt unit was kept stationary due to malfunctioning hardware. Despite this, the camera's wide FOV still allowed the robot to adequately survey its environment without moving the base, much like how cuttlefish use their broad vision to monitor their surroundings effectively. Limit switches were added around the robot to provide sensory input for detecting collisions with walls and objects. These switches function similarly to the cuttlefish's sensory organs, which detect objects in their environment through physical contact. The robot's claw was also equipped with limit switches, enabling it to distinguish between successful and unsuccessful prey captures, akin to how a cuttlefish uses its tentacles to grasp prey and sense a secure hold.

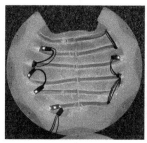 

Fig. 2. Left. LEDs embedded inside the shell. Right. Translucent shell on CuttleBot

A 3D-printed, translucent shell was designed to facilitate color-changing abilities such as hypnotizing prey and evading predators (Fig. 2). To achieve this, an array of 96 programmable LEDs were positioned internally within a translucent shell. These LEDs emitted different colors, creating the illusion of dynamically changing skin. The artistic vision for the shell was conceptualized using Blender, a 3D modeling program. Initially, the design was blueprinted with sketch studies of the cuttlefish's anatomy, where the dynamic form of its fins soon became a central theme for the shell design. While a more realistic design was considered, a minimalistic appearance was chosen to highlight its functional attributes. The wide and spherical shape of the shell created a spacious area for all the necessary electronics while maximizing the vibrancy and visibility of the colorful LED lights. The rippled fins functioned both as an aesthetic reference to the cuttlefish and a connection point to the rest of the robot. The shell was then replicated in SolidWorks for 3D printing by RapidTech at UC Irvine.

3.2 CuttleBot Behavior

Cuttlefish are adept hunters and masters of evasion. They use chromatophores and skin papillae to camouflage and effectively match the color and texture of their surroundings [4,22]. This adaptation serves both defensive and predatory purposes, allowing them to evade predators by blending into the environment and ambush prey undetected. Additionally, cuttlefish employ mesmerizing confusion signals to hypnotize prey during hunting and create opportunities for capture [23]. Their attack sequence involves three phases (Fig. 3): attention, positioning, and striking [23]. When threatened, cuttlefish blanch, a phenomenon where their skin pattern turns pale in response to a threat [22]. The Cuttlebot replicates these complex behaviors, emulating the remarkable adaptability of its natural counterpart.

Hunting Prey. Initially, the Cuttlebot identifies an object as potential prey and initiates its hypnosis sequence. Next, it focuses on the object by aligning its body and claw to directly face it. To gather depth information, the robot stores the pixel width of the object, backs up by 0.15 m, and measures the pixel width at the new position. Assuming a camera is radially aligned and centered on the object, the following equation was derived: $d = m_{rad\_in} \cdot \frac{P_1}{P_2 - P_1}$ where d is the distance the camera is from the object at the second position, m_{rad_in} is the distance the camera moves radially inward, P_1 is the pixel width at the first position, and P_2 is the pixel width at the second position. The equation allows the robot to approximate the depth of the prey, mimicking the cuttlefish's ability to extract depth information by converging its eyes onto the prey [3,23]. Finally, it disengages the hypnosis pattern, uses the computed depth to pounce toward the prey, and swiftly grabs the object with its claw. To account for any errors during the alignment or depth computation, the robot will only pounce until it is $\frac{1}{3}$ of the distance (minimum distance of 0.2 m) from the object. From then, the robot will slow down and adjust its body as needed to ensure it can grab the object. If the object is prey, the robot will then drop it and turn around to explore a different area. If the object is a predator, the robot will initiate the defense sequence. (Fig. 3)

Fig. 3. Attack Sequence. From left to right, attention, position, strike.

Avoiding Predators. Similar to the defense mechanisms of cuttlefish [19], the robot either blanches and retreats before camouflaging (Fig. 4) or directly camouflages to avoid being eaten. The decision to blanch first is dependent on the proportion an object occupies in the robot's visual field as follows: $P(blanch|obj_{prop}) = \exp\left(\frac{-3.4657}{100 \cdot obj_{prop}}\right)$ where $P(blanch|obj_{prop})$ is the probability of first blanching given the proportion the object occupies in the visual view (obj_{prop}). The function was chosen as opposed to the logistic function due to an object's proportion in a visual scene not linearly scaling with its distance. Additionally, the constant was chosen such that $P(blanch|0.05) = 0.5$. A visual scene coverage of 5% was selected to achieve a 50% blanching probability, as this setting effectively accommodated the camera's FOV and the predator's size. This increases the likelihood of the robot blanching for closer or larger objects and directly camouflaging for farther or smaller objects. By mimicking the color of the environment, the robot can avoid detection and back away before resuming its activities.

Fig. 4. Defense Sequence. From left to right, detect, avoid and blanch, camouflage.

Learning Mechanism. Q-learning, a model-free reinforcement learning algorithm, was utilized for the robot to learn between predator and prey [21], formulated as follows:

$$Q(s,a) = Q(s,a) + \alpha[r + \gamma \max_{a'} Q(s',a') - Q(s,a)] \qquad (1)$$

where s represents the current state, a the current action, r the immediate reward received, s' the new state after taking action a, and a' the possible actions in the new state. The parameters α and γ denote the learning rate and discount factor, respectively.

The CuttleBot selects its actions by drawing from a probability distribution based on the beta-modulated softmax function [8] as follows:

$$P(a_i|s) = \frac{e^{\beta Q(s,a_i)}}{\sum_j e^{\beta Q(s,a_j)}} \qquad (2)$$

where $P(a_i|s)$ is the probability of selecting action a_i in state s, $Q(s,a_i)$ is the Q-value of action a_i in state s, and β is a controlling parameter that determines how sensitive the function is to differences between Q-values. A higher β value results in more exploitation (favoring actions with higher Q-values), while a lower β value encourages exploration (choosing actions more uniformly).

States were based on the camera's detection of the color and size of objects (Fig. 5). Actions were the attack and defense phases described above. Successfully grabbing a prey resulted in a reward of 5, and approaching a predator, which entailed approaching followed by an unsuccessful attempt to grab, resulted in a penalty of -10. This contrast in reward outcomes prioritizes avoiding predators over attacking prey as attacking a predator is significantly more detrimental than missing out on a prey. In a successful grab, the limit switches in the claw were triggered. The CuttleBot's claw could grab the blue cups (prey), but could not grab green balls (predators) due to their large size. The CuttleBot was programmed to "die" if it unsuccessfully grabbed a predator three times. While this is not a natural constraint for cuttlefish, it establishes a lower bound for how rapidly the CuttleBot must adapt to its environment. If the Cuttlebot cannot learn to avoid a predator after two unsuccessful attempts, it will "die" upon its third failed attempt, preventing it from learning the correct action to take. Once this threshold is reached, the robot will stop in place and turn off any lights present, simulating its "death". This condition adds urgency to the learning process, as the robot must learn to avoid predators and eat prey before reaching this fatal threshold.

3.3 Environment and Experimental Design

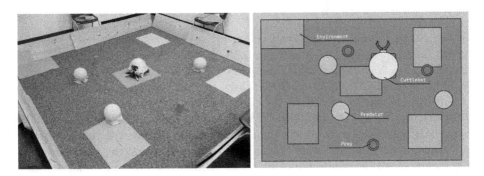

Fig. 5. Example Set Up. Left. Image of CuttleBot with predator (green objects) and prey (blue objects) with colored construction paper mimicking a coral reef. Right. Annotated drawing of environment. (Color figure online)

The CuttleBot was placed into an enclosed arena alongside variable numbers of predators (green balls) and prey (blue cups) (Fig. 5). The arena's walls confined the space for the robot's exploration, with limit switches assisting in navigation. Colored paper patches on the floor provided diverse color opportunities for camouflaging. Before each trial, the arena was reset, and the robot was positioned at the center. The environment was modified in four ways: number of predators (1, 3), number of prey (1, 3), learning rate (0.1, 0.2, 0.4), and β value (0.5, 1.0, 2.0). Each scenario was subjected to five trials, lasting for five minutes or until the CuttleBot was considered "dead".

The study includes two control groups: 1) No learning (learning rate set to zero) and 2) Only approach. In the no learning control, the robot had a 50% chance of approaching stimuli due to the action selection criteria and initial weight settings. Therefore, an only approach condition served as another control where the robot always approached stimuli. These served as baselines for comparison with learning.

4 Results

4.1 Camouflaging and Hypnotizing

Fig. 6. Camouflaging behavior on different backgrounds.

For the camouflage functionality of the robot, a color sensor located under the Sphero RVR+ was used to generate an RGB value, which was converted into the HSV (Hue, Saturation, and Value) color space and adjusted to account for the lighting effects of the robot. Specifically, the hue remained unchanged while the saturation was increased to its maximum and the value was set to 80% of the original value. The modified HSV value was converted back into an RGB color space and displayed on the shell's LED array to give the robot a better appearance when camouflaging (Fig. 6). With this method, the CuttleBot was able to match all the colors found in its experimental arena, including the blue carpet.

Fig. 7. Hypnotizing behavior. Bands of color repeatedly move from front to back.

Cuttlefish hypnotize prey with a dynamic pattern on their skin. To mimic this behavior in the CuttleBot, a wave pattern was generated by setting the background of the shell to a specified color and having two staggered rows of LEDs traverse across the shell with an alternative color (Fig. 7).

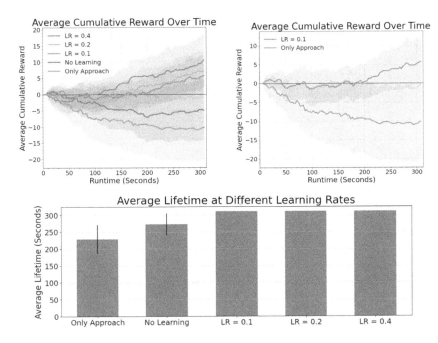

Fig. 8. Comparing different learning rates. Five trials for each condition. Top Left. Cumulative rewards over the time of all trials. Shaded region shows 95% confidence interval. Top Right. Cumulative rewards focused on two trials: learning rate of 0.1 and only approach. Bottom. Time before "dying" threshold reached. Bars show mean time and error bars show 95% confidence interval. No error bars indicate the survival of all runs. A β value of 1.0 was used in all scenarios

4.2 Adaptive Behavior

Incorporating learning into the CuttleBot increased its survival time (Fig. 8, Bottom). Runs employing the learning algorithm consistently demonstrated significantly longer survival times with the robot being able to survive the entire run. In contrast, the control groups, which lacked learning, often approached three predators and died. These findings further emphasize the importance of learning to successfully navigate and adapt to its environment. While there were clear differences between control runs and runs employing a learning rate, there was little difference between the learning rates used, all of which survived until the end of their respective runs.

To get a better understanding of the CuttleBot's learned behavior, the β values used were varied. A similar trend to varying the learning rate was observed when examining the results of different β values (Fig. 9). The lower the β value, the more the robot explored, so it reacted similarly to a low learning rate. When the β value used was greater than 0.5, the robot demonstrated rapid learning behavior and consequently had a positive cumulative reward and longer survival

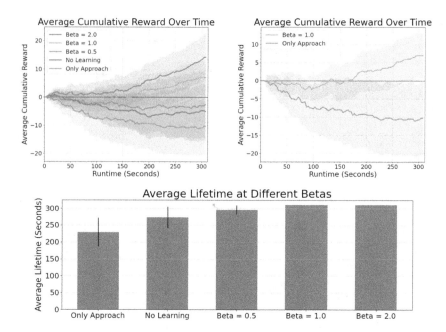

Fig. 9. Comparing different β values. Five trials for each condition. Top Left. Cumulative rewards over the time of the trial. Shaded region shows 95% confidence interval. Top Right. Cumulative rewards focused on two trials: β of 1.0 and only approach. Bottom. Time before "dying" threshold reached. Bars show mean time and error bars show 95% confidence interval. No error bars indicate the survival of all runs. A learning rate of 0.2 was used in all scenarios.

times. However, we observed that with $\beta > 1.0$ the behavior was less flexible recovering from false positives or false negatives.

As the CuttleBot learned, it changed its skin coloring similar to cuttlefish in the wild (Fig. 10). It blanched and camouflaged to avoid predators (Fig. 6), and displayed a hypnotizing traveling wave sequence before eating prey (Fig. 7). Taken together, these results show that incorporating learning and exploration/exploitation trade-offs result in appropriate responses to predators and prey. In future work, we will investigate reversal learning and other aspect of cognition [1,14].

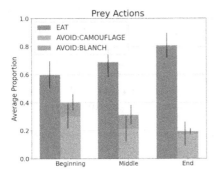 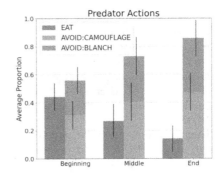

Fig. 10. Average proportion of actions taken in response to environmental stimuli during the first third, middle third, and final third of a run. β value of 1.0 alongside all positive learning rates used. Bars show mean and error bars show 95% confidence interval.

5 Discussion and Conclusions

Similar to natural cephalopods, the CuttleBot demonstrated dynamic camouflaging, behavioral signaling, and learning. Its unique shell with programmable LEDs enabled realistic behavioral responses and adaptive behavior, which emerged from reinforcement learning. Future CuttleBot iterations will incorporate more learning behaviors, such as delayed gratification and episodic learning, as well as further improvements to its embodiment like binocular vision and more movement degrees of freedom. It will be of interest to use more natural and realistic environments in the future.

Acknowledgments. The CuttleBot team was supported by the UC Irvine California Institute for Telecommunications and Information Technology (CALIT2) in collaboration with the UC Irvine Undergraduate Research Opportunities Program (UROP). The authors would like to thank members of the CuttleBot team, which included Hazel Kurien, Connor Marks, Zachary Sawdai, Erin Yi, Tiana Zhen, and Custo Yang for their valuable contributions. The authors also thank Ben Dolan of UCI RapidTech for his support and ideas on the robot design.

References

1. Birch, J., Schnell, A.K., Clayton, N.S.: Dimensions of animal consciousness. Trends Cogn. Sci. **24**(10), 789–801 (2020). https://doi.org/10.1016/j.tics.2020.07.007
2. Brown, J.W., et al.: Implementing goal-directed foraging decisions of a simpler nervous system in simulation. eNeuro **5**(1), 1–10 (2018). https://doi.org/10.1523/ENEURO.0400-17.2018
3. Feord, R.C., Sumner, M.E., Pusdekar, S., Kalra, L., Gonzalez-Bellido, P.T., Wardill, T.J.: Cuttlefish use stereopsis to strike at prey. Sci. Adv. **6**(2), eaay6036 (2020). https://doi.org/10.1126/sciadv.aay6036, https://www.science.org/doi/abs/10.1126/sciadv.aay6036

4. How, M.J., Norman, M.D., Finn, J., Chung, W.S., Marshall, N.J.: Dynamic skin patterns in cephalopods. Front. Physiol. **8**, 1–13 (2017). https://doi.org/10.3389/fphys.2017.00393
5. Hwu, T., Krichmar, J.: Neurorobotics: Connecting the Brain. Body and Environment. MIT Press, Cambridge, MA (2022)
6. Jozet-Alves, C., Bertin, M., Clayton, N.S.: Evidence of episodic-like memory in cuttlefish. Curr. Biol. **23**(23), R1033-5 (2013). https://doi.org/10.1016/j.cub.2013.10.021
7. Kim, H., Choi, J., Kim, K.K., Won, P., Hong, S., Ko, S.H.: Biomimetic chameleon soft robot with artificial crypsis and disruptive coloration skin. Nat. Commun. **12**(1), 1–11 (2021). https://doi.org/10.1038/s41467-021-24916-w
8. Kiviluoto, K., Oja, E.: Softmax-network and s-map-models for density-generating topographic mappings. In: 1998 IEEE International Joint Conference on Neural Networks Proceedings. IEEE World Congress on Computational Intelligence (Cat. No.98CH36227), vol. 3, pp. 2268–2272 (1998). https://doi.org/10.1109/IJCNN.1998.687214
9. Kotak, P., Maxson, S., Weerakkody, T., Cichella, V., Lamuta, C.: Octopus-inspired muscular hydrostats powered by twisted and coiled artificial muscles. Soft Rob. **11**(3), 432–443 (2024). https://doi.org/10.1089/soro.2023.0069
10. Mazzolai, B., Laschi, C.: Octopus-inspired robotics. preface. Bioinspir. Biomim. **10**(3), 1–2 (2015). https://doi.org/10.1088/1748-3190/10/3/030301
11. Morin, S.A., Shepherd, R.F., Kwok, S.W., Stokes, A.A., Nemiroski, A., Whitesides, G.M.: Camouflage and display for soft machines. Science **337**(6096), 828–832 (2012). https://doi.org/10.1126/science.1222149
12. Pikul, J.H., Li, S., Bai, H., Hanlon, R.T., Cohen, I., Shepherd, R.F.: Stretchable surfaces with programmable 3d texture morphing for synthetic camouflaging skins. Science **358**(6360), 210–214 (2017). https://doi.org/10.1126/science.aan5627
13. Pratakshya, P., et al.: Octopus-inspired deception and signaling systems from an exceptionally-stable acene variant. Nat. Commun. **14**(1), 1–11 (2023). https://doi.org/10.1038/s41467-023-40163-7
14. Schnell, A.K., Clayton, N.S.: Cephalopods: Ambassadors for rethinking cognition. Biochem. Biophys. Res. Commun. **564**, 27–36 (2021). https://doi.org/10.1016/j.bbrc.2020.12.062
15. Schnell, A.K., Clayton, N.S., Hanlon, R.T.: Cuttlefish exert self-control in a delay of gratification task. Roy. Soc. Biol. Sci. **288**, 1–9 (2021). https://doi.org/10.1098/rspb.2020.3161
16. Shih, C.H., Naughton, N., Halder, U., Chang, H.S., Kim, S.H., Gillette, R., Mehta, P.G., Gazzola, M.: Hierarchical control and learning of a foraging cyberoctopus. Adv. Intell. Syst. **5**(9), 1–13 (2023). https://doi.org/10.1002/aisy.202300088
17. Shook, E.N., Barlow, G.T., Garcia-Rosales, D., Gibbons, C.J., Montague, T.G.: Dynamic skin behaviors in cephalopods. Curr. Opin. Neurobiol. **86**, 102876 (2024) https://doi.org/10.1016/j.conb.2024.102876, https://www.sciencedirect.com/science/article/pii/S0959438824000382
18. Sivitilli, D.M., Smith, J.R., Gire, D.H.: Lessons for robotics from the control architecture of the octopus. Front. Robot. AI **9** (2022). https://doi.org/10.3389/frobt.2022.862391
19. Staudinger, M.D., Buresch, K.C., Mathger, L.M., Fry, C., McAnulty, S., Ulmer, K.M., Hanlon, R.T.: Defensive responses of cuttlefish to different teleost predators. Biol. Bull. **225**(3), 161–74 (2013)

20. Watanuki, N., Kawamura, G., Kaneuchi, S., Iwashita, T.: Role of vision in behavior, visual field, and visual acuity of cuttlefish sepia esculenta. Fish. Sci. **66**(3), 417–423 (2000). https://doi.org/10.1046/j.1444-2906.2000.00068.x
21. Watkins, C.J., Dayan, P.: Q-learning. Mach. Learn. **8**, 279–292 (1992)
22. Woo, T., Liang, X., Evans, D.A., Fernandez, O., Kretschmer, F., Reiter, S., Laurent, G.: The dynamics of pattern matching in camouflaging cuttlefish. Nature **619**(7968), 122–128 (2023). https://doi.org/10.1038/s41586-023-06259-2
23. Wu, J.J., Hung, A., Lin, Y.C., Chiao, C.C.: Visual attack on the moving prey by cuttlefish. Front. Physiol. **11**(648), 1–11 (2020). https://doi.org/10.3389/fphys.2020.00648
24. Xu, C., Stiubianu, G.T., Gorodetsky, A.A.: Adaptive infrared-reflecting systems inspired by cephalopods. Science **359**(6383), 1495–1500 (2018). https://doi.org/10.1126/science.aar5191

The Emergence of a Complex Representation of Touch Through Interaction with a Robot

Louis L'Haridon[1](✉), Raphaël Bergoin[1,2], Baljinder Singh Bal[1], Mehdi Abdelwahed[1], and Lola Cañamero[1]

[1] ETIS Lab (UMR 8051), CY Cergy Paris University/ENSEA/CNRS, Cergy-Pontoise, France
{louis.lharidon,raphael.bergoin,lola.canamero}@cyu.fr
[2] Center for Brain and Cognition, Department of Information and Communications Technologies, Pompeu Fabra University, Barcelona, Spain

Abstract. In this paper, we present a novel robot model of touch, and its representation in an artificial cortex, that aims to capture some of the complexity of human touch. In particular, our approach integrates artificial mechanoception and nociception in an adaptive sensory field (the robot's "sensory body"), allowing for a more comprehensive simulation of tactile sensations. The robot's sensory field is then processed by a biologically plausible neural network in a way akin to sensory processing in the somatosensory and anterior cingulate cortex. Findings from our experimental results show our model's ability to integrate complex data from infrared sensors, leading to the emergence of a spatial sensory body representation in our neural network, with potentially significant implications for robot perception and interaction.

Keywords: Adaptive Model of Touch and Pain · Bio-inspired Robotics · Computational Neuroscience · Neural network · Neurorobotics

1 Introduction

Touch is a sensory modality that allows organisms to perceive and respond to physical contact with their environment through specialized receptors in skin and other tissues. It is essential for constructing internal representations of the world, enabling the detection of noxious stimuli that signal harm. It also supports spatialization of the body helping to understand its position and movement in space. Touch perception involves sensory inputs such as mechanoception and nociception.

Mechanoception is the biological process through which the body perceives and interprets mechanical stimuli, including touch, pressure and vibration [27]. This process is mediated by sensory cells called mechanoreceptors [8] which are distributed through the skin, muscles and other tissues. These receptors respond to mechanical change, converting physical force into electrical signals transmitted through $A\beta$ fibers to the brain—specifically to the somatosensory cortex. The cortex processes and interprets the tactile sensation, enabling the construction of a physical body representation [9] (see, e.g., the *cortical homonculus* [28]).

Nociception is a biological process crucial for detecting potential or actual tissue damage [16]. It uses specific sensors, the nociceptors located in the skin, muscles, and organs, that detect harmful stimuli, triggering signals through nerve fibers, including $A\delta$ fibers for sharp pain, to the brain. Research in robotics has explored nociception models, incorporating nociceptors in robot designs for simulating pain detection mechanisms [18,19].

This predefined pathway from sensory receptors to precise areas of the cortex could be at the origin of the brain's spatial and functional modular organization, where neurons and regions associated with common modalities or functions are more strongly connected [26]. More precisely, plasticity seems to shape these neural assemblies associated with specific sensory modalities or features within a modality, under the action of co-activation zones [12].

Building on Louis L'Haridon's PhD thesis on pain modeling in robots, and on Raphaël Bergoin's thesis on inhibitory plasticity in neural memory formation [4], the initial motivation for this study was to understand and model the emergence of a complex sensory representation of touch combining a new representation of mechanoception and nociception in a "sensory body" of a mobile robot, and a neural network model of the anterior cingulate cortex and somatosensory cortex in an immature brain (i.e. not having reached its definitive organization). More precisely, the aim is to model the pathway from nociceptors (respectively mechanoreceptors) associated with the skin, to the anterior cingulate cortex (respectively somatosensory cortex).

To this end, we use a Khepera IV robot (http://www.k-team.com/khepera-iv). The robot chassis can be considered as a "human skin", a metaphor that overlooks the fact that the input that robots can process, often limited to proximity sensors like IR or ultrasonic, as in our case, is very far from the complexity and the nuanced information human skin provides, such as texture, temperature, and pressure variations. Although efforts have been made to develop sensors mimicking biological features [1,21], they may not align with the current capabilities of robots [3]. Although we use the IR sensors fitted around the robot's chassis to map the receptive field of the robot's body, we have developed a complex representation of touch (mechanoception and nociception) in the "sensory body" of the robot, in order to extract relevant tactile features to be transmitted to a biologically realistic artificial neural network. This neural network will then adapt to these external signals, shaping its overall organization.

2 The Sensory Body

This section describes the robot's "sensory body" used to model tactile sensations and the interaction between mechanoreceptors and nociceptors.

2.1 Tactile Sensory Fields

To model complex tactile sensing in our robot, we have developed a "sensory field" using proximity sensors, to better capture some of the complexity of human skin's tactile sensations. It is conceptualized through both its nominal and actual forms, where sensor readings, relative to a nominal (undisturbed) position, assess deformations triggered by environmental interactions. This analysis not only quantifies these interactions but also provides a qualitative insight into the robot's tactile experience.

We employ the Khepera-IV robot, a compact circular robot fitted with eight evenly distributed InfraRed (IR) sensors around its chassis. To allow for a more detailed spatial analysis, we first interpolate additional values between the IR— the average of the readings of two consecutive sensors is interpolated twice between each pair of consecutive IR sensors, to enhance spatial resolution from 8 to 32 values. These interpolated values are then translated into polar coordinates, offering a nuanced understanding of sensory interactions. For visualization, these polar coordinates are further transformed into Cartesian coordinates, depicting the sensory field as a "blob". This representation effectively illustrates the sensory body's "deformation" (its changing, adaptable shape) in response to external stimuli, with a nominal position established to denote its "undisturbed" state (lack of sensory stimulation).

Our model also differentiates between nociceptors and mechanoreceptors, and aims to simulate the skin's elasticity and responsiveness by employing a dual-field approach to encapsulate their distinct responses to stimuli [22]. This distinction is crucial, as mechanoreceptors and nociceptors transmit signals at different speeds: whereas β fibers relay touch and vibration quickly (30-70 m/s), δ fibers, associated with pain, which conduct at slower speeds (5-30 m/s) [24].

2.2 Mechanoceptors

In humans, mechanoreceptors are crucial for detecting a broad spectrum of tactile stimuli; they are primarily found within the dermis layer of the skin [14,15]. Our model mirrors this with the placement of IR sensor values in our defined sensory body (Fig. 1, A). They play a vital role in interpreting force, strain, and stress applied to skin, quantifiable through Hooke's Law ($F = k \cdot \Delta L$), which describes the linear relationship between the force applied and the elastic deformation of an object, and formulas for strain ($\epsilon = \frac{\Delta L}{L_0}$), stress ($\sigma = \frac{F}{A}$), and Young's Modulus ($E = \frac{\sigma}{\epsilon}$), which defines the stiffness of a material, with measurements in standard units of Newtons, Pascals, and dimensionless ratios.

The mechanotransduction pathway activates with these receptors' response to external forces, using Young's Modulus at each sensory point to determine activation levels. This data integrates with a neural network simulating the somatosensory cortex's response to tactile stimuli (Fig. 1, F). This model encapsulates the complex relationship between physical deformation and neural response, showcasing the encoding and processing of tactile information.

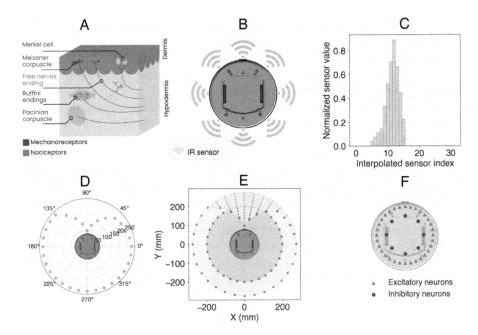

Fig. 1. (A) Shows four types of mechanoceptors (in blue) and nociceptive free nerve endings (in red) in the human hand, drawn after [16]. (B) Top view representation of a Khepera-IV robot with 8 IR sensors. (C) Normalized and interpolated IR sensor data from 8 to 32 values. (D) Polar coordinates from the data showing the robot's physical outline. (E) Representation of sensory body's deformed through pressure in Cartesian space; the red circle indicates the nominal position of mechanoceptors, the green circle marks the nominal nociceptors' position, illustrating proximity to the robot's body and a positional threshold. The yellow and green fields represent the "deformed" sensory and nociceptive layers, respectively. (F) Cortical neurons processing mechanoceptive information, with red circles for excitatory and blue for inhibitory neurons. (Color figure online)

2.3 Nociceptors

Nociceptors, recognized as the sensors of pain [10], are sensitive to noxious stimuli, playing a critical role in the body's ability to detect and respond to potentially harmful conditions. These receptors are adept at discerning various characteristics of stimuli, including intensity, duration, and even the type of the pain: whether it is sharp, throbbing, or burning.

Based on this understanding, we compute physical information on our sensory field deformation to distinguish touch events, drawing parallels between the nociceptors' functionality and our system's ability to interpret tactile data. This involves analyzing the diffuse or intense deformation of the blob, using a classification algorithm, the velocity of the touch, its duration, and the frequency of touch encounters. From our sensory body, we thus compute eight pieces of information across four physical aspects (deformation, velocity, frequency and duration), each with two classes, forming an 8-dimensional vector representing noxious stimuli for the neural network. This vector accounts for the speed difference between slower δ fibers (nociceptive) and faster β fibers (mechanoceptive). To account for this speed difference, we delay stimuli in time.

As depicted in Fig. 2, our methodology for computing nociceptive features begins by assessing the force exerted across the sensory field to evaluate deformation. This analysis helps us identify whether the contact is focused or disperse, categorized as intense or diffuse deformation respectively (Fig. 1, A & B). We then determine touch activation based on these deformations (Fig. 1, C), leading to the calculation of key tactile characteristics: frequency of touch events (Fig. 1, D & E), duration(Fig. 1, I & J), and velocity of deformation (Fig. 1, G & H).

3 Neural Network Model

To process sensory information from mechanoreceptors and nociceptors and to model the neuronal activity of the somatosensory cortex and the Anterior Cingulate Cortex (ACC), we employ a biologically inspired spiking neuronal network subject to synaptic plasticity. For more details on the model and implementation choices, see [4,6].

3.1 Spiking Neuronal Network Model

Throughout this study, we use a network of excitatory-inhibitory heterogeneous quadratic integrate and fire (QIF) neurons [11]. The network is composed of 80% of excitatory neurons and 20% of inhibitory neurons, as commonly accepted in the human cortex [20]. The inhibitory neurons are divided into two distinct populations: a population following Hebbian learning, and a population following anti-Hebbian learning (i.e. neurons that fire together, decoupled together).

Therefore, the evolution of the membrane potential V_i of each neuron ($i = 1, ..., N$) is described by the following equation:

$$\tau_m \dot{V}_i = V_i^2(t) + \eta_i + g_e S_i^e(t) + g_{hi} S_i^{hi}(t) + g_{ai} S_i^{ai}(t) + I_i(t) + \xi_i(t), \quad (1)$$

where synaptic inputs $S_i^e(t)$, $S_i^{hi}(t)$, and $S_i^{ai}(t)$ (excitatory, Hebbian inhibitory, and anti-Hebbian inhibitory, respectively) for neuron i are defined by:

$$\tau_d^{e(i)} \dot{S}_i^{e(hi,ai)} = -S_i^{e(hi,ai)} + \frac{\tau_d^{e(i)}}{N_{e(hi,ai)}} \sum_j^{N_{e(hi,ai)}} w_{ij} \delta(t - t_j), \quad (2)$$

where $\tau_m = 0.02$s is the membrane time constant, $\tau_d^e = 0.002$s and $\tau_d^i = 0.005$s the time decay of excitatory and inhibitory neurons, $\eta_i \sim \mathcal{N}(0.0, (\pi \tau_m)^2)$ the excitability parameter, $N = N_e + N_{hi} + N_{ai} = 100$, $N_e = 80$, $N_{hi} = 10$ and $N_{ai} = 10$ respectively the number of excitatory and Hebbian and anti-Hebbian inhibitory neurons, $g_e = 100$, $g_{hi} = 400$ and $g_{ai} = 200$ the global coupling strength for the excitatory neurons and Hebbian and anti-Hebbian inhibitory neurons. The coupling weights from neuron j to i is depicted by w_{ij}, t_j is the time of spike of the j-th neuron, and $\delta(t)$ is the Dirac delta function. Finally, $I_i(t) = \{0, (50\pi \tau_m)^2\}$ is an external input current and $\xi_i(t) \sim \mathcal{N}(0.0, (4\pi \tau_m)^2)$ is a Gaussian noise. We consider a fully connected network without self-connections.

3.2 Plasticity Functions

Regarding the adaptation of the weights w_{ij}, we use spike-timing-dependent plasticity (STDP) rules that depend on the time difference $\Delta t = t_i - t_j$ between the last spikes of the post-synaptic neuron i and pre-synaptic neuron j. The plasticity functions $\Lambda^+(\Delta t)$ and $\Lambda^-(\Delta t)$ from Eqs. 3 for potentiation and depression respectively, depend on the nature of the pre-synaptic neuron.

$$\Lambda^+(\Delta t) = \begin{cases} \Lambda(\Delta t), & \text{if } \Lambda(\Delta t) \geq 0, \\ 0, & \text{if } \Lambda(\Delta t) < 0, \end{cases} \quad \Lambda^-(\Delta t) = \begin{cases} 0, & \text{if } \Lambda(\Delta t) \geq 0, \\ \Lambda(\Delta t), & \text{if } \Lambda(\Delta t) < 0. \end{cases} \quad (3)$$

For excitatory neurons we use a Hebbian STDP asymmetric function commonly used in the literature [7] described by Eq. 4.

$$\Lambda(\Delta t) = \begin{cases} A_+ e^{-\frac{\Delta t}{\tau_+}} - A_- e^{-\frac{4\Delta t}{\tau_+}} - f, & \text{for } \Delta t \geq 0, \\ A_+ e^{\frac{4\Delta t}{\tau_-}} - A_- e^{\frac{\Delta t}{\tau_-}} - f, & \text{for } \Delta t < 0, \end{cases} \quad (4)$$

with the time constants $\tau_+ = 0.02$s and $\tau_- = 0.05$s, the amplitudes $A_+ = 5.296$ and $A_- = 2.949$. The forgetting term $f = 0.1$ allows to have a constant small depression of the weights whatever the spike timing difference. It models the natural, constant and slow forgetting of memories [13].

For Hebbian (anti-Hebbian) inhibitory neurons we use a Hebbian (anti-Hebbian) STDP symmetric function [17,23] described by Eq. 5.

$$\Lambda(\Delta t) = \pm A(1 - (\frac{\Delta t}{\tau})^2)e^{\frac{-\Delta t^2}{2\tau^2}} \mp f, \quad (5)$$

with time constant $\tau = 0.1$s, amplitude $A = 3$ and forgetting term $f = 0.1$.

3.3 Adaptation of Synaptic Weights

The evolution of the synaptic weights, which remain continually subject to adaptation, unlike more conventional learning systems, follows this ordinary differential equation:

$$\tau_l \dot{w}_{ij} = (-1)^{a_q}[\tanh(\lambda(w_q^l - w_{ij})) * \Lambda_q^+(\Delta t) + \tanh(\lambda(w_{ij} + w_q^u)) * \Lambda_q^-(\Delta t)] \quad (6)$$

where q denotes if the pre-synaptic neuron is excitatory $q = e$ or Hebbian (anti-Hebbian) inhibitory $q = hi$ ($q = ai$), for excitatory (inhibitory) neurons we set $w_q^l = 1$ ($w_q^l = 0$) and $w_q^u = 0$ ($w_q^u = 1$), thus ensuring that the excitatory (inhibitory) couplings are defined in the following interval $w_{ij} \in [0:1]$ ($w_{ij} \in [-1:0]$). Moreover, $a_e = 2$ and $a_{hi} = a_{ai} = 1$, thus for inhibitory synapses, the plasticity functions $\Lambda^+(\Delta t)$ and $\Lambda^-(\Delta t)$ are inverted and multiplied by -1 since potentiation (depression) of inhibitory weights makes them converge towards -1 (0). Finally, $\tau_l = 0.2$s is the learning time scale for the adaptation.

4 Experiments and Results

4.1 Experimental Setup

A Khepera IV robot was placed on a table in the robotics laboratory. A human interacted with the robot, stimulating its IR sensors by inducing contact within the sensory field area with his hand (as shown in 2) in the following experimental conditions. Three repetitions of touch tests focusing on four metrics were carried out: Speed (Slow, Fast), Frequency (Low, High), Duration (Short, Long), and Intensity (Diffuse, Intense). For each metric, sensors underwent individual testing under each condition, resulting in a total of 64 tests (8 sensors x 4 metrics x 2 conditions). To mitigate bias, both the order of conditions and the sensors were randomized. Each touch event was separated by 1 s, with sensor data captured at a frequency of 100 Hz, ensuring a comprehensive dataset. The inclusion of randomization in sensor and condition order aimed to prevent any sequential bias and ensure the robustness of the experimental results. The experiment lasted 20 min, and data were collected throughout. The experiment was run 5 times, with similar results. Below we discuss a representative run.

4.2 Results

Nociceptors and Mechanoceptors Output. In Fig. 2, we can observe the output of the nociceptive vector and of the physical information we used to compute it in a specific time windows between 125 and 150 s. Mechanoceptor vector has been computed as described in 2.2. These data are sent to the neuronal model.

Dynamics During Learning. We first describe the dynamics of the network during learning of external sensory stimuli in Fig. 3, A. First, we observe that neurons associated with mechanoreceptors (excitatory neurons 64 to 127 and inhibitory neurons 144 to 159) respond to tactile input slightly earlier than neurons associated with nociceptors (excitatory neurons 0 to 63 and inhibitory neurons 128 to 143). This time delay between the information from the two types of sensor stems from the beta (for mechanoreceptors) and delta (for nociceptors) fibers, which transmit sensory information to the two cortical areas at different speeds.

In the somatosensory area (see light spikes), a touch on the robot is characterized by an increase in the firing rate of neurons (excitatory and inhibitory) associated with a specific location on the robot's body. Thus, a more spread contact will activate more neurons than a more targeted contact. In addition, a deeper touch will increase neuron activity more than a softer touch.

On the side of the anterior cingulate area (see dark spikes), neurons (excitatory and inhibitory) increase their activity in the presence of particular touch features such as frequency bands, duration bands, deformation shapes and velocity bands. In this way, certain neurons are never activated at the same time, given the opposite nature of the features they encode (e.g. intense versus diffuse deformation). Conversely, certain features can be activated at the same time

Emergence of Complex Representation of Touch 113

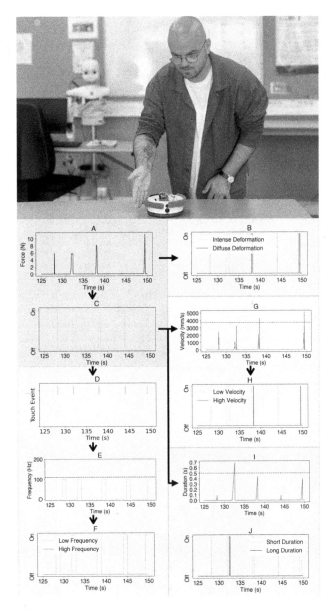

Fig. 2. Picture of our experimental setup (top), and Dynamic Interactions and Time-Based Analysis of Tactile Sensory Events in a Nociceptive Model. Black arrows indicate the directional interactions between different tactile sensory events within the same time window. (A) Mean force applied to the nociceptive field over time. (B) Partitioning of deformation on the nociceptive blob over time. (C) Activation of touch on the blob over time, true if a certain amount of force is applied to the blob. (D) Touch event over time. (E) Frequency of touch in Hz over time. (F) Separation of frequency bands over time. (G) Velocity of touch in mm/s over time. (H) Segregation of velocity bands over time. (I) Duration of touch over time. (J) Division of duration bands over time.

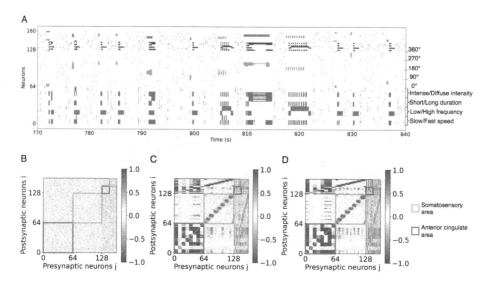

Fig. 3. Neuronal simulation of touch stimulation. (A) The raster plot displays the firing times of excitatory (red dots) and inhibitory (blue dots) neurons during the simulations. Dark red (blue) dots represent spikes in the anterior cingulate cortex (nociceptors), while light dots represent spikes in the somatosensory cortex (mechanoreceptors). (B,C,D) The matrices represent the connection weights between neurons at the beginning, middle and end of the simulation. The color denotes if the connection is excitatory (red) or inhibitory (blue) or absent (white). The magenta area represents the ACC with touch features, while the green area represents the somatosensory cortex with locations of touch. (Color figure online)

when they are not incompatible (e.g. a diffuse, long and low-frequency touch). In addition, it should be noted that some feature neurons activate only after other features have been activated. For example, a touch must be perceived as short before it can be considered long.

Resulting Weight Connectivity. This learning leads to the formation of particular structures in the weights connectivity of Fig. 3, B to D. Firstly, at the global level, we observe the formation of two modular structures, where excitatory neurons associated with the same sensory receptors (either nociceptors in magenta or mechanoreceptors in green) share strong connections, while connections between the two sensory areas are sparse and weak. This segregation between cortical areas is made possible by the beta and delta fibers described above, which prevent temporal correlations between information from nociceptors and mechanoreceptors, and hence their structural reinforcement.

In the area associated with mechanoreceptors (in green), we find that neurons associated with physically close sensors are strongly connected, while connections between distant areas are essentially suppressed. In other words, we obtain the formation of a kind of ring connectivity representing the robot's body.

Concerning the nociceptors area (in magenta), we can see that neurons coding for contradictory information (e.g. low and high frequencies) are totally decoupled. Nevertheless, some distinct feature neurons are strongly connected, showing that certain types of touch are characterized by different features. In particular, we find that some feature neurons, such as those associated with low frequency touch, share connections with all other feature neurons. Indeed, these features almost always remain active, which also explains the few weak connections between the two cortical areas.

5 Discussion and Conclusion

In this paper, we have presented a sensory body model for a mobile robot able to capture complex information about touch, including painful touch, from few data. We further investigated the coding of nociceptive features in an artificial neural model of Anterior Cingulate Cortex, a brain region associated with processing emotions and pain [25]. In the representation that emerged, we observed correlations and decorrelations between some nociceptive features extracted from different types of noxious stimuli, suggesting specific coding mechanisms.

Using a Khepera IV robot, we modeled a specific sensory body and tested various noxious stimuli designed to elicit distinct responses. The observed correlations and decorrelations between specific nociceptive inputs imply that some features co-activate, potentially encoding and discriminating different types of noxious stimuli. For example, in skin, pressure, scratch, and pinch stimuli activate distinct features. These observations support Acuña's [2] description of nociceptive coding in the ACC.

We also investigated how tactile information from nociceptors and mechanoreceptors can be learned by an artificial neural network to form two distinct areas, comparable to what can be observed in biology with the anterior cingulate cortex and the somatosensory cortex. This highlights the segregation of information, with the specialization of brain regions for specific tasks or modalities [29]. These results echo those obtained by Bergoin et al. in [5, 6] with simpler stimuli.

Further, the ACC and somatosensory cortex individually provide information on the characteristics of the touch (what and how) and on its location (where). More precisely, in our somatosensory network, we found that neurons coding for physically close robot body parts were more strongly connected than those for distant ones. This reminds us of the concept of semantic memory, where we find an association between mental representations and topology [29].

Our model contributes to the explainability of robot behavior, since, given that the neural network is able to react to particular features and the location of touch, we could read these neural activities directly and associate them with particular behaviors or reflexes.

In future work we could learn these associations and teach the robot to link certain features with types of pain or pleasure and particular movements. Finally, the neural network used would allow us to carry out experiments in more complex and changing environments, and assess the ability of the model to maintain continual learning without catastrophic forgetting of memories.

Acknowledgement. LLH and this research are supported by an INEX Chair in Neuroscience and Robotics to LC.

References

1. Abdelwahed, M., Zerioul, L., Pitti, A., Romain, O.: Using novel multi-frequency analysis methods to retrieve material and temperature information in tactile sensing areas. Sensors **22**(22), 8876 (2022)
2. Acuña Miranda, M.A., Kasanetz, F., De Luna, P., Falkowska, M., Nevian, T.: Principles of nociceptive coding in the anterior cingulate cortex. Proc, Nat. Acad. Sci. USA-PNAS **120**(23) (2023)
3. Bagnato, C., Takagi, A., Burdet, E.: Artificial nociception and motor responses to pain, for humans and robots. In: 2015 37th Annual International Conference of the IEEE Engineering in Medicine and Biology Society (EMBC), pp. 7402–7405. IEEE (2015)
4. Bergoin, R.: The role of inhibitory plasticity in the formation and the long-term maintenance of neural assemblies and memories. Ph.D. thesis, CY Cergy Paris Université; Universitat Pompeu Fabra (2023)
5. Bergoin, R., Torcini, A., Deco, G., Quoy, M., Zamora-Lopez, G.: Inhibitory neurons control the consolidation of neural assemblies via adaptation to selective stimuli. Sci. Rep. **13**(1), 6949 (2023)
6. Bergoin, R., Torcini, A., Deco, G., Quoy, M., Zamora-López, G.: Emergence and long-term maintenance of modularity in spiking neural networks with plasticity. arXiv preprint arXiv:2405.18587 (2024)
7. Bi, G.Q., Poo, M.M.: Synaptic modifications in cultured hippocampal neurons: dependence on spike timing, synaptic strength, and postsynaptic cell type. J. Neurosci. **18**(24), 10464–10472 (1998)
8. Christensen, A., Corey, D.: Trp channels in mechanosensation: direct or indirect activation? Nat. Rev. Neurosci. **8**, 510–521 (2007)
9. Cobo, R., et al.: Peripheral mechanobiology of touch-studies on vertebrate cutaneous sensory corpuscles. Inter. J. Molecular Sci. **21**(17) (2020). https://doi.org/10.3390/ijms21176221
10. Dubin, A.E., Patapoutian, A., et al.: Nociceptors: the sensors of the pain pathway. J. Clin. Investig. **120**(11), 3760–3772 (2010)
11. Ermentrout, B.: Type i membranes, phase resetting curves, and synchrony. Neural Comput. **8**(5), 979–1001 (1996)
12. Gilson, M., et al.: Effective connectivity inferred from fmri transition dynamics during movie viewing points to a balanced reconfiguration of cortical interactions. Neuroimage **180**, 534–546 (2018)
13. Hardt, O., Nader, K., Nadel, L.: Decay happens: the role of active forgetting in memory. Trends Cogn. Sci. **17**(3), 111–120 (2013)
14. Johnson, K.O.: The roles and functions of cutaneous mechanoreceptors. Curr. Opin. Neurobiol. **11**(4), 455–461 (2001)
15. Julius, D., Basbaum, A.: Molecular mechanisms of nociception. Nature (2001)
16. Kandel, E.: Principles of Neural Science. McGraw-Hill, New York (2013)

17. Lamsa, K.P., Heeroma, J.H., Somogyi, P., Rusakov, D.A., Kullmann, D.M.: Anti-hebbian long-term potentiation in the hippocampal feedback inhibitory circuit. Science **315**(5816), 1262–1266 (2007)
18. L'Haridon, L., Cañamero, L.: The effects of stress and predation on pain perception in robots. In: 2023 11th International Conference on Affective Computing and Intelligent Interaction (ACII) pp. 1–8. IEEE (2023)
19. Maniscalco, U., Infantino, I.: An artificial pain model for a humanoid robot. In: De Pietro, G., Gallo, L., Howlett, R.J., Jain, L.C. (eds.) KES-IIMSS 2017. SIST, vol. 76, pp. 161–170. Springer, Cham (2018). https://doi.org/10.1007/978-3-319-59480-4_17
20. Markram, H., Toledo-Rodriguez, M., Wang, Y., Gupta, A., Silberberg, G., Wu, C.: Interneurons of the neocortical inhibitory system. Nat. Rev. Neurosci. **5**(10), 793–807 (2004)
21. Parvizi-Fard, A., Salimi-Nezhad, N., Amiri, M., Falotico, E., Laschi, C.: Sharpness recognition based on synergy between bio-inspired nociceptors and tactile mechanoreceptors. Sci. Rep. **11**(1), 2109 (2021)
22. Pawlaczyk, M., Lelonkiewicz, M., Wieczorowski, M.: Age-dependent biomechanical properties of the skin. Adv. Dermatol. Allergology/Postępy Dermatologii i Alergologii **30**(5), 302–306 (2013)
23. Perez, Y., Morin, F., Lacaille, J.C.: A hebbian form of long-term potentiation dependent on mglur1a in hippocampal inhibitory interneurons. Proc. Nat. Acad. Sci. **98**(16), 9401–9406 (2001)
24. Perl, E.: Myelinated afferent fibers innervating the primate skin and their response to noxious stimuli. J. Physiol. **197**, 593–615 (1968)
25. Rainville, P., Duncan, G.H., Price, D.D., Carrier, B., Bushnell, M.C.: Pain affect encoded in human anterior cingulate but not somatosensory cortex. Science **277**(5328), 968–971 (1997)
26. Scannell, J.W., Blakemore, C., Young, M.P.: Analysis of connectivity in the cat cerebral cortex. J. Neurosci. **15**(2), 1463–1483 (1995)
27. Vallbo, A.B., Johansson, R.S., et al.: Properties of cutaneous mechanoreceptors in the human hand related to touch sensation. Hum. Neurobiol. **3**(1), 3–14 (1984)
28. Wang, L., Ma, L., Yang, J., Wu, J.: Human somatosensory processing and artificial somatosensation. Cyborg Bionic Syst. (2021)
29. Zamora-López, G., Zhou, C., Kurths, J.: Exploring brain function from anatomical connectivity. Front. Neurosci. **5**, 83 (2011)

Collective Behavior

Analyzing Multi-robot Leader-Follower Formations in Obstacle-Laden Environments

Zachary Hinnen[✉] and Alfredo Weitzenfeld

University of South Florida, Tampa, FL 33620, USA
{zhinnen,aweitzenfeld}@usf.edu

Abstract. Observations in biologically inspired swarm formations from nature, like flocks of birds, herds of mammals, and packs of wolves, have inspired the innovation of various multi-robotic architectures. This work presents a robotic system that mimics leader-follower behaviors in the navigation and formation of sparse and dense environments. This work extends the original work by Weitzenfeld et al. to evaluate new swarm-based multi-robot architectures with obstacle avoidance and variations in group formations. The multiple robot architecture is based on a wolf pack with a defined 'alpha wolf,' which acts as the leader, and defines 'betas wolves,' which act as followers. The 'alpha wolf' leads multiple 'beta wolves' that follow in formation behind the lead wolf, keeping track of a group member and maintaining a set angle and distance while performing obstacle avoidance, staying in formation, and performing speed adjustment. Variations in swarm formation behaviors being analyzed with robots include (1) beta robots following the alpha robot, (2) beta robots following the closest neighboring robot, and (3) robots following the same robot identified since the beginning. Experiments are performed in simulation, using Webots, to analyze robot formations.

Keywords: Biologically Inspired · Multi-Robot · Swarm

1 Introduction

Animal swarms and pack formations have served as inspiration for multi-robot swarm formations in complex environments. Nature's portrayal of collaborative and synchronized movement in schools of fish and flocks of birds has sparked significant interest in understanding the underlying principles of collective intelligence. This concept, rooted in the interactions of simple individuals forming substantial groups, enables swarms to achieve objectives through cumulative interactions [1]. These insights have advanced biological sciences and fostered transformative applications in robotics, especially in multi-robot system development, while creating multiple challenges when navigating in obstacle-laden environments [2,3], and adaptability requirements amidst uncertainties [3, 4].

Inspired by previous work, this study focuses on crafting a leader-follower swarm formation algorithm sustaining cohesion while navigating obstacle-laden terrain. Experiment outcomes revolve around environment layout, formation patterns, leader-follower velocities, robot quantity, and adopted robot-tracking behaviors [2,3]. Throughout experiments, constants include the maintenance strategy, responses to obstacles or peers, and methods for reestablishing formation.

2 Literature Review

This work is inspired by visually guiding animals, many using optic flow, for navigation and collision avoidance [4,5]. Swarming, flocking, and schooling behaviors in fish schools and bird flocks serve as basis for self-organization driven by predation risk minimization [6]. Cohesive motions within these groups adjust based on proximity, maintaining shapes like spheres or oblongs [6,7]. The cohesion and stability of formations rely on velocity adjustments, emphasizing the significance of proximity in dictating movement patterns. These collective behaviors highlight the complex interplay of fluid dynamics, visual perception, and social interactions in shaping organism motion and organization. Swarm robotics, inspired by natural collective behaviors, emphasizes the emergence of collective behavior from local interactions among agents [8]. Craig Reynolds' pioneering work on bird flocking behavior laid the foundation for simulated environments like Boids, which simulate cohesive yet diverse motion in virtual bird flocks [3]. The principles of swarm robotics have been applied to various domains, including multicopter flight and robotic schools of fish, drawing inspiration from natural swarming behaviors [9,10]. Underwater robotics systems have replicated the seamless coordination seen in fish schools using local implicit vision-based coordination [11] Bionic visual perception techniques inspired by natural visual acuity have been explored to enhance environmental perception and recognition [12]. Additionally, bird swarm algorithms now incorporate simulated social behaviors to mimic foraging and vigilance behaviors [13]. Drawing inspiration from efficient animal herding, bio-herding proposes employing unmanned aerial vehicles to monitor and herd animal groups [14]. These approaches present promise and challenges in understanding animal responses, developing suitable algorithms, and integrating them into software and hardware architecture for practical applications.

3 Behavior Model

This research aims to enhance robot formations' scalability and obstacle avoidance, using prior wolf pack models as inspiration, Weitzenfeld, et al. [2], and Reynolds' Boids model for bird flocking [3]. The current study expands on cohesion, separation, alignment, and obstacle avoidance.

3.1 Alpha-Beta Model

The alpha-beta wolf pack model comprises an alpha leader and multiple beta followers. The alpha leads the pack's navigation, while the betas adopt one of three behaviors: (1) following the alpha, (2) the closest robot, or (3) a specific robot set initially as the closest robot. The Leader-Follower formation model operates under the following assumptions:

1. Groups consist of 5 to 9 green-colored followers and one red-colored leader
2. Followers maintain an initial formation while autonomously adjusting positions to avoid obstacles and other robots throughout the navigation process.
3. Each robot uses local visual information and sensor data to assess its environment and adjust velocity.
4. Dynamic movement speeds are based on the relative positions of robots relative to their target, with a baseline velocity maintained.
5. Individual robot identification within the group is available for robots to always follow the same other robot.
6. Experiments are completed with static obstacles in a controlled environment.

3.2 Formation Behaviors

The leader behavior is defined as wandering, i.e. random, navigation. while three distinct formation-following behaviors are developed:

1. Follow-the-Leader: This pack-following behavior is based on [2], where all followers trail the leader.
2. Follow-the-Nearest-Neighbor: Each follower tracks and follows the nearest robot.
3. Follow-the-Same-Initial-Neighbor: A robot ID is assigned to the closest initial robot, enabling subsequent navigation by following this identified robot. This system simplifies recognition in realistic scenarios, such as distinguishing animals based on unique sensory information like color patterns.

Follow the Leader. Follow the leader behavior expands upon traditional formations of three followers and one leader while evaluating outcomes as the number of robots involved increases, follower behavior can be see in Fig. 1.

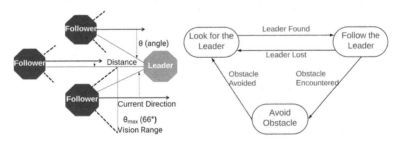

Fig. 1. The figure (Left) illustrates the follow the leader formation configuration, with a leader in red and followers in green. The state diagram in the figure (Right) describes the behavior. (Color figure online)

Follow the Nearest Neighbor. The Follow Nearest Neighbor behavior focuses on tracking the closest neighboring robot within visual range. If multiple robots are visible, the behavior selects the nearest one to follow; if only one is visible, it follows that one. The closest robot may be either the leader or another follower. The following behavior process can be observed in Fig. 2.

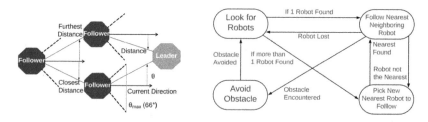

Fig. 2. The figure (Left) illustrates the nearest neighbor formation configuration with a leader in red and followers in green. The state diagram in the figure (Right) describes the behavior. (Color figure online)

Follow the Initial Nearest Neighbor. The Follow Initial Neighbor behavior entails identifying the nearest neighboring robot within visual range (uniquely identified by an "ID") and maintaining a constant relative angle and distance, as seen in Fig. 3 (left); while Fig. 3 (right) illustrates state-based decision making. Due to ID tracking, this behavior does not differentiate between following a leader or a follower. However, an initial formation ensures that at least one follower tracks the leader.

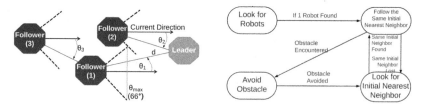

Fig. 3. The figure (Left) shows the follow-initial-nearest-neighbor formation configuration with the leader in red and followers in green. The state diagram in the figure (Right) describes the behavior. (Color figure online)

4 Experiments and Results

4.1 Robot Velocity and Orientation

To adjust robot velocity and orientation, the model uses a proportional feedback controller to maintain a constant distance and angle with respect to the robot being followed. The controller error function e(t) is set as the difference between the desired constant distance and angle with respect to their current values, where the control signal adjusts robot velocity (speed and orientation), u(t), calculated by taking the error function and multiplying it with a proportional gain K_p (set as 1.2). This control is then adjusted using a saturation motor velocity function $u_r(t)$) after applying a saturation f_{sat} that limits the maximum and minimum motor velocity adjustments (maximum is 6.28 and min is 0). Control functions are described by Eqs. 1–3.

$$e(t) = r(t) - y(t) \tag{1}$$

$$u(t) = K_p * e(t) \tag{2}$$

$$u_r(t) = f_{\text{sat}} * (u(t)) \tag{3}$$

Figure 4 (Left) illustrates the relative distance and angle are indirectly controlled by velocity. The figure depicts the maximum ($\theta_{\max}$), minimum ($\theta_{\min}$), and center (θ_{center}) angle values, along with the closest (d_{close}) and furthest (d_{far}) distance values, also showing initial target angle (θ_0) and distance (d_0). The PID feedback control loop is described in Fig. 4 (Right).

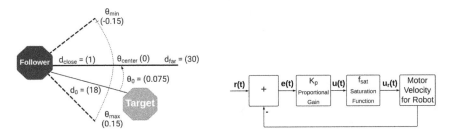

Fig. 4. Follower Tracking Diagram with Max/Min Distance and Angle Values (left) and Proportion Control PID Diagram (right)

4.2 Robot Simulation Environment

Webots [15], an open-source physically realistic robot system simulation software was used that provided built-in visual object color recognition to distinguish between different robots within the environment, with alpha robots in red and beta robots numbered in green, as shown in Fig. 5 (Left). Leader and closest neighbor robot are computed using relative position and orientation using the follower robot camera. An internal robot ID tracked a specific robot being followed, while GPS recorded the followers' global positions; Fig. 5 (right).

Diverse behaviors were tested within an obstacle-laden environment, assessing adaptability and performance across different obstacle positions and sizes. Small obstacles, scaled down and positioned in various locations, were used to

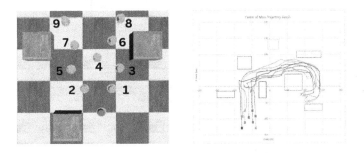

Fig. 5. (Left)Multiple robot formation in Webots simulator. (Right) illustrates the follow-the-nearest-neighbor path in a large obstacle environment.

evaluate maneuverability and responsiveness; an enlarged version can be seen in Fig. 5 (Left). Large obstacles were used to test sustained obstacle avoidance behaviors while maintaining formation integrity. The study aimed to understand how the robot formations navigate through different obstacle configurations.

4.3 Initial Formation Configurations

Various robot formations are evaluated, each with a single leader and different numbers of followers (5, 7, and 9) with initial formation configurations. Figure 6 shows the initial formation configuration of seven and nine robots.

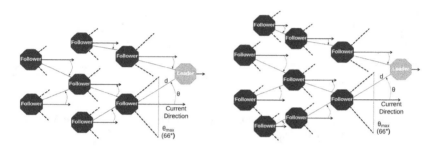

Fig. 6. Formation setup with 7 followers (left) and 9 followers (right). For the purpose of these experiments, the 9 robot formation will be primarily used.

4.4 Metrics

The experiment metrics are based primarily on center of mass (COM) of the formation, i.e. formation's center computed from each robot individual x and y locations. For example, for a 9-robot swarm with robots numbered 1-9 (see Fig. 5), $(x_i(t), y_i(t))$ represent the position of follower i at time t. Note that we use Webots GPS coordinate to obtain x and y robot positions).

COM for the formation system is calculated by averaging the $x_i(t)$ and $y_i(t)$ position data for all followers. This results in the COM for the x and y coordinate plane $(C_x(t), C_y(t))$ as described in Eq. 4 and 5:

$$C_x(t) = \frac{1}{N} \sum_{i=1}^{N} x_i(t) \tag{4}$$

$$C_y(t) = \frac{1}{N} \sum_{i=1}^{N} y_i(t) \tag{5}$$

We then compute the distances between each follower and the COM at time t. For each follower i, the distance $(d_i(t))$ to the $(C_x(t), C_y(t))$ at time t is calculated using the distance formula shown in Eq. 6.

$$d_i(t) = \sqrt{(x_i(t) - C_x(t))^2 + (y_i(t) - C_y(t))^2} \tag{6}$$

After establishing the distances between the followers and the COM, we track changes over time by recording the difference between each follower's current distance to the COM and its initial distance $d_i(0)$. Denoted as $\Delta d_i(t)$, this difference indicates whether a follower has moved closer to or farther away from the COM compared to its initial position, see Eq. 7.

$$\Delta d_i(t) = d_i(t) - d_i(0) \tag{7}$$

To further analyze formation dynamics, we compute the average $\bar{\Delta d_i}(t)$ (Eq. 8) and standard deviation s_i (Eq. 9) of $\Delta d_i(t)$ over time for all followers. The average distance $\bar{\Delta d_i}(t)$ is calculated by summing the distances traveled by follower i at each of the T time points and dividing by the total number of time points T. This metric offers insight into the typical distance covered by follower i throughout the entire runtime of the analysis and provides context for understanding follower behavior within the formation.

$$\bar{\Delta d_i}(t) = \frac{1}{T} \sum_{t=1}^{T} \Delta d_i(t) \tag{8}$$

$$s_i = \sqrt{\frac{1}{T-1} \sum_{t=1}^{T} (\Delta d_i(t) - \bar{\Delta d_i}(t))^2} \tag{9}$$

The normal distribution's probability density function (PDF) computes differences over time between a follower's current and initial distances to the COM. Denoted as $P(\Delta d_i)$, it uses the average $\bar{\Delta d_i}$ (Eq. 8) and standard deviation s_i (Eq. 9) of these differences. The formula assigns higher probabilities to differences closer to the average, while s_i quantifies their dispersion. By applying normal distribution, formation's stability is analyzed and outliers are noted.

$$P(\Delta d_i(t)) = \frac{1}{s_i \sqrt{2\pi}} \exp\left(-\frac{(\Delta d_i(t) - \bar{\Delta d_i}(t))^2}{2s_i^2}\right) \tag{10}$$

4.5 Results

Experiments include four behaviors: Follow the Leader, Follow Nearest Neighbor, Follow Initial Neighbor, and Varying Leader Speed. The first three maintain constant speed along random paths, while the last varies the leader's speed in 0.5 rad/s increments. Linear speeds are calculated as follows: 2 rad/s: 0.0796 m/s (constant speed), 2.5 rad/s: 0.0995 m/s, 3 rad/s: 0.1194 m/s, 3.5 rad/s: 0.1393 m/s, and 4 rad/s: 0.1592 m/s, resetting after 4 rad/s. Detailed results are in the figures.[1]

Follow the Leader Results. The experiment lasted 60 s due to formations becoming unusable beyond this time. 9 followers were used, but leader tracking was very inconsistent throughout follower groups of 5, 7, and 9. Results are shown in Fig. 7.

---

[1] See Hinnen MS thesis [16] for further configurations and results.

 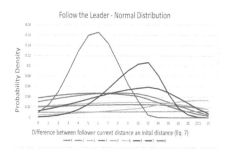

Fig. 7. The figure (left) illustrates the variation between each follower's current and initial distance to the COM (computed using Eq. 7) over the experiments runtime while (right) presents the normal distribution of the difference in distance over time. The x-axis is the distance difference (Eq. 7), and the y-axis is the probability density (Eq. 10); probability density is calculated using standard deviation (Eq. 9) and average (Eq. 8).

Follow Nearest Neighbor Results. The Follow the Nearest Neighbor experiment ran for 120 s, longer than follow the leader but shorter than initial-nearest-neighbor with Varying Leader Speed. However, the formation consistently broke apart after this duration, rendering formation following unusable. Comparatively, using 9 followers proved just as effective as seven for nearest neighbor following. Results are shown in Fig. 8.

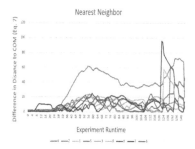 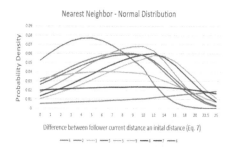

Fig. 8. The distance over time (left) and Normal Distribution for nearest neighbor has the same parameters and equations as Fig. 7 (right).

Follow Initial Neighbor Results. The Follow the Nearest Neighbor experiment ran for 180 s, the longest duration possible for optimal mapping and data recording. The experiment was deliberately halted at 180 s to ensure precision. 9 followers were used due to the consistent success among follower counts of 5, 7, and 9. Results are shown in Fig. 9.

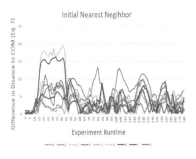
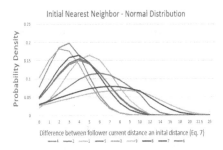

Fig. 9. The distance over time (left) and Normal Distribution for initial nearest neighbor has the same parameters and equations as Fig. 7 (right)

Varying Leader Speed Results. The application of leader speed changed was only applied for initial nearest behavior and the results in Fig. 10 reflect that. 9 followers were used due to the consistent success among follower counts of 5, 7, and 9.

Fig. 10. The distance over time (left) and Normal Distribution for Varying Leader Speed has the same parameters and equations as Fig. 7 (right)

5 Discussion

Table 1 shows the distance averages and standard deviation for behavior experiments with constant leader speed: Follow the Leader (C-FTL), Follow Nearest Neighbor (C-NN), and Follow Initial Nearest Neighbor (C-INN). Follow Initial Nearest Neighbor had a secondary test with Varying Leader Speed (V-INN).

Examining average distances between followers' initial and current positions relative to the center of mass reveals distinct patterns. C-INN behavior is the most successful, with the lowest average distance (4.78) and standard deviation (3.18), indicating consistent and close follower clustering around the COM. V-INN demonstrates moderate metrics (average distance: 6.55, standard deviation: 6.15), balancing cohesion and adaptability. In contrast, C-FTL and C-NN behaviors show higher average distances (C-FTL: 12.17, C-NN: 11.62) and standard

Table 1. The tables list the (left) average $\overline{\Delta d_i}(t)$ (Eq. 8) and (right) standard deviation s_i (Eq. 9) of $\Delta d_i(t)$ over time for all followers.

Follower	C-FTL	C-NN	C-INN	V-INN
1	10.7	12.63	4.98	9.52
2	7.98	8.0	4.05	8.34
3	22.4	9.65	2.35	3.91
4	10.67	32.63	2.75	6.17
5	22.4	4.54	5.93	5.43
6	5.7	8.64	4.02	6.11
7	11.21	10.49	7.06	4.95
8	23.32	11.44	3.72	8.36
9	11.79	6.61	8.16	6.19
Average	12.17	11.62	4.78	6.55

Follower	C-FTL	C-NN	C-INN	V-INN
1	13.33	6.84	2.42	8.77
2	8.35	6.58	2.66	8.07
3	32.18	5.85	2.16	3.33
4	15.69	19.84	2.00	5.81
5	8.43	5.17	3.47	4.09
6	2.39	6.73	2.58	4.82
7	3.64	16.85	5.07	6.62
8	11.7	6.61	2.40	6.90
9	6.73	9.89	5.85	6.97
Average	11.38	9.37	3.18	6.15

deviations (C-FTL: 11.38, C-NN: 9.37), suggesting greater follower variability and dispersion tendencies. This emphasizes the pivotal role of follower behavior selection in achieving desired outcomes within multi-agent systems. INN is the optimal strategy for tight follower clustering around a designated focal point while maintaining an initial formation.

6 Conclusions and Future Works

6.1 Conclusions

The paper presents a multi-robot navigation model inspired by animal behavior, exploring various formations and behaviors including Follow-the-Leader, Follow-Nearest-Neighbor, and Follow-Initial-Neighbor. Follow-Initial-Neighbor consistently outperformed the others, while Follow-the-Leader exhibited the poorest outcome. The paper's contributions include extending existing formation models to enhance obstacle avoidance and accommodate a large number of follower robots, analyzing distinct following behaviors in multi-robot formation algorithms, assessing the impact of different environments and formations on formation-based navigation efficacy, and providing a deployable code environment to support further research in formation-based navigation.

6.2 Future Work

The model's limitations stem from constraints within the visual processing system and initial behavior development. Incorporating bilateral vision or lidar would enhance obstacle avoidance capabilities and allow for better environmental perception. Distance-based velocity control and a multi-level follower behavior model could improve follower performance while distinguishing between robot-based and obstacle-based avoidance, enhancing overall performance and adding the potential for mapping algorithms. Overcoming these limitations would allow for better mimicry of predator-prey behavior and swarm leadership, aligning with natural formations like bird flocks and fish schools. Future work will focus on integrating dynamic obstacle handling and reinforcement learning to develop the model further, aiming for better adaptability to spatial configurations and advanced visual techniques resembling biological perception and behaviors.

Finally, our results can be seen as mirroring those found in bird flocks and fish schools, where following the initial nearest neighbor provide stability akin to a fixed reference point, while following the nearest neighbor show shifting formation integrity akin to dynamic interactions in bird flocks. Research by Ballerini et al. [17] illustrates that birds interact with six or seven nearest neighbors, and studies by Berlinger et al. [11] reflect fish schools relying on visual cues from nearby neighbors. Although we haven't yet tracked multiple nearest neighbors for a centralized position, future research will explore this direction, including their implementation with both wheeled robots and drones.

References

1. Liu, Y., Passino, K.M.: Swarm intelligence: Literature overview. Department of electrical engineering, the Ohio State University (2000)
2. Weitzenfeld, A., Vallesa, A., Flores, H.: A biologically-inspired wolf pack multiple robot hunting model. In: 2006 IEEE 3rd Latin American Robotics Symposium, pp. 120–127. IEEE (2006)
3. Reynolds, C.W.: Flocks, herds and schools: A distributed behavioral model. In: Proceedings of the 14th Annual Conference on Computer Graphics and Interactive Techniques, pp. 25–34 (1987)
4. Niehorster, D.C.: Optivc flow: a history. i-Perception **12**(6), 20416695211055770 (2021)
5. Bhagavatula, P.S., Claudianos, C., Ibbotson, M.R., Srinivasan, M.V.: Optic flow cues guide flight in birds. Curr. Biol. **21**(21), 1794–1799 (2011)
6. Hemelrijk, C.K., Hildenbrandt, H.: Schools of fish and flocks of birds: their shape and internal structure by self-organization. Interface focus **2**(6), 726–737 (2012)
7. Goodenough, A.E., Little, N., Carpenter, W.S., Hart, A.G.: Birds of a feather flock together: insights into starling murmuration behaviour revealed using citizen science. PLoS ONE **12**(6), e0179277 (2017)
8. Sahin, E., Labella, T.H., et al.: Swarm-bot: pattern formation in a swarm of self-assembling mobile robots. In: IEEE International Conference on Systems, Man and Cybernetics, vol. 4, pp. 6–pp. IEEE (2002)
9. Vásárhelyi, G., et al.: Outdoor flocking and formation flight with autonomous aerial robots. In: 2014 IEEE/RSJ International Conference on Intelligent Robots and Systems, pp. 3866–3873. IEEE (2014)
10. Swain, D.T., Couzin, I.D., Leonard, N.E.: Real-time feedback-controlled robotic fish for behavioral experiments with fish schools. Proc. IEEE **100**(1), 150–163 (2011)
11. Berlinger, F., Gauci, M., Nagpal, R.: Implicit coordination for 3d underwater collective behaviors in a fish-inspired robot swarm. Sci. Robot. **6**(50), eabd8668 (2021)
12. Duan, H., Xu, X.: Create machine vision inspired by eagle eye. Research (2022)

13. Meng, X.B., Gao, X.Z., Lu, L., Liu, Y., Zhang, H.: A new bio-inspired optimisation algorithm: bird swarm algorithm. J. Exper. Theoretical Artifi. Intell. **28**(4), 673–687 (2016)
14. King, A.J., et al.: Biologically inspired herding of animal groups by robots. Methods Ecol. Evol. **14**(2), 478–486 (2023)
15. https://www.cyberbotics.com/
16. Hinnen, Z.J.: Analyzing Multi-Robot Leader-Follower Formations in Obstacle-Laden Environments. Master's thesis, University of South Florida (2023)
17. Ballerini, M., et al.: Interaction ruling animal collective behavior depends on topological rather than metric distance: evidence from a field study. Proc. Nat. Acad. Sci. **105**(4), 1232–1237 (2008)

Spatio-Temporal Dynamics of Social Contagion in Bio-inspired Interaction Networks

Yunus Sevinchan[1,2(✉)], Carla Vollmoeller[1,2], Korbinian Pacher[3,4], David Bierbach[1,3,4], Lenin Arias-Rodriguez[5], Jens Krause[1,3,4], and Pawel Romanczuk[1,2]

[1] Research Cluster of Excellence 'Science of Intelligence', Technical University Berlin, Berlin, Germany
[2] Institute for Theoretical Biology, Department of Biology, Humboldt University of Berlin, Berlin, Germany
yunus.sevinchan@hu-berlin.de
[3] Department of Biology and Ecology of Fishes, Leibniz-Institute of Freshwater Ecology and Inland Fisheries, Berlin, Germany
[4] Faculty of Life Sciences, Albrecht Daniel Thaer-Institute, Humboldt University of Berlin, Berlin, Germany
[5] División Académica de Ciencias Biológicas, Universidad Juárez Autónoma de Tabasco, Villahermosa, Mexico

Abstract. In collective biological systems, social contagion processes play an important role in evaluating and processing information on the level of the collective. These group-level abilities typically arise from individual-level mechanisms and through local interactions. We are interested in the role of these mechanisms and their effect on the system's response to environmental inputs.

In this paper, we present a spatially embedded network model that is inspired by large fish shoals performing collective action in response to predation. We compare the observations of spatio-temporal dynamics in the model simulations with empirical observations, studying specifically the effect of spatial heterogeneities on system activity. The model demonstrates how already simple mechanisms suffice to represent key characteristics of the study system, and highlights the importance of taking into account the spatial embedding for understanding group-level processes in animal collectives.

Keywords: Collective Behavior · Social Contagion · Spatially Embedded Networks · Agent-based Modeling · Animal Collectives

1 Introduction

In animal collectives, social contagion behavior within groups and its spatial embedding are strongly intertwined: Individuals can typically interact only with group neighbors in their immediate surroundings; yet, social contagion and self-organization processes cause group-level responses that manifests on a much larger spatial scale [12,15]. This is in contrast to neuronal systems where dentrites and axons add long-range links that lead to networks with a small-world topology [2], rather than being restricted entirely to local interactions. With the localization of agents in space being a fundamental property of animal collectives, we are interested in the nature of individual-level social contagion mechanisms and their effect on the resulting spatio-temporal collective dynamics.

To that end, we looked at a predator-prey system in southern Mexico consisting of large fish shoals of sulphur mollies (*Poecilia sulphuraria*) and birds attacking these shoals [16]. Besides toxic H_2S in the water and high water temperatures [9], dissolved oxygen levels in the water can be so low that fish need to stay near the surface and perform aquatic surface respiration [7]. While at the water surface, they are preyed upon by fish-eating birds; as a defense mechanism, the shoals perform synchronous and repeated diving behavior, causing ripples on the water surface ("surface waves", see videos), effectively reducing the frequency of predator attacks [3].

Often, thousands of fish can take part in these surface waves, making this system a suitable test-bed for studying collective dynamics in general, and in particular to investigate how information spread is realized and evolutionarily optimized. For instance, a small number of fish may dive down as a reaction to a predator attack; this activity may then spread throughout the network and cause the whole shoal to partake in a sequence of surface waves, sometimes lasting for minutes [3]. These surface waves have certain spatio-temporal characteristics – activity patterns propagating in space and time – which can be compared to model observations, allowing to study the effect of individual-level mechanisms on the resulting wave properties. Thus, empirical observations can motivate mechanisms in a synthetic model of such a system; in turn, the model can be used to generate hypotheses and inspire empirical experiments.

Specifically, in this paper, we investigate the mechanisms by which the fish system is able to produce the observed wave sequences, despite the stimulus no longer being detectable. Furthermore, we ask how spatial heterogeneities like density fluctuations affect the collective response to a stimulus.

We present a spatially embedded network model with agents that individually aggregate information from neighboring agents and their environment. Mediated by a social contagion process, activity can spread throughout the network and thus in space, representing the diving response of individual fish in the wild. The model is inspired by observed behavior of individual fish in the study system, but makes a number of abstractions to make it conceptually tractable and computationally feasible. In contrast to previously studied models employing similar dynamics [4,5], the spatial embedding of the interaction network and controlled manipulation of the network topology allows to investigate the effect these system properties have on the collective-level dynamics of the system.

2 Methods

2.1 Model Description

The fish (or: agents) are represented as nodes in an interaction network, where each node is additionally placed with coordinates in a 2D square domain – the spatial embedding. Weighted links between the nodes represent possible interaction partners, with weights determining the strength of social interaction. For simplicity, links are undirected, meaning that interactions are symmetric. Values for link weights w_{ij} depend on the distance d_{ij} between two nodes (see below).

The interaction network is static: neither node positions nor links change. The assumption underlying this choice is that individual fish in the shoal are interchangeable and the macroscopic structure of the fish shoal remains similar over time, despite fish changing their position within the shoal.

The contagion dynamics are implemented as a three-state model, akin to SIR-type contagion models that have also previously been used to study large human and animal collectives [4,5,14]. Each agent can be in one of three states: at the surface ($\mathcal{S}$), currently diving ($\mathcal{D}$), or underwater ($\mathcal{U}$). At the surface, a fish is susceptible to diving. Diving represents the active state and induces social input to neighbors. The underwater state is a passive refractory state, before the fish resurface and are again susceptible to dive. Essentially, susceptible agents are activated by diving neighbors or if they receive sufficiently strong input from their environment, see Fig. 1.

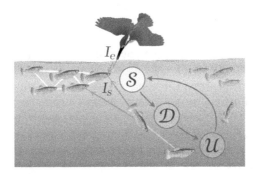

Fig. 1. Schematic depiction of the agent-based computer model and corresponding aspects of the biological study system. White lines between fish depict edges of their interaction network. Agents at the surface (state $\mathcal{S}$) can receive environmental input I_e, for instance from bird attacks, and social input I_s from diving agents (state $\mathcal{D}$) in their neighborhood (magenta edges). After a refractory period (state $\mathcal{U}$), agents resurface.

Overall, the chosen abstraction as a spatially embedded interaction network with fixed positions allows to study the effect of the interaction network properties, while being much simpler than spatially explicit 2D or 3D models with changing positions, which would need many additional assumptions for the movement dynamics of the fish and the cohesion dynamics of the shoal.

Network Construction. Networks are generated through a multi-step process: First, node positions for N agents are determined using a 2D Poisson Disk algorithm [1], creating a macroscopically homogeneous but locally unordered distribution of positions in the domain. To re-introduce local density fluctuations, positional noise is applied by drawing offset vectors from a 2D normal distribution with variance σ_{xy}. The generated positions are re-scaled and shifted such that they fit into a square domain of area L^2. The size of the domain is chosen such that a density of $\rho = 1000\,\mathrm{m}^{-2}$ is reached.

In the next step, links are determined depending on the node positions. In this paper, agents are linked to their first-shell Voronoi neighbors, thus approximating the visual field of the fish [10].

Finally, distance-dependent link weights are computed using $w_{ij} = w_{ji} = \beta_1/(1 + \beta_2 \cdot d_{ij}^{\beta_3})$, with $d_{ij} = d_{ji}$ denoting the metric distance between agents i and j, scaling parameters β_1 and β_2, and power-law exponent β_3.

Dynamics. The temporal dynamics of the model are primarily determined by the state transition rates of the agents. To represent that agents are accumulating cues from their surroundings, we employ a stochastic integrator model. Here, we use a second-order model with a single decision boundary, making it behave similarly to spiking dynamics in models of neuronal systems.

We define an agent-specific excitation $E_i(t)$ which increases depending on the drift $D_i(t)$:

$$\dot{E}_i(t) = -\gamma_E E_i(t) + D_i(t) + \alpha_\xi \xi_i(t) \qquad (1)$$

$$\dot{D}_i(t) = -\gamma_D D_i(t) + \alpha_s I_{s,i}(t) + \alpha_e I_{e,i}(t) \qquad (2)$$

The drift $D_i(t)$ integrates inputs from social interactions (I_s) and from environmental cues (I_e) like predator attacks or artificial stimuli; the factors α_s and α_e are respective interaction strengths. There is an agent-specific noise term, $\xi_i(t) \geq 0$, representing stochastic processes that are not explicitly modeled. Here, the decay rates γ_E and γ_D (chosen as $\gamma_E = 4\,\mathrm{s}^{-1}$ and $\gamma_D = 1\,\mathrm{s}^{-1}$) introduce two time scales at which the excitation and aggregated cues decay.

Social interactions are defined as $I_{s,i}(t) = \sum_{j \in \mathcal{D}_i} w_{ij}$, with $\mathcal{D}_i(t)$ being the set of neighbors of i that are in the diving state at time t and interaction weights w_{ij}. Alternative interaction modes (like quorum decisions) can also be chosen, but are not studied here.

Activation ($\mathcal{S} \to \mathcal{D}$) occurs once E_i passes a threshold $\theta_i = 1$. Subsequently, E_i is reset to zero and social input to neighboring agents is induced. Transition to the refractory underwater state $\mathcal{U}$ happens deterministically after a short time. Agents resurface to $\mathcal{S}$ stochastically after a biologically plausible time. The $\mathcal{D} \to \mathcal{U}$ transition happens deterministically after an agent was in the diving state for 0.1 s. During this time, the excitation E_i is reset to and kept at zero and no inputs ($I_{s,i}$, $I_{e,i}$) are received. Finally, for the $\mathcal{U} \to \mathcal{S}$ transition back to the susceptible state, the time in the $\mathcal{U}$ state is drawn from a normal distribution centered around a mean underwater time of $(2.0 \pm 0.1)\,\mathrm{s}$, corresponding to a sigmoidally-shaped resurfacing probability function.

The model parameters are formulated to correspond to natural units (seconds, meters), allowing to choose biologically plausible parameter values where possible and compare model outputs to observations. For instance, the mean time the fish spend underwater has previously been measured to be 2–4 s [3]. While we report specific values for parameters here, we check model behavior for robustness by also using other values from a plausible range.

Implementation. All numerical simulations and their analyses are carried out using the *Utopia* modeling framework [11,13], which handles simulation configuration, parallelized parameter sweeps as well as efficiently reading, writing, and evaluating high-dimensional simulation output. The model itself is implemented in Python, using NetworkX for representing the spatially embedded networks, and vectorized numpy operations for efficiently determining the social input and step-wise integration of agent-specific excitation and drift levels.

2.2 Observation of Study System

For comparison to the biological study system, we recorded videos of sulphur molly shoals in the wild. We then manually marked the coordinates of bird attacks (by Green kingfisher individuals) on the shoals as well as the resulting starting points of waves. Using a calibration object of known size and a perspective transformation of the video frames, we attained coordinates in the 2D top-down view onto the shoal, where a quantitative geometric analysis becomes possible. The outline of the shoal was estimated from disturbances in the water surface.

3 Results

3.1 Spatio-Temporal Model Activity Resembling Biological Behavior

First, we look at the spread of diving activity in the network after a stimulus is introduced at the center of the domain. In Fig. 2, the network state at two times is shown after a stimulus I_e was introduced. After such a stimulus, the agents that received environmental input cross their excitation threshold θ_i and transition to the diving state $\mathcal{D}$; this activity produces social input I_s to their neighbors, increasing their excitation and causing their activation, which eventually spreads throughout the whole domain. The observed waves are typically circular, only slightly distorted depending on local heterogeneities, and terminate at the boundary of the domain due to no susceptible agents being available to carry on the activity. Given the choice of $\gamma_D = 1\,\mathrm{s}^{-1}$, the effect of the initial stimulus on D_i is negligible already after a short time period.

After an initial wave of activity triggered by the stimulus, subsequent waves emerge at varying locations within the domain, sometimes also at multiple locations, as shown in Fig. 2(b). This is caused by some agents' drift level D_i still

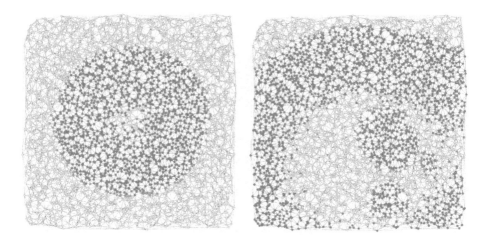

Fig. 2. Snapshots of the network state at different times for a representative simulation with $\alpha_s = 1.1$. Colors denote the agent states $\mathcal{S}$ (gray), $\mathcal{D}$ (orange), and $\mathcal{U}$ (blue). A stimulus was applied at $t = 2.0$ s, initiating system activity. The full animation for this realization and others can be found online. Note that the long-range links along the edge of the domain typically have vanishing link weights and effectively do not contribute to the contagion dynamics. **(a)** At $t = 4.3$ s, diving activity is spreading throughout the whole network. The agents that have initially received the stimulus in the center of the domain have already transitioned back to the susceptible state $\mathcal{S}$ (gray). **(b)** At $t = 12.2$ s, new waves are seen to emerge independently at different locations in the domain and in the wake of a previous wave.

being sufficiently high such that it drives the excitation E_i beyond the threshold. The timing of these new waves is irregular, sometimes allowing the whole system to be in the susceptible state before wave activity recommences from one or more individual agents. In other cases, drift values are not sufficiently high to cause subsequent diving activity, thus leading to system activity ending.

Notably, while the dynamics include a noise term $\alpha_\xi \xi_i$, the diving behavior is not a result of the individual noise input and also occurs with $\alpha_\xi = 0$. Instead, the source of stochasticity in wave locations and timing is a result of the previous interactions, topological heterogeneities, and variations in resurfacing time. The noise input merely causes an effective reduction of the distance to the excitation threshold because of the mean input of α_ξ and thus an overall higher system activity.

Different qualitative regimes of waving behavior can be distinguished in this model, controlled mainly by the social coupling parameter α_s. For instance, for low coupling $\alpha_s \approx 0.3$, a stimulus only elicits very slow-traveling and fragmented activity clusters that quickly cease. For an intermediate coupling strength of $\alpha_s \approx 0.5$, spiral waves can be observed [17]: wave fronts curving in on themselves with a timing that allows the previously active agents to be susceptible again. Such a spiral core dominates the system and sustains its activity,

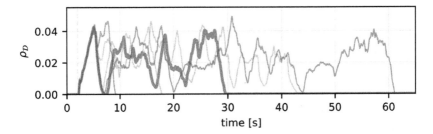

Fig. 3. Representative time series of system diving activity as measured in the fraction of agents in state $\mathcal{D}$. The thick line corresponds to the scenario shown in Fig. 2, other lines are from realizations with identical parameters but varying random number sequences. The dashed line denotes the time a subset of nodes received the stimulus.

often indefinitely. While this behavior is interesting, we do not focus on it here, because it cannot be linked to any observations of fish in the wild.

For interaction strengths beyond $\alpha_s \approx 0.8$, wave fronts travel too quickly for spiral waves to be possible. With increasing social coupling, overall system excitation $\sum_i E_i$ is higher and activity may persist also after the initial stimulus response in form of a circular wave has ceased, subsequently leading to additional waves emerging in the system. With higher α_s, more waves are produced, prolonging the activity period of the system. For very high coupling (exceeding $\alpha_s \approx 1.3$), waving activity does not stop within the simulation time. We call the parameter range producing waving activity that *terminates* within a biologically motivated simulation time "repeat wave regime". During that time, the wave sequences exhibit a variability in wave sizes, duration and inter-wave intervals, manifesting in irregularities in diving activity time series, examples of which are shown in Fig. 3.

Repeat wave behavior can also be observed for different choices of γ_E, γ_D and other system parameters; the specific choice mostly determines the time scale at which a wave emerges and travels through the domain. However, choosing $\gamma_E/\gamma_D \approx 4$ seems to set the appropriate time scale ratio producing irregular waving activity, i.e. locally correlated behavior that spreads throughout the domain *without* leading to global synchronization.

These qualitative observations of the model's behavior allow a comparison to the behavior of the biological study system. In particular, the aforementioned repeat wave regime shows similarities to observations in the field in that waves seem to emerge from locations other than the stimulus location.

To compare this in more detail, we quantified the locations at which new waves emerge in the wild. Figure 4 shows the approximate outline of a fish shoal, ten bird attack locations and the starting points of 250 resulting surface waves. As can be seen there and in the distance distribution, waves do indeed originate from various locations, with roughly half the waves starting more than 2 m away from the attack; given shoal widths of roughly 3–4 m and fish body lengths of 3–4 cm, this can be considered a rather large distance. At the same time, a clustering around the initial attack location is apparent.

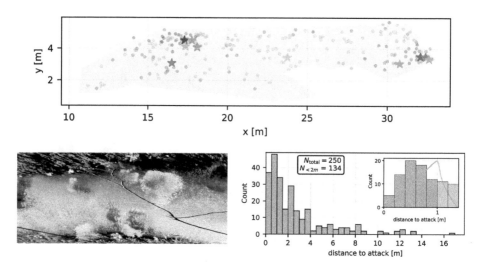

Fig. 4. Empirical observations. (**a**) Manually annotated locations of ten bird attacks (star symbols) and the beginnings of 250 resulting waves (corresponding circles) within the shoal (orange polygon). (**b**) Rectified view of part of the fish shoal with multiple surface waves propagating through the system. (**c**) Distribution of observed distances between wave starting points and the corresponding attack location. (**c, inset**) Distance distribution of waves that started $\geq 5\,\mathrm{s}$ after an attack and originated within a short distance (1.4 m) from the attack location. The orange line shows the KDE of a distance distribution (in arbitrary units) resulting from wave locations that are uniformly distributed within a $L = 2\,\mathrm{m}$ finite domain; the drop-off beyond 1 m is the result of boundary effects, thus a comparison should only be made up to that distance.

Given that domain size of the model is $L = 2\,\mathrm{m}$, a comparison should focus on the distance distribution of waves originating within a short distance of the attack and not on the whole domain; such a distribution is shown in Fig. 4(c, inset). In the model, wave locations are randomly distributed throughout the whole domain: there is no mechanism that would systematically differentiate one location over another, except the boundary layer where input weights are effectively lower. Hence, the distance distribution resulting from the model (orange line in inset) is peaking at 1 m and is then dropping off due to the finite size of the system and the reduced area of the ring-shaped spatial histogram bins.

Comparing the empirically observed distribution to that of these spatially uniformly originating waves suggests that the clustering near the impact location is not a result of a geometrical effect alone, but that it is indeed more likely that waves originate near the location of the bird attack. This may be indicative of fish that have originally witnessed the attack triggering new waves. In the model, in contrast, no such clustering near the impact location can be seen because the information from the initial stimulus is typically lost after the first wave.

3.2 Spatial Heterogeneity Increases System Responsiveness

Next, we studied the effect of spatial heterogeneities on system activity. The distribution of nodes in the domain affects system dynamics via the non-linearly distance-dependent interaction weights w_{ij}. The Poisson Disk algorithm itself produces a macroscopically homogeneous distribution of node positions, prescribing a minimum distance between nodes, while not creating local order. By adding positional noise from a 2D normal distribution with variance σ_{xy}, local density fluctuations are re-introduced, affecting the distribution of link weights, see Fig. 5(a). Specifically, higher positional noise shifts the maximum of the distribution to lower values while increasing the number of higher-valued noise weights from agents that are very close to each other. As the overall weights $\sum_{ij} w_{ij}$ may change as a result of this procedure, we rescaled weights such that the mean weight was fixed at 1.0, allowing a comparison between the various noise levels.

We find that for higher positional noise σ_{xy}, the onset of system activity in reaction to a stimulus moves to smaller social interaction strengths α_s, see Fig. 5(b). In other words: systems with more local heterogeneity maintain activity for a longer time. The underlying reason for increased system activity lies in the increased variance of weights, allowing for local clusters that have a larger mean weight than the global mean weight and thus more social input I_s. Subsequently, drift rates D_i on these agents are larger and they will dive sooner than other local clusters that have lower mean interaction weights.

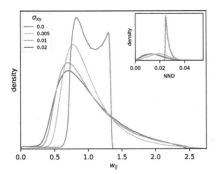
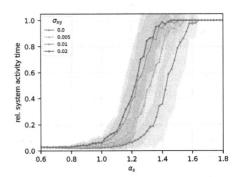

Fig. 5. Weight distributions and system activity depend on spatial heterogeneity induced by positional noise σ_{xy}. For each parameter combination, 120 realizations of spatial networks with 4000 agents were simulated for a simulation time of 240 s. (a) Distribution of interaction weights w_{ij} and nearest-neighbor distance (inset). The abrupt cutoff and bi-modality in the weights distribution for $\sigma_{xy} = 0$ is a result of the minimum distance imposed by the Poisson Disk algorithm (see inset) and corresponding geometric constraints for the distance to other neighbors. (b) System activity depending on spatial noise level σ_{xy} and interaction strength. Higher noise, inducing higher local variations in weights, facilitates strong system responses also for weaker coupling between agents.

Qualitatively, these observations are robust against the specific choice of weighting parameters $\beta_{1,2,3}$ and also other weighting methods like $w_{ij} \propto \exp(-d_{ij})$ or using the log-metric-distance [12], as long as weights are distance-dependent. If constructing the network starting from a hexagonal or square grid where neighbor distances have no variation, the required spatial noise level or interaction strength needs to be higher overall to elicit the same response.

4 Discussion

Using simple assumptions, we show that a spatial interaction network model can produce dynamics that resemble aspects of the behavior observed in sulphur molly shoals, specifically: irregular, self-terminating wave sequences with varying starting locations. By further investigating the role of the spatial embedding of the agents, we found that local heterogeneities in the interaction can play an important role in the system's responsiveness to external stimuli, highlighting the importance of studying these systems with explicit spatial embedding.

The key mechanisms to achieve this behavior in the model are: first, individual-level information aggregation with threshold-like activation; second, agent localization in space, allowing distance-dependent link weights and restricting interactions to the local scale; finally, stochasticity introduced by spatial distribution of agents. Interestingly, no mechanisms with agent-specific adaptations were required to elicit repeated diving and variation in wave locations.

At the same time, the comparison to empirical observations highlighted that some aspects are not reproducible with the chosen model mechanisms. For instance, there are qualitative differences in the wave geometry: In the model, waves are always circular, which is owed to the undirected interaction network and agents being assumed point-like, thus not distinguishing agent orientation and possibly asymmetric interactions. In the wild, waves can be observed to propagate in a directed, non-circular fashion.

Furthermore, waves do not terminate within the model domain but always propagate to its edge or until they cancel out by running into another wave. The network can thus be thought of as representing only a small part of a larger shoal, with the additional assumption that connectivity is sufficiently high to form a quasi-continuous aggregation of fish; fluctuations like the one studied in Fig. 5 are only on a local scale. In contrast, we presume that fish shoals like the one observed in Fig. 4 have larger-scale density fluctuations, where low-density areas may reduce or altogether stop wave propagation due to lower interaction strengths. Along the same line, wave termination could also be an effect of asymmetric interaction weights, e.g. as caused by abrupt local changes in fish orientation and thus reduced linking between two parts of a network. Thus, for the spatial scale the model represents and the selected interaction topology, differences in the distance distributions can be expected.

An intriguing path for further investigation of wave termination would include agent-specific adaptive mechanisms like an increased dampening in the social response as a result of successive diving events. Similar processes have been shown to play a role in the termination of agitation waves of starling flocks [6].

More generally, adaptive mechanisms could also be key in exploring further behavioral phenomena in animal collectives like the sulphur molly shoals. For instance, an agent-specific adaptive diving threshold could represent a temporarily heightened alertness after a stimulus, with the adaptive threshold following a different mechanism than the information aggregation. Such adaptive mechanisms may also be important for better understanding how animal collectives optimize their information-processing capabilities by self-organizing to the critical point of a phase transition [8]. It was previously found that the sulphur molly shoals operate near a phase transition between high individual diving activity and low overall diving activity [5]; however, it remains an open question by which individual-level mechanisms the collective adapts to changes in their environment.

Future work should also aim to more quantitatively compare the behavior in the fish system to the model simulations. Video recordings of the fish waves could be used to extract wave activity time series or isolate individual waves, allowing to compare time series and distribution of wave properties like inter-wave-intervals or wave sizes. Determining geometric properties of individual waves (like their location, speed, or direction) would allow to better understand whether and how the fish that have witnessed an attack respond differently than those that are farther away.

In summary, our results demonstrate that core aspects of the observed collective evasion of sulphur mollies can be explained by a rather simple generative model. However, they also suggest potential limitations of the model in terms of representing biologically observed behavior. Overall, our agent-based approach provides a promising framework for future investigation of the specific mechanisms underlying the adaptive collective behavior of sulphur mollies, as well as, more generally, a whole class of spatially embedded collective systems exhibiting social contagion dynamics.

Acknowledgments. YS, CV, DB, JK, and PR are part of the *Science of Intelligence* Cluster of Excellence, funded by the Deutsche Forschungsgemeinschaft under Germany's Excellence Strategy (EXC-2002/1, project number 390523135). KP acknowledges funding by the federal state of Berlin via the Elsa-Neumann-Fellowship.

References

1. Bridson, R.: Fast poisson disk sampling in arbitrary dimensions. In: ACM SIGGRAPH 2007 sketches. SIGGRAPH07. ACM (Aug 2007). https://doi.org/10.1145/1278780.1278807
2. Bullmore, E., Sporns, O.: Complex brain networks: graph theoretical analysis of structural and functional systems. Nat. Rev. Neurosci. **10**(3), 186–198 (2009). https://doi.org/10.1038/nrn2575
3. Doran, C., et al.: Fish waves as emergent collective antipredator behavior. Current Biol. **32**(3), 708–714.e4 (feb 2022). https://doi.org/10.1016/j.cub.2021.11.068
4. Farkas, I., Helbing, D., Vicsek, T.: Mexican waves in an excitable medium. Nature **419**(6903), 131–132 (2002). https://doi.org/10.1038/419131a

5. Gómez-Nava, L., et al.: Fish shoals resemble a stochastic excitable system driven by environmental perturbations. Nat. Physi (2023). https://doi.org/10.1038/s41567-022-01916-1
6. Hemelrijk, C.K., Costanzo, A., Hildenbrandt, H., Carere, C.: Damping of waves of agitation in starling flocks. Behav. Ecol. Sociobiol. **73**(9) (2019). https://doi.org/10.1007/s00265-019-2734-4
7. Lukas, J., et al.: Diurnal changes in hypoxia shape predator-prey interaction in a bird-fish system. Front. Ecol. Evolu. **9** (2021). https://doi.org/10.3389/fevo.2021.619193
8. Muñoz, M.A.: Colloquium: criticality and dynamical scaling in living systems. Rev. Mod. Phys. **90**(3) (2018). https://doi.org/10.1103/revmodphys.90.031001
9. Pacher, K., et al.: Thermal tolerance in an extremophile fish from mexico is not affected by environmental hypoxia. Biol. Open **13**(2) (2024). https://doi.org/10.1242/bio.060223
10. Poel, W., Winklmayr, C., Romanczuk, P.: Spatial structure and information transfer in visual networks. Front. Phys. **9** (2021). https://doi.org/10.3389/fphy.2021.716576
11. Riedel, L., Herdeanu, B., Mack, H., Sevinchan, Y., Weninger, J.: Utopia: a comprehensive and collaborative modeling framework for complex and evolving systems. J. Open Source Softw. **5**(53), 2165 (2020). https://doi.org/10.21105/joss.02165, https://utopia-project.org/
12. Rosenthal, S.B., Twomey, C.R., Hartnett, A.T., Wu, H.S., Couzin, I.D.: Revealing the hidden networks of interaction in mobile animal groups allows prediction of complex behavioral contagion. Proc. Nat. Acad. Sci. **112**(15), 4690–4695 (2015). https://doi.org/10.1073/pnas.1420068112
13. Sevinchan, Y., Herdeanu, B., Traub, J.: dantro: a Python package for handling, transforming, and visualizing hierarchically structured data. J. Open Source Softw. **5**(52), 2316 (2020). https://doi.org/10.21105/joss.02316
14. Sosna, M.M.G., et al.: Individual and collective encoding of risk in animal groups. Proc. Nat. Acad. Sci. **116**(41), 20556–20561 (2019). https://doi.org/10.1073/pnas.1905585116
15. Sumpter, D.: The principles of collective animal behaviour. Philos. Trans. Royal Soc. B: Biol. Sci. **361**(1465), 5–22 (2005). https://doi.org/10.1098/rstb.2005.1733
16. Tobler, M., Kelley, J.L., Plath, M., Riesch, R.: Extreme environments and the origins of biodiversity: adaptation and speciation in sulphide spring fishes. Mol. Ecol. **27**(4), 843–859 (2018). https://doi.org/10.1111/mec.14497
17. Zykov, V.S.: Spiral wave initiation in excitable media. Philos. Trans. Royal Soc. A: Mathem. Phys. Eng. Sci. **376**(2135), 20170379 (2018). https://doi.org/10.1098/rsta.2017.0379

Behavioural Contagion in Human and Artificial Multi-agent Systems: A Computational Modeling Approach

Maryam Karimian[1,2(✉)], Fabio Reeh[1,2], Asieh Daneshi[3], Marcel Brass[3], and Pawel Romanczuk[1,2,4]

[1] Institute for Theoretical Biology, Department of Biology, Humboldt-Universität zu Berlin, 10115 Berlin, Germany
[2] Science of Intelligence, Research Cluster of Excellence, 10587 Berlin, Germany
m.karimian@scioi.de
[3] Berlin School of Mind and Brain, Department of Psychology, Humboldt-Universität zu Berlin, 10099 Berlin, Germany
[4] Bernstein Center for Computational Neuroscience, Berlin, 10115 Berlin, Germany

Abstract. This study employs a modified formulation of the second-order Drift-Diffusion model to investigate how the interaction between environmental and social cues influences individual decision-making in a binary choice scenario. Environmental information is represented as stochastic cues, often biased towards one of the choices, while social information is conveyed through signals from a group of identical agents making random decisions. The model incorporates simplified human perceptual characteristics via a visual network of social interactions, which considers perceptual limitations due to physical distances and visual occlusions. Model parameters and assumptions are informed by an ongoing behavioural experiment on behavioural contagion, conducted in human and artificial multi-agent systems using virtual reality. The stochastic evolution of decision states in response to environmental and social input mirrors the behavioural choices of human participants, who respond to stimuli presented in the virtual reality environment and social cues from a group of virtual agents. Manipulating the size and density of the group revealed that larger group sizes and lower densities lead to greater alignment of individual decisions with social cues, accompanied by shorter and more homogeneous response times and reduced accuracy. These findings afford preliminary insights into the behavioural experiment. With reciprocal informative exchange from experimental findings, this study would contribute to enhanced realism in future steps.

Keywords: behavioural contagion · second-order Drift-Diffusion model · Social and environmental information · Visual network

1 Introduction

Behavioural contagion refers to the spread of behaviour among individuals within a group, akin to the spread of infection [14]. Examples include spread of escape maneuvers in animal groups in response to a threatening stimulus [11], or spread of violence and aggression during protests [10]. On the one hand, cognitive science has studied individual decision-making within groups in research on social conformity, from a first-person perspective [4,7]. However, this approach overlooks dynamic social interactions and the impact of collective dynamics on individual decisions. On the other hand, in the field of collective behaviour, research has focused on simplified models of social interactions [3,8,9], while neglecting the cognitive complexity of individual agents. Consequently, there is a research gap in studying behavioural contagion, which incorporates not only the intricate decision processes of individuals, but also the concurrent social interactions that influence and are influenced by decisions of individuals within the group. This study aims to establish the groundwork for a comprehensive investigation into behavioural contagion, which is designed to bridge this gap. That is by integrating an ongoing virtual reality (VR) experiment focusing on the cognitive aspects, with the computational modeling study presented in this paper, which incorporates a stochastic integrator model (a variation of the second-order Drift-Diffusion Model) [12,15] to account for the collective dynamical aspects.

In VR experiments, a human participant perceives a group of virtual human agents performing a task in response to visual stimuli (environmental cues) presented in the virtual environment. The visual stimuli consist of a random pattern of yellow and blue pixels. Each agent signals the color majority by raising their right hand for mostly yellow and their left hand otherwise (see Fig. 1.a). Human participants within this VR environment perceive both the environmental cues provided by the screen and the social information conveyed by the virtual agents. In the computational model, the decision process of the participant is modeled using a second-order Drift-Diffusion model (DDM). Our modeling approach explicitly accounts for the temporal dynamics of incoming environmental and social information, and allows for real-time updates and adaptations of decision states in response to changing environmental and social signals. Our purpose was to develop a computational model that transcends experimental limitations, enabling the identification of effective parameter ranges and the accurate prediction of experimental findings. VR setups offer unique opportunities to study behavioural contagion mediated by vision in controlled laboratory settings. Simultaneously, our collective behaviour model incorporates local interactions between individuals using visual networks that consider the sensory constraints of visual perception. This approach presents a novel perspective in the study of collective behaviour by explicitly accounting for the first-person perspective [2,5,9,11,13]. That is while, traditional collective behaviour models often assume oversimplified local interactions based on an omniscient third-person perspective [1,3,8].

Davidson and colleagues (2021) [6] reported that the detectability of visual cues among individual group members depends on both the group size and the spatial arrangements of its individuals. Specifically, they highlighted that in larger groups, visual occlusions may constrain the detection capability, a limitation influenced by the spatial arrangement of individuals. Expanding on these insights and findings from other relevant studies (e.g. [7]), it becomes clear that group size and density, shaping spatial arrangements and visual occlusions, are pivotal determinants of behavioural contagion dynamics. Thus, we designed this study to systematically explore the role of these group characteristics in shaping the dynamics of behavioural contagion.

2 Methods

2.1 Overview of the Behavioural Experiment

The behavioral experiment involves a two-alternative forced choice perceptual decision-making task conducted within a virtual reality (VR) environment, where human participants are exposed to a screen displaying blue and yellow pixels (Fig. 1a). Specifically, participants are required to determine whether the screen predominantly displays yellow or blue pixels. Additionally, within the VR environment, a group of artificial agents also engage in the same task. They signal their decisions by raising their right hands to indicate the dominance of blue pixels or their left hands for the dominance of yellow pixels. As such, participants receive two types of cues to inform their decisions: environmental cues provided by the screen and social cues derived from the behaviour of virtual agents. The experiment consists of multiple trials, each incorporating variations in the number of virtual agents, number of responding agents, density of their group, and the blue-yellow ratio on the screen. Following each trial, participants' choices (if any) and their response times will be recorded for subsequent analysis and comparison with model predictions.

2.2 Computational Model Design

We introduce a computational model to simulate contagious decision-making processes occurring in the behavioural experiment. Specifically, the model is tailored to predict the results of the behavioural experiment by replicating key aspects of the decision-making task employed therein. To that end, we simulated VR experiment sessions with consistent setup and parameterization including the duration of trials, as well as environmental and social input variables. Specifically, the model is designed to capture necessary geometric attributes of the VR environment, such as the size and placement of virtual agents. We utilized a variation of second-order Drift-Diffusion model (DDM), referred to as the second-order DDM to represent decision processes in response to the environmental and social cues. The model is implemented using Python 3.10.9.

Second-Order Drift-Diffusion Model: The second-order Drift-Diffusion model (DDM) consists of four key components: 1) Accumulation of environmental and social cues within each trial; 2) Decaying mechanism to reflect individuals' limited memory; 3) A white Gaussian noise to accommodate fluctuations in decision processes; and 4) Setting decision boundaries corresponding to the two response options. With these components, the second-order DDM is formulated as the following stochastic differential equations:

$$\dot{s}(t) = -\gamma_s s(t) + \beta(t) + \sigma\xi(t)$$
$$\dot{\beta}(t) = -\gamma_\beta \beta(t) + \mu_e I_{\text{env}}(t) + \mu_s I_{\text{soc}}(t)$$
(1)

The first equation governs the temporal evolution of the state variable, $s(t)$, which represents the evidence accumulated by the focal individual (participant) from environmental and social cues at time t. For each trial with duration t_{tr}, if s reaches the thresholds of $+\theta$ or $-\theta$ (i.e. either of the decision boundaries), a decision is made. Otherwise, the trial remains undecided. Each of the decision boundaries represents the perceived dominance of either of the two colors (blue or yellow) displayed on the virtual screen in VR experiments. The state rate, $\dot{s}(t)$, is the result of an interplay between a dissipation factor (γ_s), a time-dependent, adaptive drift variable ($\beta(t)$), and a temporarily uncorrelated Gaussian white noise ($\xi(t)$) with intensity σ, zero mean and amplitude ϵ: $\langle\xi(t)\xi(t')\rangle = \epsilon^2\delta(t-t')$. The second equation defines the evolution of the drift variable, $\beta(t)$, which indicates the bias towards one of the choices over time, with respect to the evidence accumulation. Accordingly, $\dot{\beta}(t)$, specifies the speed of evidence accumulation. Based on the second equation, drift rate is derived from a stochastic process of accumulating environmental input ($I_{\text{env}}(t)$) and social input ($I_{\text{soc}}(t)$) with specific strengths, μ_e and μ_s, respectively. Similar to the state rate, drift rate is subject to a dissipation factor (γ_β). The environmental input ($I_{\text{env}}(t)$) is derived from a random flow of environmental cues occurring with a constant probability rate (p). In fact, p determines the extent of attention to environmental cues. Accumulation of positive values of $I_{\text{env}}(t)$ favour the choice of blue dominance, while accumulation of negative values of $I_{\text{env}}(t)$ leads the state towards the choice of yellow dominance. In addition, a reliability index (q) determines the likelihood of positive cues, indicating the blue-yellow ratio on the screen. $q = 1$ indicates complete coverage by blue pixels, while $q = 0$ represents an entirely yellow screen. Thus, $q = 0.5$ implies an equal distribution of both colours, resembling a random noise.

At each time-step, $I_{\text{soc}}(t)$ represents the cumulative cues received from virtual agents. These cues are influenced by two key factors: 1) the decision made by each agent and 2) the degree to which that decision affects the focal individual's decision. Agents' decisions are stochastic, characterized by an instantaneous decision signal of either +1 or -1, corresponding to the choice of blue or yellow dominance, respectively. In our simulations, social signals occur randomly across trials but remain consistent within the virtual group and within a predefined interval during each simulation trial. These signals are scaled by the influence of the respective agent on the focal individual's decision. The scaling factor is determined by the extent to which the agent is visible to the focal individual.

Projection of Social Interactions in Visual Networks: We created interaction networks where each node signifies a virtual agent, except for the one representing the focal individual. Nodes representing virtual agents are linked to the focal individual via outgoing, weighted edges. The weight of each edge reflects the visibility of the corresponding agent to the focal individual, thus the strength of its input. This structure is commonly known as a visual network [5]. We constructed synthetic visual fields from the perspective of the focal individual, with each visual field containing dummy representations of human agents positioned identically to those in the VR environment (Fig. 1b). These dummies were color-coded with distinct RGB values. The number of pixels allocated to each dummy correlates with the three-dimensional angle in the focal individual's visual field for the corresponding agent. This measure serves as the weight of edges in our visual networks (see Fig. 1c). Given that all virtual agents are avatars with identical shapes, visibility of agents is influenced by their spatial proximity to the focal individual (according to perspective rules), their centrality in the VR environment, and whether they are (partially) occluded by other agents.

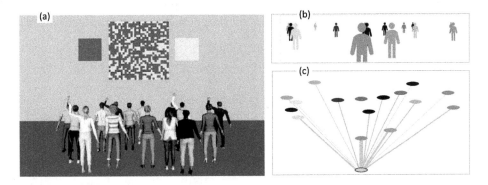

Fig. 1. Panel (a): An example visual scene of the VR experiment (https://github.com/Asieh-Daneshi/SciOI-images-and-Videos). **Panel (b):** Simplified illustration of VR visual scenes featuring color-coded human dummies. **Panel (c):** Display of an interaction network with a single focal individual, where connection weight with each avatar is determined by the portion of the visual field occupied by that avatar. The thickness of the link with the focal individual represents this weight. Connection weights are normalized by a constant value, representing an avatar positioned 1 m away from the focal individual.

Predictors and Dependent Measures: We manipulated the size and density parameters within the modelled network of virtual agents. The network structure, resembling the arrangement of avatars' positions in the VR experiment, is represented by an irregular grid extended radially from the centre of the focal individual's visual field, consisting of $N = n^2$ agents, corresponding to the number of avatars, and with a density denoted by $\rho = \frac{n}{r\sqrt{n-1}}$. Here r represents the averaged distance between grid nodes. Specifically, we considered four distinct values for n (4, 9, 16, 25) and four values for ρ (0.64, 1, 1.78, 4), resulting in a total of 16 combinations of group size and density. We conducted 200 simulations for each combination using unique random seeds over 2000 steps (equivalent to 4 s). These simulations generated time series data for state (s), drift (β), and distributions of response times. Furthermore, correlation between averaged final states and environmental/social signals across all combinations was quantified. Additionally, we examined how these measures are affected by variations in the reliability of environmental signals (biased signals). In the VR experiment, this is achieved by adjusting the ratio between blue and yellow colors displayed on the screen, thereby manipulating task difficulty. For simulations with biased environmental signal we also evaluated accuracy of decisions. Results of simulations conducted with this setup (scenario I) were compared to those of a similar simulation with a single difference (scenario II). This difference was that instead of varying the group size, we maintained a constant size ($N = 25$) and adjusted the number of agents expressing their choices ($M = 4, 9, 16, 25$). These agents were randomly selected from within the group. It's important to note that in this scenario, the suppressive effect of non-responding agents is accounted for by scaling the social information based on the ratio of responding agents to the total number of agents in the group.

3 Results

Here, decision processes are analyzed based on dependent measures of the second-order DDM, namely, "state" (s) and "drift" (β). The former signifies the decision state, while the latter indicates the inclination of the decision toward one of the options. We employed this model to explore how an individual's decision state and drift are influenced by three key determinants: 1) the presence or absence of social cues, 2) the structure of the group of virtual agents, and 3) the reliability of environmental cues. Additionally, we investigated how these factors affect the response time of the focal individual.

In Fig. 2, we visually compare the results of two decision processes modeled using the second-order DDM. Figure 2a displays these modeling outcomes for a decision process influenced solely by environmental signals, generated by a noisy screen with a 70%-30% blue-yellow ratio. Notably, these simulations do not consider social cues. As anticipated, the majority of simulations show a preference for the blue majority choice, indicated by state trajectories converging towards the upper decision boundary. This preference aligns with the 70%-30% blue-yellow ratio of the environmental signal. The introduction of social signals enhances the stochasticity of the decision process. Figure 2b illustrates this dynamic, where social signals initially favour the lower decision boundary by chance. As a result, many decision trajectories tend towards this boundary, leading to an average state close to zero. Furthermore, incorporating social signals reduces average reaction times and their variance. Here, the additional social signals lead to overall larger drifts, and thus faster decisions.

We propose that visual interactions with VR agents and, consequently, the dynamics of behavioural contagion are influenced by fundamental spatial characteristics of the virtual group such as size and density, which determine the

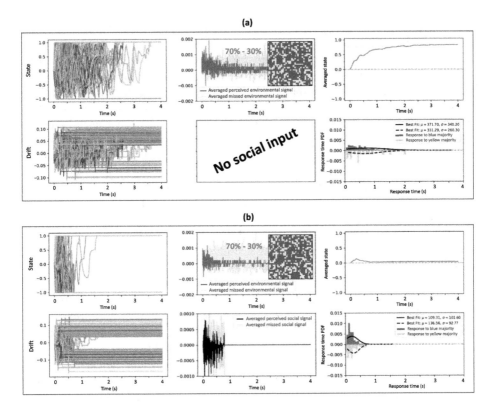

Fig. 2. Comparison of two example simulations using the second-order DDM, where a focal individual receives environmental signals with an increment of $\mu_e = 0.04$ and a rate of $10/s$, generated by a noisy screen with a 70%-30% blue-yellow ratio. Additionally, white Gaussian noise with intensity $\sigma = 0.01$, zero mean and amplitude $\epsilon = 0.001$ is included, along with dissipation factors (γ_s and γ_β) both set to 0.03. Panel (a) displays state and drift trajectories over a 2000-step trial (equivalent to 4 s) for 200 simulations, environmental signal dynamics (averaged at each time step over the 200 simulations), time series of states averaged over the simulations, and distributions of response times for both blue and yellow choices. Notably, this simulation lacks social signals. Panel (b) depicts the same visualizations as in panel (a), with the inclusion of social signals from a group of 25 agents, a density of $1/m$, and $\mu_s = 0.02$, distributed within the first quarter of the trial. (Color figure online)

interaction network. To investigate this, we manipulated these network characteristics in two scenarios I and II (refer to the Methods section). Additionally, we compared our findings across various environmental signal distributions.

Figure 3 highlights the influence of social information on the decision-making process of an unbiased focal individual, in conjunction with both noisy and biased environmental cues. It underscores how exposure to social signals affects the consistency between the individual's final state and their perceived social cues, contingent upon various factors such as the social group structure and model parameters including dissipation factors, strength of environmental and social signals, and noise intensity.

Furthermore, consistent with observations from example simulations in Fig. 2, the inclusion of social signals leads to shorter response times with reduced variance. Comparing Fig. 3a and 3b, where the environmental signal consists solely of pure noise, the incorporation of social signals in both scenarios alters decision states. Specifically, the averaged correlation between the individual's final state (s_{final}) and the perceived environmental signal (I_{env}) decreases. The color map displaying averaged correlation of s_{final} with perceived social signals (I_{soc}) in the space defined by group size and density exhibits an inverse relationship with the color map of $s_{\text{final}} - I_{\text{env}}$ correlations. The $s_{\text{final}} - I_{\text{soc}}$ correlation increases, particularly for group sizes larger than $N = 4$, while it diminishes with higher group densities. The significance of group size is notably more pronounced compared to group density, as demonstrated in Fig. 3b. Note that in Scenario I, there exists a threshold for group density ($\rho = 4/m$), beyond which the effect of density becomes more pronounced in reducing the $s_{\text{final}} - I_{\text{soc}}$ correlation, but group size exhibits less influence in intensifying the impact of social signals on an individual's final state. Additionally, it's important to highlight that the effect of group size is noticeably more significant in Scenario II compared to Scenario I. Interestingly, the pattern observed in color maps of response time medians and variances mirrors the pattern observed in the correlation color maps. This implies that more influential social signals prompt quicker responses with reduced variance across various simulations. When environmental signals are biased, favouring one choice over the other (e.g., with a 70%-30% blue-yellow ratio, as depicted in Fig. 3c), despite the increased reliability of environmental signals and their stronger impact on individual's state trajectories, social signals persist in shaping these trajectories as well. Comparing Figs. 3b and 3c reveals that the $s_{\text{final}} - I_{\text{soc}}$ correlation is overall lower in Fig. 3c. Additionally, in Fig. 3c for scenario I, the significance of the density effect diminishes at a lower threshold ($\rho = 1.8/m$) compared to the corresponding threshold observed in Fig. 3b. Similar to Fig. 3b, the relationship between response time metrics (medians and variances) and group structure characteristics (size and density) mirrors the relationship observed between the $s_{\text{final}} - I_{\text{soc}}$ correlations and group structure characteristics. In this configuration, given the biased environmental signals, we could delineate correct and incorrect behavioural choices. Consequently, color maps of response evaluation for each scenario were included in Fig. 3c. Interestingly, these color maps also reveal a pattern consistent with other color maps in their respective scenarios, suggesting a reduced likelihood of correct responses when social signals exert greater influence on state trajectories.

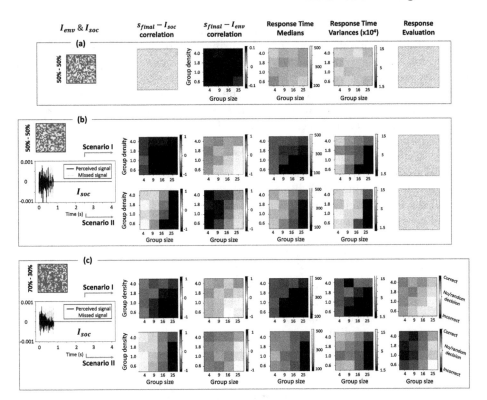

Fig. 3. Visualization of simulation results with different environmental and social signal parametrizations. **Panel (a)** displays simulations with random environmental cues (50%-50% blue-yellow ratio) without social information. **Panel (b)** illustrates results with random environmental signals as well as social cues for both scenarios I and II. **Panel (c)** depicts simulations with environmental signals corresponding to 70%-30% blue-yellow ratio and social cues akin to panel (b) for both scenarios. color maps illustrate the correlation between the final state (averaged over the final 200 time steps) and perceived social/environmental signals (first and second columns), response time medians (third column), response time variances (fourth column), and response evaluations (last column) across various group sizes and densities. All color map elements are averaged over 200 simulations. In scenario I, group size refers to the number of avatars present in the VR environment, whereas in scenario II, it represents the number of responding avatars among the 25 present avatars. Response evaluations assign a value of 1 for choosing the blue majority (considered correct) and -1 for selecting the yellow majority (deemed incorrect). These values range from -1 to 1 when averaged over the 200 simulations, with a value of 0 indicating either entirely random or indecisive choices. All parameters are the same as in Fig. 2, unless otherwise stated.

4 Discussion

We employed a second-order Drift-Diffusion Model (DDM) to simulate decision-making processes within a virtual reality (VR) environment, where a focal individual interacts with environmental and social cues. The VR setup provides stimuli for a behavioural experiment, featuring environmental cues displayed on a screen for a two-alternative-forced choice task, alongside social signals from a group of virtual human agents performing the same task. Our research aimed to investigate the impact of virtual agents' behaviours (social input) on the focal individual's performance, particularly in scenarios where there is discrepancy between social and environmental cues, and when environmental cues reliably inform individuals of the correct performance. We defined social interactions between virtual agents and the focal individual using visual networks. Visual networks used in this study, enable each individual to receive information from all neighbors occupying a two-dimensional angular area within their visual field. This allows for the adoption of a first-person viewpoint in our visual interaction networks. We proposed that social interaction in this network would be influenced by fundamental group characteristics, such as group size, and group density via presence and absence of visual occlusions, as well as the perceived size of avatars [5].

Our modeling outcomes and observations of this behavioural experimentation have concurrent exchanges of insights for better adjustment of modeling assumptions and experimental setup. For instance, preliminary results from online 3D experiments conducted by our collaborators as prelude to VR experiments suggest that increasing the ratio of signaling agents to the total number of agents has a pronounced effect on the alignment between participants' decisions and social signals. In addition, Rosenthal and colleagues (2015) [11] have reported that the transmission of behaviour within their fish schools is optimally described by a fractional contagion process. In this framework, the response probability of fish depends on the fraction of active neighbours observed, as inhibition from non-responding neighbours diminishes the probability of response. Consequently, we devised two scenarios of adjusting the influx of social signals, one involving alteration of the total number of agents (scenario I), and the other targeting the fraction of responding agents (scenario II). It is noteworthy that in the second-order DDM, the distinction between these scenarios is represented by a term we introduced to the social signal increment, which reflects the ratio of responding agents to the total number of agents. This approach allows for the incorporation of the inhibitory effect of non-responding agents into the model.

With the specified assumptions, our computational model predicts that when individuals receive social signals from larger groups with densities below a certain threshold, there is an increased alignment observed between the perceived social cues and the final decision states. This effect is observed for both scenarios defined for adjusting the influx of social signals. However, in scenario II, increasing the number of responding agents leads to a considerably more pronounced effect on this consistency. In addition, an increase in group density when it is below the specified threshold, specifically for larger group sizes, was found to attenuate the influence of social signals on individual's decision states. However, this effect is less pronounced than that of group size in both scenarios. Remarkably, social signals maintain their influence with a qualitatively similar dependency on group size and density in both scenarios, even in simpler tasks where individuals are more confident in their decisions based on environmental cues. Finally, our results imply that enhanced social influences correspond to shorter and more homogeneous response times as well as to less accuracy in behavioural choices. This might reflect a speed-accuracy trade-off. However, this also stems from the randomness of social cues. Therefore, the next phase of this study involves implementing a multi-agent DDM. In this model, agents' responses will be driven from an interaction between their perceived environmental signals and social cues from other agents within their visual field, as opposed to random responses. According to these findings and given that scenario II receives more substantial support from relevant studies [11], future endeavors in this research may concentrate solely on scenario II. It would be intriguing to manipulate the centrality and proximity of responding agents in relation to the focal individual, while keeping the group structure and number of responding agents constant. This manipulation aims to assess the respective impacts of centrality (given that the source of environmental signals is at the center of the focal individual's visual field) and proximity of signalling agents on the alignment of individuals' decisions with social cues. It is noteworthy that our model partially addresses attentional restrictions in signal perception by incorporating sporadic environmental and social signals over time. However, the current treatment of attentional effects in the model remains simplistic and warrants further refinement to achieve a more comprehensive understanding. This may affect the alignment between individual's decision states and social/environmental signals, especially when the size of the virtual group increases and/or the density is low (i.e. attention is more distributed). Moreover, in cognitive-behavioural research, the suggested mechanism for social influence includes two levels. First, bottom-up imitative processes evoke a reflexive inclination towards imitation, which is then followed by top-down strategic processes that govern the decision to either enact or inhibit this reflexive response [4]. A limitation of our model is that it does not assume mechanisms to distinguish between these two processes and the extent to which they contribute to contagion processes. Therefore, addressing this limitation could also serve as a focus for future steps.

In summary, the model employed in this study demonstrates the capacity to generate intuitive predictions regarding some aspects of behavioural contagion, such as the trade-off between social and environmental impacts in relation to the structure of social groups. These predictions align with the results of several experimental and modeling studies on collective behaviour (e.g. [6,7,9,11]). Furthermore, these predictions will undergo further scrutiny against forthcoming virtual reality (VR) experimental data. This comparison aims to refine the model parameters, capturing additional realistic factors such as attentional effects and perceptual hierarchies.

Acknowledgements. We acknowledge funding by the Deutsche Forschungsgemeinschaft (DFG, German Research Foundation) under Germany's Excellence Strategy - EXC 2002/1 "Science of Intelligence" - project number 390523135. We extend our heartfelt appreciation to Prof. Dr. Jens Krause and Prof. Dr. Anne Nassauer for their insightful guidance, which enriched our understanding throughout the development of our model. We also thank Anant Kulkarni, our diligent student assistant, for his assistance in this project.

References

1. Ballerini, M., et al.: Empirical investigation of starling flocks: a benchmark study in collective animal behaviour. Anim. Behav. **76**(1), 201–215 (2008)
2. Bastien, R., Romanczuk, P.: A model of collective behavior based purely on vision. Sci. Adv. **6**(6), eaay0792 (2020)
3. Couzin, I.D., Krause, J., Franks, N.R., Levin, S.A.: Effective leadership and decision-making in animal groups on the move. Nature **433**(7025), 513–516 (2005)
4. Cracco, E., et al.: Evidence for a two-step model of social group influence. Iscience **25**(9) (2022)
5. Dachner, G.C., Wirth, T.D., Richmond, E., Warren, W.H.: The visual coupling between neighbours explains local interactions underlying human 'flocking'. Proc. R. Soc. B **289**(1970), 20212089 (2022)
6. Davidson, J.D., Sosna, M.M., Twomey, C.R., Sridhar, V.H., Leblanc, S.P., Couzin, I.D.: Collective detection based on visual information in animal groups. J. R. Soc. Interface **18**(180), 20210142 (2021)
7. Gallup, A.C., et al.: Visual attention and the acquisition of information in human crowds. Proc. Natl. Acad. Sci. **109**(19), 7245–7250 (2012)
8. Ginelli, F., Chaté, H.: Relevance of metric-free interactions in flocking phenomena. Phys. Rev. Lett. **105**(16), 168103 (2010)
9. Poel, W., Winklmayr, C., Romanczuk, P.: Spatial structure and information transfer in visual networks. Front. Phys. **9**, 716576 (2021)
10. Reicher, S.: Collective protest, rioting, and aggression. In: Oxford Research Encyclopedia of Communication (2017)
11. Rosenthal, S.B., Twomey, C.R., Hartnett, A.T., Wu, H.S., Couzin, I.D.: Revealing the hidden networks of interaction in mobile animal groups allows prediction of complex behavioral contagion. Proc. Natl. Acad. Sci. **112**(15), 4690–4695 (2015)
12. Sosna, M.M.G., et al.: Individual and collective encoding of risk in animal groups. Proc. Natl. Acad. Sci. **116**(41), 20556–20561 (2019). https://doi.org/10.1073/pnas.1905585116
13. Strandburg-Peshkin, A., et al.: Visual sensory networks and effective information transfer in animal groups. Curr. Biol. **23**(17), R709–R711 (2013)
14. Thorpe, W.H.: Learning and instinct in animals (1956)
15. Tump, A.N., Pleskac, T.J., Kurvers, R.H.: Wise or mad crowds? The cognitive mechanisms underlying information cascades. Sci. Adv. **6**(29), eabb0266 (2020)

Transient Milling Dynamics in Collective Motion with Visual Occlusions

Palina Bartashevich[1,2](✉) ⓘ, Lars Knopf[1,2], and Pawel Romanczuk[1,2] ⓘ

[1] Institute for Theoretical Biology, Department of Biology,
Humboldt-Universität zu Berlin, Berlin, Germany
{palina.bartashevich,pawel.romanczuk}@hu-berlin.de
[2] Research Cluster of Excellence "Science of Intelligence",
Technische Universität Berlin, Berlin, Germany

Abstract. The coordinated circular motion of individuals within a group, known as milling, is a widely observed collective motion pattern across biological and artificial systems. However, existing models focused on achieving stable, albeit unnatural, patterns, while overlooking the embodiment aspect of real-world agents. Here, we employ a spatially explicit agent-based model with visual occlusions and a collision avoidance mechanism to address this gap and investigate the emergence of temporary milling states over time. We show that short yet frequent milling dynamics are prevalent in a distinct parameter region, characterised by a qualitative shift in group behaviour between dynamical regimes, suggesting adaptability. We also show that such milling states require a minimal field of vision that not only promotes their occurrence but also matches the typical field of view observed in biological systems.

Keywords: milling · collective motion · dynamic states · criticality

1 Introduction

Models of collective motion based on interactions such as local repulsion, directional alignment, and longer-range attraction predict three primary collective behavioural states: disordered swarms, directed groups, and milling (rotating in circles, see Fig. 1A) [8]. In nature, collective behaviour typically serves some biological function, such as protection from predators or searching for food [13]. Milling remains one of the least understood and perhaps even controversial among collective self-organised patterns in terms of its functional role [10,14], despite being observed across a wide variety of animal species (like fish, ants, sheep, reindeer, and horses). It poses a significant risk to group members, as once got into the milling state, animals can stay moving in circles for prolonged durations, leading to exhaustion and eventual death [11]. Such stable milling can arise spontaneously or be induced by external factors such as confined space [15,24] or external perturbations due to predators [19]. In the latter, it can also be seen as a collective defence pattern, making it almost impossible to target a single individual for attack. To be potentially effective in an anti-predator context, it is plausible to assume that milling has to be a transient pattern, allowing individuals to easily transition in and out of it, avoiding getting stuck in a "deadly" trap. However, little attention has been paid to the conditions necessary for such *temporary milling states* to occur, which would allow for a broader understanding of milling behaviour in nature [10].

Evolutionary modelling in [14] has shown that milling emerged as the second most common and effective strategy against a confusable predator, after directed cohesive motion, if prey are unable to perform evasion manoeuvres. In the same context, among stationary patterns, milling behaviour was identified as the most effective compared to swarming or staying compact [14]. Taken together, alternating between cohesive forward motion and stationary milling states can be considered a plausible anti-predatory strategy. The reinforcement learning method in [17] also indicated that the groups trained to perform milling demonstrated extreme resistance to external perturbations, which could pose risks to animals (e.g., strong wind or currents). Meanwhile, it has been suggested that groups may be the most effective at evading predators when they operate at or near critical points, i.e., phase transition, where the system is, on the contrary, the most susceptible to its environment and external forces [12,20]. There, dynamic changes in the group's structure can give rise to different spatio-temporal collective patterns over time. The question of whether *temporary milling* can be one of these emerging patterns has remained unexplored.

The main goal of this paper is to analyse collective rotational dynamics across different parameter regimes and offer insights into how these dynamics can relate to transient milling states and group behaviour near and far from criticality. For this, we use the definition of criticality as in [12], defined as the directional order-disorder transition, where an increase of alignment strength beyond a critical point causes a spontaneous directed movement in a disordered swarm. We employ a spatially explicit generic agent-based model of collective motion [3,12] but consider agents not as particles but rather as entities with "physical bodies" together with visibility constraints which they introduce. In particular, we analyse how occlusion-based perception of the environment differs from Voronoi-based interactions, known as an approximation of the vision field [22], in their impact on the emergence of temporary milling states. We also propose an additional explicit collision avoidance mechanism that does not qualitatively alter the collective dynamics of the underlying swarm model [3,12]. It ensures no collisions between agents, regardless of the parameter regime in which they operate. Besides correspondence with the inherent biological nature of living systems, this feature is of crucial importance for future applications in swarm robotics.

The paper is organised as follows. In Sect. 2, we provide an overview of the literature that illustrated sufficient conditions for the emergence of milling behaviour in simulations. Afterwards, we describe the vision-based collective motion model of the current study, including the introduced collision avoidance mechanism and performance metrics in Sects. 3 and 4 respectively. Section 5 presents the analysis of the results from the simulated experiments, with a subsequent summary of the findings and discussion in Sect. 6.

2 Related Work

Milling has been observed in models of collective motion based on one or a combination of several social forces between agents [8,10]. Notably, single-force models involved either local attraction alone [2,23] or velocity alignment mechanism only [1,7], with agents moving at a constant speed. Regardless of the terms used to describe social interactions, an introduction of a blind angle (i.e., the region behind an individual where others remain undetected) was shown to significantly impact milling behaviour. In particular, large blind angles (about half of a perception range) have been found to amplify its occurrence and create more complex shapes of mills [1,2,7,8,23]. Newman et al. [16] suggested that this holds only if organisms can keep moving at a near-constant pace. In their variable speed model, with attraction-repulsion force without alignment, a blind zone of just 10% of the perception range was sufficient to destroy milling patterns [16]. Although all these models considered a limited field of view, social interactions were metric-based, i.e., within a particular radius, and did not account for the possibility of visual occlusions. This aspect is especially crucial in densely packed groups and also considering that milling was suggested to be density-dependent [6].

The importance of visual networks in collective animal motion has been demonstrated both empirically and in simulations [9,22]. Bastien et al. [4] proposed a minimal vision-based model of collective behaviour without spatial representation but with steering governed by visual perception field accounting for visual occlusions. There, the emergence of uncoordinated stable milling states was observed, with particles rotating in both directions in the same mill. While their model has regions of parameter space where collisions between disc-like agents can be avoided, the region with milling was one of those where collisions were always observed. However, real-world entities with solid bodies can not pass through each other, regardless of the parameter regime. Therefore, integrating physical constraints into future simulations is vital for further investigations.

Previous studies (e.g., [2,8]) mainly focused on understanding social interactions that create a certain stable collective state, in which transitions to other states are triggered by changing parameters. Milling behaviour seems to be special in this regard, as it was observed to co-exist together with directed (polarised) motion, with reoccurring transitions between the two states, for the same combinations of individual parameters [5,8,18]. This makes it challenging to disentangle in natural groups whether the given milling event occurred in response to a threat or rather spontaneously due to its bistable nature. However, both of these possibilities may not be necessarily mutually exclusive. Our goal in this paper is to provide a systematic analysis of the emergence of dynamic milling states over time in a generic model of collective motion with visual occlusions and to study how the frequency of transitions into milling states varies with certain parameter regimes that may be associated with an anti-predator context.

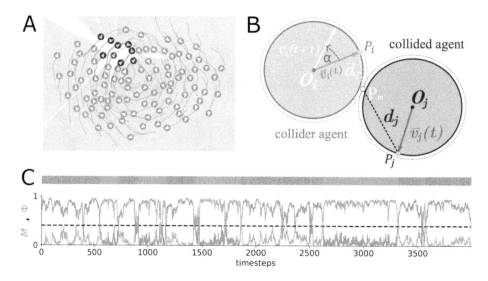

Fig. 1. A) A swarm in the progress of rotating, with an example of neighbour selection for one focal agent (red). Blue agents are perceived by the focal agent, grey ones are not. White areas mark the focal agent's vision area, gray areas mark areas the focal agents can not perceive due to occlusion. B) An example for collision avoidance: Agent i at O_i is identified as a "collider" due to the shorter distance d_i between its heading point P_i and collision point P_m. Agent i therefore changes its velocity $\vec{v}_i(t)$ to $\vec{v}_i(t+1)$ by angle α. C) Timeseries of rotational order M and polarization Φ for a swarm with Voronoi-based neighbour selection. The dashed line marks a threshold of $M = 0.4$, above which the swarm is considered rotating. The horizontal bar designates consecutive rotational events in blue and straight movement in red. (Color figure online)

3 Model of Collective Motion

We consider a spatially explicit agent-based model of coordinated motion, presented in [12] and further motivated by empirical observations in [3]. Each agent $i \in \{1, ..., N\}$ moves on a two-dimensional plane with a constant velocity $\vec{v}_i$, to which a directional Gaussian white noise with a strength of D_φ is added at each discrete time-step Δt. Agent's relative position $\vec{r}_{ij}$ to each other is noted by $\vec{r}_{ji} = \vec{r}_j - \vec{r}_i$, where $\vec{r}_i$ is the position vector of an agent i. Social interactions are modulated by a combined force $\vec{F}_i$, being the sum of the alignment $\vec{F}_{i,alg}$ and distance regulating $\vec{F}_{i,d}$ forces. This way, each agent i strives to align itself to the velocity of its interaction partners $j \in \mathcal{N}_i$ computed as the mean velocity difference $\vec{v}_j - \vec{v}_i$, multiplied by the alignment strength μ_{alg}. The distance between interacting agents is regulated based on their relative distance $r_{ji} := |\vec{r}_{ji}|$ to each other by linearly increasing attraction and repulsion forces with magnitude k, such that the total social force $\vec{F}_i$ acting on an agent i is computed as follows:

$$\vec{F}_i = \frac{1}{|\mathcal{N}_i|} \sum_{j \in \mathcal{N}_i} \mu_{alg} \cdot (\vec{v}_j - \vec{v}_i) + \frac{1}{|\mathcal{N}_i|} \sum_{j \in \mathcal{N}_i} k \cdot (\vec{r}_{ji} - r_d \hat{r}_{ji}), \qquad (1)$$

where r_d being the preferred distance between agents and $\hat{r}_{ji} := \vec{r}_{ji}/r_{ji}$.

Different from previous work [12], we assume that agents have circular rigid bodies of a radius r_a. Interaction partners are selected based on the agent's field of vision (FOV) Θ, within which others that are not blocked by closer ones are perceived (see Fig. 1A). To implement such a visual network, we use Python package *vmodel* developed in [21] to simulate visual-detection-based swarms. There, agents are considered invisible to the focal individual if they are occluded to any degree by others, i.e., only the closest agents with an uninterrupted line of sight are visible to the focal one. To ensure the possibility of swarm formation, the minimal FOV is set to $\Theta = 30°$ in all simulations.

Collision Avoidance Mechanism. Depending on the parameter values, the repulsion force used in Eq. 1 to keep the agents away from each other at the preferred distance does not necessarily ensure that the agents will move collision-free. As a result, real-world agents employing such behavioural rules can experience physical collisions. To simulate this, we introduce an additional collision avoidance mechanism activated as soon as $r_{ij} \leq 2r_a + \epsilon$. We consider each collision event pair-wise and assume that only the *collider* agent alters its course of motion to resolve the collision, while the *collided* agent remains unaffected by collision force. To determine which agent is a collider and which is a collided one, the point of collision P_m has to be detected. Following this, the distances d_i and d_j from each agent's velocity heading point to the collision point P_m are computed (see Fig. 1B). The agent i with the smaller distance $d_i < d_j$ to P_m is considered to be the *collider* agent, while the agent j with the larger distance d_j is considered to be the *collided* one. In a rare case of head-on-head collision when $d_i = d_j$, both agents are assumed to be *colliders*. The *collider* agent i then computes an avoidance turning angle α_i such that both agents i and j are no longer colliding when they move at $t+1$, while minimising the angular difference to its current velocity $\vec{v}_i(t)$. As a result, the avoidance angle α_i determines the *collider* agent's velocity $\vec{v}_i(t+1)$ for the next time step, while discarding all other forces which are acting on the agent i. In this way, the collision force leads to a drastic change in the agent's velocity $\vec{v}_i(t)$, therefore it is paramount that it is only applied when a collision is imminent. Since the *collider* agent can not anticipate the exact position of the *collided* agent at $t+1$, it uses the currently available velocity vector $\vec{v}_j(t)$ of the *collided* agent j as an approximation of its next movement direction at $t+1$. The turning angle α_i of the *collider* agent i is gradually increased until a new movement direction is found, which will presumably allow the agent to move collision-free at $t+1$.

4 Experimental Setup

To identify the emergence of temporary milling states, we compute the angular momentum order parameter M over time, which quantifies the rotation of the group around its centre of mass $\vec{r}_{com}$:

$$M = \left| \frac{\sum_{i=1}^{N} \vec{r}_{i,c} \times \vec{v}_i}{\sum_{i=1}^{N} \|\vec{r}_{i,c}\| \|\vec{v}_i\|} \right|, \tag{2}$$

where $\vec{r}_{i,c} = \vec{r}_i - \vec{r}_{com}$. We define the group as being in a milling (rotating) state when $M > 0.4$ (value also used in previous studies [15]; see Fig. 1C), and denote the fraction of the time of a single simulation T the group spends in this state, averaged further across 40 simulations, as the average rotational order parameter Ψ. Additionally, the average rotation duration τ over these 40 runs is calculated as the mean of the durations of consecutive rotation events which persist longer than $100 \cdot \Delta t$ during a single simulation. Alongside, we also quantify the amount of orientational order in the system using polarization Φ as the mean value of agents velocities, i.e., $\Phi = \sum_{i=1}^{N} \vec{v}_i / N$. In a directed group, all agents will be aligned by having a similar velocity, resulting in a polarization value of 1, whereas in a chaotic (disorganised) swarm this value will be closer to zero.

To analyse changes in the group's structure, we estimate the swarm diameter as the maximum distance $d_s := \max(r_{ij})$ observed between any two agents i and j during simulation time T, which serves as an approximation for the spatial size of the swarm. As a proxy of swarm density, we use the average number of perceived neighbours, denoted as NN. This metric shows the average number of interaction partners each agent has over time T, reflecting the tightness of the swarm structure.

Each simulation starts with $N = 100$ agents located in circles, with positional variation and initial velocity vectors being set orthogonal to the individuals' radius vector to the group's centre of mass. Preliminary experiments showed that the latter affects only the first milling transition, but does not significantly impact subsequent transitions compared to random. The main control parameters of the current study are the alignment strength μ_{alg}, orientational noise D_φ, and the agent's field of view Θ. Other model's parameters are set unchanged as follows: the agent's radius $r_a = 0.5$, the attraction-repulsion strength $k = 3$, the preferred neighbour distance $r_d = 1.5$, the individual's constant speed $v = 1$, and the discrete-time step $\Delta t = 0.02$. In the following, we analyse collective dynamics using the described metrics in the model with visual occlusions and further compare them with Voronoi tessellation, known as an approximation of the field of vision [22].

5 Results

In the occlusion-based model with the full FOV $\Theta = 360°$ (Fig. 2A), for $D_\varphi > 0$, we observe the average rotational order Ψ is the strongest along the directional order-disorder transition (i.e., criticality line) defined by group polarisation Φ. However, the duration of the milling events τ at the phase transition tends to be short. Given the body length of the agents (i.e., BL $:= 2r_a$) and their speed $v = 1$, for the agent located at 1 BL from the group's centre of mass to make a perfect full circle move will take approx. 3 iterations. Meanwhile,

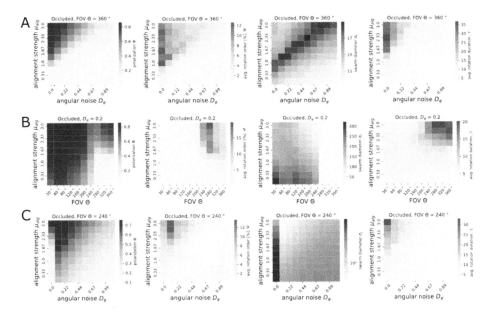

Fig. 2. Heatmaps for polarization Φ, average rotational order Ψ and swarm diameter d_s in the presence of occlusion based neighbour selection. A) Alignment strength μ_{alg} and noise intensity D_φ are being varied while keeping a constant field of vision $\Theta = 360°$. B) Alignment strength μ_{alg} and field of vision Θ are being varied while keeping a constant noise intensity $D_\varphi = 0.2$. C) Alignment strength μ_{alg} and noise intensity D_φ are being varied while keeping a constant field of vision $\Theta = 240°$.

the swarm diameter d_s is at its maximum at the phase transition, with the cohesive structure being preserved (no fractured swarm). Taken together, this indicates that the swarm does not make a single full mill but frequently gets into a temporal rotational state. With an increase of the alignment strength μ_{alg}, as the systems gets in a more ordered regime, the duration of the milling events becomes longer. Considered together with the rotational order parameter Ψ, this indicates that milling occurs less frequently but is more stable compared to when the system operates at or near criticality. Due to the initial swarm configuration, in the absence of orientational noise $D_\varphi = 0$, the system easily gets into a milling state and stays in it the longest compared to other noise levels, especially with the increase of μ_{alg}.

Figure 2B shows the impact of FOV Θ on the occlusion-based collective motion, depending on the alignment strength μ_{alg} with fixed $D_\varphi = 0.2$. In addition to the order-disorder phase transition at alignment strength $\mu_{alg} = 1.67$ (which corresponds to the criticality point at $D_\varphi = 0.2$ in Fig. 2A), we observe a sharp transition at $\Theta = 240°$ across all metrics. In particular, the occurrence frequency of dynamic milling states, as depicted by the average rotational order parameter Ψ, increases with decreasing FOV from the full $\Theta = 360°$ up to $\Theta = 240°$ and completely vanishes for $\Theta < 240°$. Meanwhile, the average

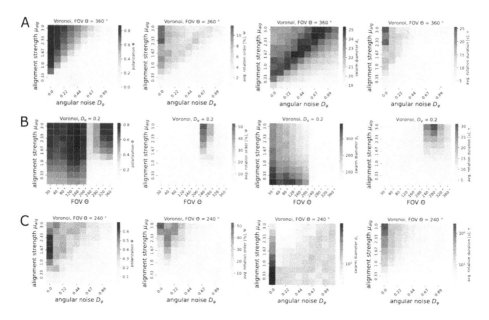

Fig. 3. Heatmaps for polarization Φ, average rotational order Ψ and swarm diameter d_s in the presence of neighbour selection via Voronoi interactions. A) Alignment strength μ_{alg} and noise intensity D_φ are being varied while keeping a constant field of vision $\Theta = 360^{circ}$. B) Alignment strength μ_{alg} and field of vision Θ are being varied while keeping a constant noise intensity $D_\varphi = 0.2$. C) Alignment strength μ_{alg} and noise intensity D_φ are being varied while keeping a constant field of vision $\Theta = 240°$.

duration of the occurring milling states τ does not change across $\Theta > 240°$ for $\mu_{alg} \geq 1.67$. In other words, this suggests that a limited FOV makes it easier for a swarm to transition into dynamic milling states, but does not significantly affect their durability. Notably, there is a sharp transition for the swarm diameter d_s at $\Theta = 240°$ regardless of the alignment strength μ_{alg}. That is, the swarm diameter d_s is significantly increased for the FOV restricted to $\Theta < 240°$ leading to highly elongated and fractured swarms. Coupled with metrics for rotational motion, this suggests that milling dynamics occur only in a cohesive swarm configuration.

Despite the shift in order-disorder phase transition observed for the occlusion-based model with $\Theta = 240°$ (Fig. 2C) compared to full FOV (Fig. 2A), the swarm retains a tendency to exhibit rotational dynamics along and above the transition. In contrast to $\Theta = 360°$, absence of noise for $\Theta = 240°$ leads to highly fractured swarms moving in opposite directions (Fig. 4A), as indicated by polarization Φ approaching zero values. It also correlates with low values of rotational order parameter Ψ for $D_\varphi = 0$, implying weak rotational dynamics primarily occurring only at the initial swarm configuration. As indicated by the duration τ of these rotational events, only high alignment strength $\mu_{alg} = 3$ allows taking over the rotational momentum from the initial velocities without noise, enabling milling dynamics and counterbalancing the impact of limited FOV.

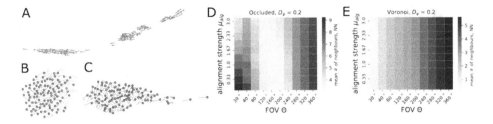

Fig. 4. A)-C) Examples of swarm shapes with occlusion-based neighbour selection for $\mu_{alg} = 2.33$, $\Theta = 240°$ and varying values of D_φ: A) $D_\varphi = 0$ B) $D_\varphi = 0.22$ C) $D_\varphi = 0.88$. D), E) Heatmaps for the mean amount of neighbours NN for occlusion and Voronoi-based neighbour selection, respectively. Alignment strength μ_{alg} and field of vision Θ are being varied while keeping a constant noise intensity $D_\varphi = 0.2$.

5.1 Voronoi-Based vs. Occlusion-Based Neighbour Selection

Comparing the average rotational order Ψ between Voronoi and occlusion-based neighbour selection models with full FOV (Fig. 3A vs. Fig. 2A), we observe that both spend a similar amount of time in rotational states along and above the criticality line. However, this changes for a restricted FOV, where the Voronoi-based model shows a significantly increased occurrence rate of dynamic milling states compared to the occlusion-based model, despite the duration τ of these states remaining similar on average (Fig. 3B vs. Fig. 2B). Notably, both models exhibit a sharp transition at $\Theta = 240°$ across the studied metrics, highlighting their common properties. However, contrary to Fig. 2C, with no noise Voronoi-based model produces elongated but less fractured swarms, i.e., $\Phi > 0.5$ at $D_\varphi = 0$ (Fig. 3C). This is also correlated with a significantly increased duration of the rotational events, especially for high alignment strength $\mu_{alg} = 3$. Overall, the swarm diameter d_s for the Voronoi-based model with $\Theta = 240°$ (Fig. 3C) suggests its ability to maintain a cohesive structure across a wider range of parameters than its occlusion-based counterpart.

The distribution of rotational state durations τ at the criticality line (order-disorder transition) for $\Theta = 240°$ shows that Voronoi-based neighbour selection provides more stable milling states, lasting longer and occurring more frequently, than in the case of the occlusion-based model (Fig. 6A,C). Above the criticality line, this trend is amplified, with milling events of longer durations but less frequency (Fig. 6B,D). For $\Theta = 360°$, both models exhibit comparable distributions, with the occlusion-based model showing more but shorter rotating events with $\tau < 3$ at criticality (Fig. 5).

Figure 4D illustrates two sharp transitions at $\Theta = 80°$ and $\Theta = 240°$ in the average number of interactions partners NN for the occlusion-based model, with the plateau of approx. $NN = 4$ in-between of the transitions. In turn, the Voronoi-based model does not have a sharp transition in NN and exhibits a linear increase in NN with the increasing FOV (Fig. 4E). Notably, both models are characterised by $NN = 4$ on average at $\Theta = 240°$, coinciding with the transition in the rotational order parameter Ψ (Fig. 2B, Fig. 3B). This suggests

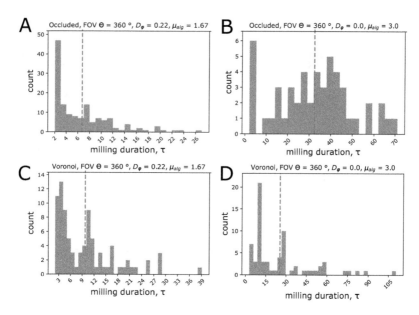

Fig. 5. Histograms of the duration of rotational events with $\Theta = 360°$ at the criticality line (A,C) and above it (B,D) for Occlusion-based (A,B) and Voronoi-based models (C,D). The dashed vertical red line shows the mean milling duration. (Color figure online)

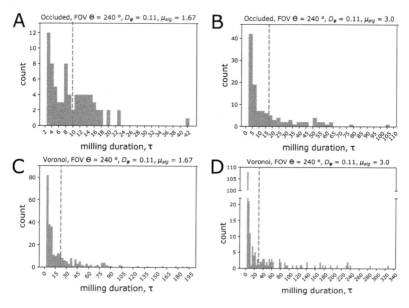

Fig. 6. Histograms of the duration of rotational events with $\Theta = 240°$ at the criticality line (A,C) and above it (B,D) for Occlusion-based (A,B) and Voronoi-based models (C,D). The dashed vertical red line shows the mean milling duration. (Color figure online)

that having at least about four interacting partners is sufficient for dynamic milling states to occur.

6 Conclusion

In this paper, we analysed the emergence of temporary milling states in a generic occlusion-based collective motion model with a collision avoidance mechanism, simulating rigid agent bodies and preventing their overlaps, making our study more relevant for future robotics implementations. Our results indicate that transient milling states are short-lasting but occur frequently along the criticality line, where the swarm undergoes a phase transition from an ordered to a disordered regime. Above the phase transition, rotation events become longer (more stable) albeit less frequent, compromising the swarm's adaptability to environmental changes. This holds especially for Voronoi-based neighbour selection which exhibits more frequent and longer rotational events at and above the criticality than occlusion-based interactions. The overall presence of transient milling states near criticality suggests that collective rotational dynamics are more likely to emerge due to the changes in the group network structure, and not necessarily in the direct response to the external perturbation (threat). Finally, we find the minimal FOV of 240° required for rotational dynamics to occur not only facilitates milling but also falls within the typical range of blind angles observed in natural biological agents (e.g., fish), along with four perceived neighbours on average, a biologically plausible number of interaction partners.

Acknowledgement. This study was funded by the Deutsche Forschungsgemeinschaft (DFG, German Research Foundation) under Germany's Excellence Strategy - EXC 2002/1 "Science of Intelligence" - project number 390523135. The funder played no role in study design, data collection, analysis and interpretation of data, or the writing of this manuscript.

References

1. Bagarti, T., Menon, S.N.: Milling and meandering: flocking dynamics of stochastically interacting agents with a field of view. Phys. Rev. E **100** (2019). https://doi.org/10.1103/PhysRevE.100.012609
2. Barberis, L., Peruani, F.: Large-scale patterns in a minimal cognitive flocking model: incidental leaders, nematic patterns, and aggregates. Phys. Rev. Lett. **117**, 248001 (2016). https://doi.org/10.1103/PhysRevLett.117.248001
3. Bartashevich, P., et al.: Collective anti-predator escape manoeuvres through optimal attack and avoidance strategies. bioRxiv (2024). https://doi.org/10.1101/2024.03.26.586812
4. Bastien, R., Romanczuk, P.: A model of collective behavior based purely on vision. Sci. Adv. **6**(6) (2020). https://doi.org/10.1126/sciadv.aay0792
5. Calovi, D.S., Lopez, U., Ngo, S., Sire, C., Chaté, H., Theraulaz, G.: Swarming, schooling, milling: phase diagram of a data-driven fish school model. New J. Phys. **16**(1) (2014). https://doi.org/10.1088/1367-2630/16/1/015026

6. Cambui, D.S., Gusken, E., Roehrs, M., Iliass, T.: The milling pattern in animal groups and its dependence on the density and on the number of particles. Phys. A **507**, 289–293 (2018). https://doi.org/10.1016/j.physa.2018.05.111
7. Costanzo, A., Hemelrijk, C.K.: Spontaneous emergence of milling (vortex state) in a Vicsek-like model. J. Phys. D Appl. Phys. **51**(13) (2018). https://doi.org/10.1088/1361-6463/aab0d4
8. Couzin, I.D., Krause, J., James, R., Ruxton, G.D., Franks, N.R.: Collective memory and spatial sorting in animal groups. J. Theor. Biol. **218**(1), 1–11 (2002). https://doi.org/10.1006/jtbi.2002.3065
9. Davidson, J.D., Sosna, M.M.G., Twomey, C.R., Sridhar, V.H., Leblanc, S.P., Couzin, I.D.: Collective detection based on visual information in animal groups. J. Royal Soc. Interface **18**(180) (2021). https://doi.org/10.1098/rsif.2021.0142
10. Delcourt, J., Bode, N.W.F., Denoël, M.: Collective vortex behaviors: diversity, proximate, and ultimate causes of circular animal group movements. Q. Rev. Biol. **91**(1), 1–24 (2016). https://doi.org/10.1086/685301
11. Delsuc, F.: Army ants trapped by their evolutionary history. PLOS Biology **1** (2003). https://doi.org/10.1371/journal.pbio.0000037
12. Klamser, P.P., Romanczuk, P.: Collective predator evasion: putting the criticality hypothesis to the test. PLoS Comput. Biol. **17**(3), 1–21 (2021). https://doi.org/10.1371/journal.pcbi.1008832
13. Krause, J., Ruxton, G.: Living in Groups. Oxford University Press, USA (2002)
14. Kunz, H., Zublin, T., Hemelrijk, C.: On prey grouping and predator confusion in artificial fish schools (2006). https://doi.org/10.5167/uzh-61728
15. Lafoux, B., Moscatelli, J., Godoy-Diana, R., Thiria, B.: Illuminance-tuned collective motion in fish. Commun. Biol. **6**(1), 585 (2023). https://doi.org/10.1038/s42003-023-04861-8
16. Newman, J.P., Sayama, H.: Effect of sensory blind zones on milling behavior in a dynamic self-propelled particle model. Phys. Rev. E **78** (2008). https://doi.org/10.1103/PhysRevE.78.011913
17. Nuzhin, E.E., Panov, M.E., Brilliantov, N.V.: Why animals swirl and how they group. Sci. Rep. **11**(1) (2021). https://doi.org/10.1038/s41598-021-99982-7
18. Paley, D., Leonard, N., Sepulchre, R., Couzin, I.: Spatial models of bistability in biological collectives, pp. 4851–4856 (2008). https://doi.org/10.1109/CDC.2007.4434523
19. Parrish, J.K., Viscido, S.V., Grünbaum, D.: Self-organized fish schools: an examination of emergent properties. Biol. Bull. **202**(3), 296–305 (2002)
20. Romanczuk, P., Daniels, B.C.: Phase transitions and criticality in the collective behavior of animals - self-organization and biological function, chap. Chapter 4, pp. 179–208 (2023). https://doi.org/10.1142/9789811260438_0004
21. Schilling, F., Soria, E., Floreano, D.: On the scalability of vision-based drone swarms in the presence of occlusions. IEEE Access **10** (2022). https://doi.org/10.1109/ACCESS.2022.3158758
22. Strandburg-Peshkin, A., et al.: Visual sensory networks and effective information transfer in animal groups. Curr. Biol. **23**(17), R709–R711 (2013). https://doi.org/10.1016/j.cub.2013.07.059
23. Strömbom, D.: Collective motion from local attraction. J. Theor. Biol. **283**(1), 145–151 (2011). https://doi.org/10.1016/j.jtbi.2011.05.019
24. Tunstrøm, K., Katz, Y., Ioannou, C., Huepe, C., Lutz, M., Couzin, I.: Collective states, multistability and transitional behavior in schooling fish. PLoS Comput. Biol. **9** (2013). https://doi.org/10.1371/journal.pcbi.1002915

Extended Swarming with Embodied Neural Computation for Human Control over Swarms

Jonas D. Rockbach[1,2](✉) and Maren Bennewitz[2]

[1] Human-Machine Systems, Fraunhofer FKIE, 53343 Wachtberg, Germany
[2] Humanoid Robots Lab, University of Bonn, 53113 Bonn, Germany
jonas.rockbach@fkie.fraunhofer.de

Abstract. With the rise of robot swarms, it has become a relevant problem how humans can control them. Extended swarming is a potential approach in which robot swarms are treated as self-organising extensions of human bodies. Swarm control takes the form of controlling the observable swarm body while robot chains connect the human operator to relevant aspects of the environment. Inspired by how natural bodies are controlled by a nervous system, we here investigate how the swarm body's self-organisation can be influenced by robot chains acting as embodied neural traces while remaining under human high-level control. Three design principles are proposed for such embodied neural computation. First, the swarm body's self-organisation is controlled both by top-down human control and bottom-up sensor inputs alike to the hierarchical control architecture of the nervous system. Second, robots participating in robot chains are treated as rate-coded neurons rendering the chains as embodied neural traces which offers intuitive control possibilities for the human. Third, neural and swarm self-organisation are integrated by utilizing the swarm's communication network as a scaffolding for neural function influencing swarm dynamics. This process is interpreted as embodied Hebbian learning. Human control over the swarm is demonstrated in a grid-based search-and-rescue simulation with the objective of selecting the most valuable subregion defined by accumulated victims in need. We evaluate how using embodied trace relevance in terms of neural activation improves completion time to finding the highest-value trace as well as how attracting units to relevant traces increases their robustness.

Keywords: Human-swarm interaction · Extended swarming · Swarm robotics · Embodied neural computation · Bio-inspiration

"A signal comes in saying, 'threat!' Something has appeared that can be detected by the changes it induces, for example alterations in the electrostatic field. At once, the flying swarm forms into this 'cloud-brain' or whatever it is..."

The Invincible, Stanisław Lem [13]

1 Introduction

In swarm robotics, minimalistic robots rely solely on local information without global leadership to promote flexibility, scalability, and robustness against node failures [4]. Therefore, swarms are especially envisioned for scenarios with large and dynamic environments such as Search-and-Rescue (SaR) missions.

However, deploying robot swarms to real-world problems requires human operators to be in control of swarms. While humans require fused information about the robot swarm and means to control it, the traditional swarm definition explicitly excludes the generation of global information and centralised leadership. Human-Swarm Interaction (HSI) therefore develops swarm control methods invariant to swarm size in order to join the natural centralised human cognition with the distributed synthetic intelligence of robot swarms [6,11,12], the latter being referred to as centralisation-decentralisation trade-off.

A potential approach to HSI is Extended Swarming (ES) which focuses on designing the joint agent as a whole [6]. The joint agent describes the combination of human and swarm as a goal-seeking unit itself. In ES, the observable robot swarm is treated as a self-organising but controllable body extending the human's sensor-act range [20,21]. Human control takes the form of influencing aspects of the extending body based on the virtual pheromone approach [19,20], e.g., by adjusting the swarm body's dispersion. The extending body is also streaked with shortest-path robot chains connecting the human operator to relevant aspects of the environment [1,5,21], such as victims in need. The underlying motivation of ES is that the embodied swarm network performs the computation itself while being under human high-level control. In comparison to fully centralised control, swarms as world-embedded, or embodied, distributed networks can compute in parallel, are more robust, can adapt to the environment, and can use the spatial location of their nodes for computation [5,19,21].

In this work, we investigate the benefit of designing robot chains as embodied neural traces given that real bodies are also controlled by the nervous system. For us, Embodied Neural Computation (ENC) refers to treating each robot participating in chains as embodied neurons that self-organise by an embodied form of Hebbian learning into embodied neural traces and has the following motivations. First, the brain provides an approach to the centralisation-decentralisation trade-off and applying neural principles to ES could be of use for HSI [6]. Second, given the brain controls the body, neural parameters could offer human control possibilities over the swarm body. Third, embodied neural activity could be used to represent chain relevance, similar to virtual pheromones in swarm robotics for site selection [1]. Fourth, the ES vision aims at integrating the extending swarm body into the human nervous system function in the long run, with a neural swarm possibly simplifying such a cyborg integration [6].

This paper provides the following contributions. In Sect. 2, the ES approach to swarm control and sensor fusion is summarised based on a SaR mission requiring the joint agent to select the most relevant subregion with victims in need. Then, ENC is introduced into the extending swarm body (Sect. 3). We first survey previous work on treating swarms as embodied neural systems and extract three design principles for ENC: Embodied hierarchical control, robots as embodied neurons, and Embodied Hebbian Learning (EHL). In embodied hierarchical control (principle 1), the swarm body's self-organisation is influenced by both bottom-up sensor inputs and top-down human operator commands [6]. If robots participating in robot chains are treated as neurons, the swarm body is streaked with embodied neural traces while neural parameters offer intuitive control over the swarm body for ES (principle 2). In order to integrate swarm and neural self-organisation, the liquid-slow swarm body communication topology is treated as a scaffolding for locking robot nodes to relevant information traces for fast-solid neural information processing [22]. This process is interpreted as EHL (principle 3) influencing the swarm body's self-organisation which also can be modulated by the human. Finally, Sect. 4 demonstrates how ES with ENC provides swarm control possibilities for HSI invariant to swarm size in the context of a grid-based SaR simulation. The completion time of establishing relevant traces is evaluated showing that taking trace relevance in terms of neural activation into account improves completion time. In addition, we demonstrate how utilizing neural activation for attracting robots increases trace robustness.

2 Extended Swarming

2.1 Model Assumptions

Consider a SaR scenario as an experimental testbed in which the joint agent's goal is to decide which subregion featuring victims in need are of the highest relevance (best-of-N problem), while we here focus on the interplay between the human operator and the swarm. The joint agent $A = h \cup S$, h being the human operator node and S the swarm, operates on the region $\mathbb{D}$, $\mathbb{D}$ here being a two-dimensional grid plane $\mathbb{N}^2$, that includes static objects of interest $o \in O$, i.e., the victims, being located at position (x_o, y_o) with relevance $\gamma \in [0, 1]$. For this work, h is a static node with position (x_h, y_h). The swarm of size N is a collection of homogenous simple robot nodes $S = \{r_1, ..., r_N\}$. A robot r with position (x_r, y_r) has the basic local state $[\theta_r, v_r]$ with $\theta_r \in [0, 2\pi]$ being the heading direction and $v_r \in \{0, 1\}$ the speed while $v_r = 1$ corresponds to $1\,\mathrm{m/s}$. Joint agent nodes $a \in A$ are able to sense other members h, r' and objects o, and estimate the relative bearings $\hat{\delta}_h, \hat{\delta}_{r'}, \hat{\delta}_o$ to them if nodes are in their bearing sensor range $d_{ij} \leq d_{\hat{\delta}max}$, d_{ij} being the Euclidean distance between two nodes a_i and a_j. In addition, A nodes can estimate the object relevance $\hat{\gamma}$ if in their state sensor range $d_{ij} \leq d_{\hat{\gamma}max}$. The A nodes can also selectively send messages at a particular angle, such as possible with directed infrared communication [19].

Robot movements are based on an object attractor force and a dispersion force being summed to θ_r. The object attraction force attracts robots towards O if $d_{ij} \leq d_{\hat{\delta}max}$, while $v_r = 0$ for all $(x_r, y_r) = (x_o, y_o)$. The dispersion force repels robots away from each other if $d_{ij} \leq d_{disp}$ resulting in a regular dispersion pattern if the swarm is initiated as aggregated.

The A nodes with positions (x_a, y_a) establish communication edges if $d_{ij} \leq d_{com}$ yielding the swarm body Euclidean graph G_{body} at the current time. In the following, G_{body} is used to superimpose virtual pheromone fields Φ and the extending shortest-path tree G_{tree}.

2.2 Extending Swarm Control

In ES, the region $\mathbb{D}$ is explored via the swarm body being controlled by the virtual pheromone approach [19] as adapted by Rockbach [20]. At predefined points in time, a source node $\phi \in A$ injects a message into G_{body} containing a hop counter $c^\phi \in \mathbb{N}$, which is distributed to connected nodes while being incremented by 1 at each hop. Here the hop counter c is used instead of estimating the Euclidean distance to nodes as this suffices [19]. Each node a only accepts the message with the lowest hop count, thereby ensuring a radial distribution of the pheromone message into the network, while also saving the bearing $\hat{\delta}^\phi$ to the transmitting node. The result of this virtual pheromone breadth-first distribution is a world-embedded potential field described by $\Phi^\phi = \{(x_a, y_a, c_a^\phi, \hat{\delta}_a^\phi) | a \in A\}$. Thus, a pheromone field is an embodied shortest-path tree with ϕ as source.

A Φ-potential field is used for robot particle allocation by defining a negatively charged field centre ϕ^- for attraction or a positively charged centre ϕ^+ for repulsion with allocations being triggered based on a hop count distance threshold τ^ϕ. An extending behaviour is given by a distance condition to the field centre ϕ such as $c^\phi < \tau^\phi$ and the resulting attraction or repulsion action to ϕ being released with probability p^ϕ. Triggered robots follow the attraction or repulsion gradient as given by their local state $(c_a^\phi, \hat{\delta}_a^\phi)$.

The extending posture is given by condition $c^h > \tau^h$ with action h^- and releaser $p^h = 1$ and is used to radially constrain the dispersion of S based on the distance to h (Fig. 1a). The human can control the extending swarm dispersion by adjusting τ^h, which is communicated via G_{body} to all connected robots, invariant to the swarm size. $\tau^h = 0$ results in a swarm aggregated at h, a "contracted pose", while a larger τ^h allows the swarm to further disperse onto $\mathbb{D}$, referred to as "extended pose". Thus, the extending posture encapsulates the robots inside the swarm body and ensures the connectivity of G_{body}. Φ^r-fields have a selected robot as source. If the above extending posture is combined with condition $c^r > \tau^r$ and action r^-, the extending posture is oriented towards (x_r, y_r) with strength p^r (Fig. 1c). Finally, if the source is a robot sensing an object, condition $c^o < \tau_o$ with action o^- "grasps" the object by allocating robots towards o with recruiting depth τ^o and allocation force p^o (Fig. 1d).

2.3 Extending Swarm Fusion

Based on Rockbach et al. [21], swarm fusion in the context of ES requires the establishment of robot chains between h and all sensor robots r^o estimating object relevance. The collection of shortest paths between h and all reachable r^o is the fusion graph G_{tree}, $G_{tree} \subset G_{body}$. As described above, swarm control is based on virtual pheromones distributed as an embodied shortest-path tree. Since each r contains the local bearings towards h, a shortest path between h and r^o is found if r^o injects a return message which is forwarded based on the local bearings. Each robot receiving such a return message is a node of the shortest path and therefore restricts its movement $v_r = 0$ for all $r \in G_{tree}$. Given the swarm dynamics, new robot positions lead to a shortening of paths with time since only the nodes participating in a shortest path stop while other nodes continue their exploration [6].

G_{tree} represents a directed radial tree topology pointing towards h extending its sensor range while each r^o injects object relevance $\hat{\gamma}_o$ into its individual chains (Fig. 1a). The fusion graph contains three types of nodes; sensor robots r^o injecting $\hat{\gamma}_o$, relay robots forwarding the injections, and fusion robots integrating information from multiple o. The fusion nodes emerge as a function of object distance resulting in a hierarchical embodied fusion topology [21]. After G_{tree} self-organisation, the human h can follow the observable path with the integrated relevance estimation leading to the highest-value subregion.

3 Embodied Neural Computation

3.1 Background

We provide a short overview of treating robot swarms as embodied neural systems given the aim of introducing neural logic to ES. Augmenting a swarm with neural logic requires an understanding of the relationship between swarms and neural systems as well as how these can be exploited for design. Swarm cognition proposes that both swarms and neural systems are decentralised computing networks that share similarities on the computational level [2,24]. However, swarms and neural systems are also fundamentally distinct; swarm agents move in space with very flexible connections whereas neurons do not [22], resulting in faster information processing for neural systems. Integrating the two systems therefore means integrating different time scales. In general, they both can be treated as graphs G; here G_{body} represents the swarm's liquid interactions and G_{tree} the solid-neuronal, while a neural graph is called a connectome [4,23].

In swarm engineering, limited work has considered how the potential overlap between swarms and neural systems could be exploited. Holland et al. [9] described the idea that the computation capabilities of drones could be linked together into an "ultraswarm"; into an artificial nervous system. An approach called "Mergeable Nervous System" to reconfigurable robotics was proposed by Mathews et al. [14]. Their proposal augments decentralised intelligence with a semi-centralised control unit (brain robot) and is similar to ES where the human operator constitutes the brain [6]. Otte [18] formalized and implemented a robot swarm as a distributed neural network that can classify scattered stimuli such as pixels making up an image. A hippocampus-inspired swarm model for navigation was also discussed where robots were treated as neurons and groups of robots formed reciprocal connected networks [16]. Finally, Hasbach and Bennewitz [6] proposed constraining robot positions based on superimposed neural activity in the context of ES. In the following, three principles for Embodied Neural Computation (ENC) are extracted and elaborated; embodied hierarchical control, embodied neurons, and Embodied Hebbian Learning (EHL).

3.2 Principle 1: Embodied Hierarchical Control

The nervous system has been abstracted as a hierarchical fusion and control architecture with fast, low-level, bottom-up, sensor-driven loops being modulated by slow, high-level, top-down, cognition-driven loops [6]. Neural fusion, more commonly called convergence, is illustrated by Hubel and Wiesel [10] who showed how the vision system first encodes simple local visual features of a visual scene that are sequentially fused over layers to more complex global representations. The self-organising hierarchical fusion topology G_{tree} represents embodied neural convergence if the plane $\mathbb{D}^2$ is seen as a visual scene with stimuli O. The swarm not only fuses local objects hierarchically into a global percept at h but the emergence of fusion nodes as a function of object distance [21] also represents an embodied instance of the Gestalt law of proximity [3], stating that stimuli close in space are more likely being grouped together in the visual percept.

In turn, neural hierarchical distributed control refers to top-down modulatable low-level sensor-motor loops [6] which is best exemplified for ES by decentralised octopus arms acting semi-autonomously [8]. In ES, local sensor information at sensor and fusion nodes is used to directly trigger robot allocations via pheromone fields close to objects, economising time by not relying on human control inputs [20]. For example, an object "seen" at the low level can trigger the swarm body "reflex" of "reaching out" to that object via using Φ^o for robot attraction. By utilizing a repulsion pheromone field, the swarm body can also be triggered to retract from an object. Such a reaching out or retraction behaviour represents the swarm's fine motor skills, alike to limbs or tentacles. Gross motor activity is achieved by orienting the pose as a function of Φ^r or Φ^o, or by changing the extension, such as if o is a threatening stimulus for the A leading to a contracted pose [20]. Finally, h controls the swarm bodies' self-organisation top-down via τ and p for extending posture control as well as by the neural parameters discussed in the following.

3.3 Principle 2: Robots as Embodied Neurons

Each $r_i \in G_{tree}$ is treated as a single world-embedded neuron [6,16] given the embodied computation viewpoint. Thus, robot chains now represent embodied neural traces. r_i is a neural rate-coding unit [15] receiving excitatory inputs $u_{ji}^{in} \in [0,1]$ from other nodes j computing an output activation $u_i^{out} \in [0,1]$ via

$$u_i^{out} = f(W * \mathbf{u}^{in} + u^{bias}) \tag{1}$$

where $\mathbf{u}^{in}$ is the input activation vector and W the weight matrix with $w_{ji} \in [0,1]$, $u^{bias} \in \mathbb{Z}$ is the bias of the neuron defining its resting activity, and f is the neuron's activation function. Here, $f(u)$ is a saturating linear unit with an adjustable activation threshold $M_u \in [0,1]$; $f(u) = 0$ for $u < 0$, u for $0 \leq u \leq M_u$, and 1 for $M_u > u$. Given the intended scalability in swarm robotics, the same virtual weight w, bias u^{bias}, activation function f, and activation threshold M_u are assumed for all robots so that the complexity for storing the neural parameters is 1 rather than $N^2 - N$ (assuming no self-connections), while w, M_u, u^{bias} are under human influence as long as $r_i \cup c \in G_{tree}$. By using directed communication at a particular angle, h can also selectively update the neural parameters for a particular trace. New meanings are given to these neural parameters in the context of ES. Assuming the same weight w, the weight becomes the swarm body's distance sensitivity [21] with $w < 1$ decreasing forwarded activity u^{out} with each hop. M_u in turn defines the swarm body's activation sensitivity, ignoring activities below its threshold. The bias u^{bias} is used to excite or suppress specific neural traces, e.g., if h pays attention to a particular trace. In sum, G_{tree} is a simplified connectome over which bottom-up neural activity spreads while being under human top-down control via w (distance sensitivity), M_u (activation sensitivity), and u^{bias} (trace attention).

3.4 Principle 3: Embodied Hebbian Learning

Neural learning is a form of self-organisation. In the neural sciences, it is well known that learning is a result of the interdependence between neural structures and neural activities [23]. More specifically, Hebbian learning [7] is described as "neurons that fire together, wire together, and neurons that do not fire together, do not wire together.", referring to trace stabilization and forgetting by updating W based on neuronal activity u^{out}.

In its embodied version, $G_{tree} \subset G_{body}$ represents the embodied connectome based on which neural activity spreads. To implement EHL, the connectome is updated by influencing the positions of robots since they in turn define G_{tree} [6]. From a swarm robotics perspective, this is a special case of an exploration-exploitation trade-off; should a robot participate in a current chain ("neural trace stabilisation" via $v_r = 0$) or rather continue exploration ("neural trace forgetting" via $v_r = 1$) [17]? In contrast to the virtual weight w, these embodied connection weights $\omega \in \{0, 1\}$ are binary since robots can be either connected or not, while node distance may be utilized for some applications [16]. EHL is implemented via

$$P(v_i = 0) = l(u_i^{out}) \quad (2)$$

where $P(v_i = 0)$ is the robot's stopping probability and l is the embodied learning function. Here, $l(u) = 1$ for $u > M_l$ and $\epsilon u + (1-\epsilon)$ for $u \leq M_l$ with parameter $M_l \in [0, 1]$ being the embodied one-shot learning threshold immediately stabilising relevant traces and $\epsilon \in [0, 1]$ being the embodied learning rate defining the stability of flexible traces for adjusting exploration-exploitation dynamics. Note that trace stability depends on the joint probability of the individual robot stopping probabilities. If $M_l = 0$ or $\epsilon = 0$, all traces are stabilized.

The above EHL locks robots to relevant traces over time while trace brittleness represents forgetting. However, brittleness can also result from undesired factors, such as robot failures. Therefore, $u_i^{out} > M_l$ traces are protected from undesired brittleness by attracting further units for redundancy based on trace pheromone fields with condition $c^T < \tau^T$ and robot particle attraction T^-, T being the nodes of G_{tree},

$$\tau^T = \alpha u_i^{out} \quad (3)$$

$$p^T = \beta u_i^{out} \qquad (4)$$

with the recruiting depth τ^T and allocation probability p^T being coupled to neural importance via scaling factors $\alpha \in \mathbb{R}^{\geq 0}$ and $\beta = [0, 1]$. Thus, the density around traces inside the swarm body is updated as a function of neural activity.

To conclude, EHL describes swarm pose self-organisation as a function of neural activation while the swarm pose enables specific neural activity in the first place. The human can control EHL by adjusting the embodied neural learning parameters M_l (one-shot learning), ϵ (learning rate), α (recruiting depth), and β (recruiting probability).

4 Computational Investigation

4.1 Simulation Setup

In order to investigate human control over the self-organising swarm body and to evaluate the establishment of embodied neural traces in ES with ENC, we implemented a simulation in MATLAB 2020b with a $\mathbb{D}^2 = 100\,\text{m} \times 100\,\text{m}$ grid. h is placed at $(50\,\text{m}, 50\,\text{m})$ and N robots are initiated inside $10\,\text{m}$ around h. The simulation parameters are $d_{com}, d_{\hat{\delta}max} = 10\,\text{m}$, $d_{\hat{\gamma}max} = 0.5\,\text{m}$, and $d_{disp} = 6\,\text{m}$.

4.2 Demonstration of Swarm Body Control

The top-down controlled bottom-up self-organisation in ES with ENC is demonstrated through the following examples. Assume a SaR unit A being deployed in area $\mathbb{D}^2$ with unknown victims O, $\gamma = 1$ being a high need of attention according to triage. Victim pairs are placed in all four directions (Fig. 1), three with equal distance to h but different criticality (east $\gamma = 1$, west $\gamma = 0.25$, south $\gamma = 0.1$), and one pair ($\gamma = 1$) further away to the north. h must chose which direction to attend to based on the enactive exploration of $\mathbb{D}^2$ with the swarm body. Neural parameters are initialised with $w = 0$, $M_u = 0.25$, $u^{bias} = 0$ and EHL parameters with $M_l = 0.5$, $\epsilon = 0$, $\alpha = 0.8$, and $\beta = 1$.

The swarm body with $N = 150$ robots disperses onto $\mathbb{D}^2$ until the commanded stretching limit is reached ($\tau^h = 3$ hops), leading to trace stabilisation for the three close victim pairs (Fig. 1a). Given the threshold $M_u = 0.25$, the swarm body assigns no relevance to the southern perception at h. Based on the EHL setup, all traces are stabilised, but only the high-value trace to the east is strengthened by the swarm bodies' bottom-up dynamics. Instead of following the provided swarm body estimation, h however becomes attentive of the western region instead. Selective top-down attention to the west is implemented by inducing $u^{bias} = 1$ into the western direction while suppressing the other traces by $u^{bias} = -2$. The bottom-up dynamics now strengthen the western trace while some units remain locked to the relevant eastern victims on standby (Fig. 1b).

The victims located at the peripheral of the northern region are not perceived via $\tau^h = 3$. h explores the region by orientating the swarm body towards north via a robot source field with $\tau^r = 5$, $p^r = 1$, and the robot source position at the border of the dispersion while stretching to $\tau^h = 5$ and deactivating trace robustness via $\alpha = 0$ (Fig. 1c). The swarm body is now pushed towards the north and the two victims are perceived while h remains connected to the previously found victims. By activating distance sensitivity $w = 0.6$ and switching off the bias $u^{bias} = 0$, the north is perceived as irrelevant given the distance, although both eastern and northern victims are in critical states. In addition, the western direction is now also filtered by the activation threshold given its relevance in regard to its distance. The perception of the global scene has been adapted by enactive top-down parameter selection influencing the swarm body's bottom-up self-organisation. h now feels that the global percept is reliable and releases a full bottom-up grasping reaction via Φ^o for first-responder action while oneself being guided by the eastern trace towards the victims in need (Fig. 1d). The traces to the other sites remain intact, enabling h to monitor the dynamics of the other victims without delay and adapt to new situations as necessary.

4.3 Evaluation of Embodied Hebbian Learning

In order to evaluate trace self-organisation via EHL, four objects $\{o_1, o_2, o_3, o_4\}$ are randomly initiated, each in a 20 m × 20 m area located in one of the four corners of $\mathbb{D}^2$. One randomly chosen o_i is relevant ($\gamma = 1$) while the other three are of low relevance ($\gamma \leq 0.5$). The swarm body disperses onto the grid without stretching constraint τ^h and a correlated random walk force is added so that robots sometimes break the dispersion to explore remote regions. The neural parameters are $w = 1$, $M_u, u^{bias} = 0$ with EHL parameters $M_l = 0.5$, and $\alpha, \beta = 0$. The task for S is to establish a connection to the relevant object.

Figure 2 shows completion time until a connection to the relevant object o_i is found for $\epsilon = 0$ (no forgetting) and $\epsilon = 0.1$ (forgetting) with 100 samples each. For limited resources, forgetting irrelevant traces increases the performance to find the relevant object [17]. However, this increased performance comes with the cost of increased brittleness of low-value traces. Specifically, connection probability, defined as the ratio of the number of connected trace time steps to the total number of time steps during the search period, is 0.38 for $\epsilon = 0$ and 0.06 for $\epsilon = 0.1$ at $N = 80$.

To investigate trace robustness, robot failures are introduced with an individual failure probability of 0.1 at each time step. Figure 3 shows the connection probabilities to a relevant o_i with $\gamma = 1$, randomly placed with a maximal distance of 20 m to the arena border and observed over 100 time steps after a connection was first established with 100 samples. Both attraction depth and probability increase the robustness of the relevant trace.

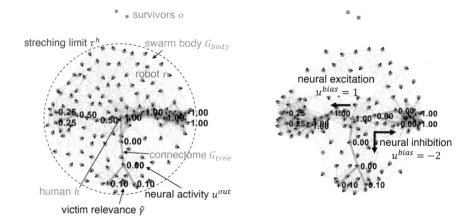

(a) The swarm body strengthens the eastern trace based on bottom-up dynamics.

(b) The western trace is strengthened based on injecting neural biases top-down.

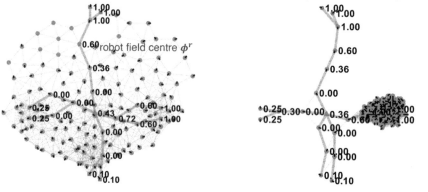

(c) The swarm body is oriented north based on top-down robot field Φ^r control.

(d) The eastern objects are "grasped" by top-down releasing the fine motor skill.

Fig. 1. Examples showing human control over the swarm body's self-organisation.

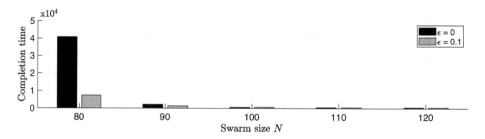

Fig. 2. Speed to relevant trace establishment improves with trace forgetting.

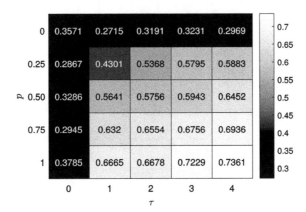

Fig. 3. Connection probability of a relevant trace for $N = 100$ and $\epsilon = 0$ over different recruiting depths τ^T and probabilities p^T.

5 Conclusion

In Extended Swarming (ES) as an approach to Human-Swarm Interaction (HSI), the swarm's embodied nature is utilised for hierarchically fusing local estimates about the world while the swarm itself embodies a potential field that is used for swarm control. This work proposed three design principles for Embodied Neural Computation (ENC) in the context of ES. First, the swarm body's self-organisation is controlled by both bottom-up sensor estimates and top-down human control inputs. Second, robots participating in robot chains are treated as embodied neurons offering intuitive control possibilities in the context of ES. Third, swarm and neural level are integrated by using the swarm's communication topology as a scaffolding for neural function that in turn influences robot movements which is interpreted as Embodied Hebbian Learning (EHL). We demonstrated how the approach establishes human control over the swarm extension in a search-and-rescue simulation. It was also shown how EHL improves completion time for finding the highest-value robot chain and how chain robustness can be strengthened. A parameter sensitivity analysis should be conducted next to explore the model in greater detail. Currently, our approach remains theoretical and requires real-world validation. However, we have demonstrated how robot swarms can not only be envisioned as autonomous agents, but also as controllable extensions of humans with embodied neural-like computation.

References

1. Campo, A., et al.: Artificial pheromone for path selection by a foraging swarm of robots. Biol. Cybern. **103**(5), 339–352 (2010)
2. Gershenson, C.: Computing networks: a general framework to contrast neural and swarm cognitions. Paladyn, J. Behav. Robot. **1**(2), 147–153 (2010)
3. Goldstein, E.B., Cacciamani, L.: Sensation and perception. Cengage Learning (2009)
4. Hamann, H.: Swarm robotics: a formal approach. Springer (2018). https://doi.org/10.1007/9783319745282
5. Hamann, H., Wörn, H.: Embodied computation. Parallel Process. Lett. **17**(03), 287–298 (2007)
6. Hasbach, J.D., Bennewitz, M.: The design of self-organizing human-swarm intelligence. Adapt. Behav. **30**(4), 361–386 (2022)
7. Hebb, D.O.: The organization of behavior: a neuropsychological theory. J. Wiley; Chapman & Hall (1949)
8. Hochner, B.: An embodied view of octopus neurobiology. Curr. Biol. **22**(20), R887–R892 (2012)
9. Holland, O., Woods, J., De Nardi, R., Clark, A.: Beyond swarm intelligence: the ultraswarm. In: Proceedings 2005 IEEE Swarm Intelligence Symposium, 2005. SIS 2005, pp. 217–224. IEEE (2005)
10. Hubel, D.H., Wiesel, T.N.: Receptive fields, binocular interaction and functional architecture in the cat's visual cortex. J. Physiol. **160**(1), 106–154 (1962)
11. Kaduk, J., Cavdan, M., Drewing, K., Vatakis, A., Hamann, H.: Effects of human-swarm interaction on subjective time perception: swarm size and speed. In: HRI, pp. 456–465 (2023)
12. Kolling, A., Walker, P., Chakraborty, N., Sycara, K., Lewis, M.: Human interaction with robot swarms: a survey. IEEE Trans. Hum.-Mach. Syst. **46**(1), 9–26 (2016). https://doi.org/10.1109/THMS.2015.2480801
13. Lem, S.: The Invincible. MIT Press (2020)
14. Mathews, N., Christensen, A.L., O'Grady, R., Mondada, F., Dorigo, M.: Mergeable nervous systems for robots. Nat. Commun. **8**(1), 1–7 (2017). https://doi.org/10.1038/s41467-017-00109-2
15. McCulloch, W.S., Pitts, W.: A logical calculus of the ideas immanent in nervous activity. Bull. Math. Biophys. **5**(4), 115–133 (1943). https://doi.org/10.1007/BF02478259
16. Monaco, J.D., Hwang, G.M., Schultz, K.M., Zhang, K.: Cognitive swarming in complex environments with attractor dynamics and oscillatory computing. Biol. Cybern. **114**(2), 269–284 (2020). https://doi.org/10.1007/s00422-020-00823-z
17. Nouyan, S., Campo, A., Dorigo, M.: Path formation in a robot swarm. Swarm Intell. **2**(1), 1–23 (2008)
18. Otte, M.: An emergent group mind across a swarm of robots: collective cognition and distributed sensing via a shared wireless neural network. Int. J. Robot. Res. **37**(9), 1017–1061 (2018)
19. Payton, D., Estkowski, R., Howard, M.: Pheromone robotics and the logic of virtual pheromones, pp. 45–57 (2005). https://doi.org/10.1007/978-3-540-30552-1_5
20. Rockbach, J.D.: Enhancing human self-regulation with controllable robot swarms acting as extended bodies. In: 2023 IEEE International Conference on Systems, Man, and Cybernetics (SMC) (2023)

21. Rockbach, J.D., Schlangen, I., Bennewitz, M.: Self-organising distributed sensor fusion networks for hierarchical swarm control and supervision. In: 2023 IEEE Symposium Sensor Data Fusion and International Conference on Multisensor Fusion and Integration (SDF-MFI), pp. 1–6. IEEE (2023)
22. Solé, R., Moses, M., Forrest, S.: Liquid brains, solid brains (2019)
23. Sporns, O.: Networks of the Brain. MIT press (2016)
24. Trianni, V., Tuci, E., Passino, K.M., Marshall, J.A.: Swarm Cognition: an interdisciplinary approach to the study of self-organising biological collectives. Swarm Intell. **5**(1), 3–18 (2011). https://doi.org/10.1007/s11721-010-0050-8

Bio-Inspired Agent-Based Model for Collective Shepherding

Yating Zheng[1,2](✉) and Pawel Romanczuk[1,2]

[1] Institute for Theoretical Biology, Department of Biology, Humboldt Universität zu Berlin, Berlin, Germany
{yating.zheng,pawel.romanczuk}@hu-berlin.de
[2] Science of intelligence, Technische Universität Berlin, Berlin, Germany

Abstract. Collective shepherding is a complex problem with potentially a broad range of applications. Its complexity arises from the interaction of two collectives: 'sheep' and 'shepherds', with the latter attempting to control and guide the 'sheep'. Here, we combine an agent-based model for the 'sheep'-flock with a heuristic algorithm for the adaptive behavior of shepherds with two different behavioral modes: *collecting*, i.e. keeping the sheep flock together, and *driving* the sheep towards the target. We show that this algorithm can achieve self-organized coordination among multiple shepherds without direct communication, and investigate how the shepherding performance depends on selected parameters of the system such as sheep flock size, number of shepherds, or parameters governing the switching between the shepherd behavioral modes. We demonstrate that the algorithm can also be applied to more challenging scenarios like controlling non-cohesive or passive agents without self-propulsion. Besides extending our understanding of collective shepherding, our model provides a starting point for future research into unexplored aspects of this complex dynamical behavior.

Keywords: Collective shepherding · agent-based model · coordination

1 Introduction

Collective shepherding is a general control method for a swarm of intelligent agents to control other types of self-organized agents [13]. Due to the asymmetric heterogeneous interaction of two different collectives, it may exhibit complex self-organized dynamics, which makes it an interesting, yet challenging, research target. Shepherding behavior can be also linked to ecologically highly relevant predator-prey dynamics [11], including the scenario of multiple predators hunting together collective prey [1,6]. The research of the shepherding behavior can be extended to other applications such as guiding human crowds [7], herding livestock [3], or collaborative transport in swarm robotics [5,20].

This research is supported by Deutsche Forschungsgemeinschaft (DFG, German Research Foundation) under Germany's Excellence Strategy-EXC 2002/1 'Science of Intelligence', Project 390523135.

Different efforts have been dedicated to researching on how a single shepherd agent can manipulate the cohesion and coordinated movement of a sheep flock [2, 8]. King et al. [10] have found that sheep exhibit a strong attraction toward the center of the flock under the threat of external attack of a shepherd. Based on these findings, Strömbom et al. [19] proposed a heuristic algorithm for one shepherd agent to collect the dispersed agents to the center of the flock and drive the well-aggregated sheep flock toward the expected location. They also found that their result resembles quantitatively experimental data of a farm dog herding a flock of sheep. However, in the proposed algorithm, all the sheep agents are attracted toward the center of the mass. This implicitly assumes access to global information by each sheep-like agent in the flock at all times, which is not achievable in the real sheep flock.

Relatively few studies have investigated collective shepherding, i.e. the cooperation of multi-shepherds to control a sheep collective. Pierson presents a control strategy for multiple dog-like agents to herd a group of sheep-like agents [14]. The shepherds can coordinate with each other into a partial circle around the sheep-like agents and herd them into a target place. The coordination between the shepherds relies on the predefined controller, which constantly requires the position information of other shepherds as feedback. Recently, King et al. [9] discuss a roadmap on how to embrace the gap between the bio-herding algorithm in theory and the hardware application of several UAVs, which is aimed at reducing the danger of using human resources.

On the swarm robotics side, there has been some interest in the herding of passive agents or objects [4, 17, 18]. The work of Gopesh et al. [4] has introduced an optimization method for a swarm of moving robots to collect a mixture of active and passive objects to a target location. This algorithm has the advantage of limited memory requirements for the robot and the flexibility of moving mixed objects. However, analogously, David et al. [17] come up with an automatic design method and neuroevolution to generate control software for shepherd robots to coordinate sheep robots.

In this paper, we consider a distributed agent-based model for the sheep flock with purely local interaction rules, combined with a heuristic algorithm for the behavior of multiple shepherds, featuring simple state transitions between two behavioral modes ("collecting" and "driving") inspired by the model of Strömbom et al. [19]. The simulation result shows that the self-organized coordination emerges with multiple shepherds and it can be improved by modifying the minimum preferred distance between shepherd agents. We also test the flexibility of the algorithm in two different scenarios: one for easily dispersing sheep agents without sheep-sheep attraction, and the other for passive agents without self-propulsion. The results show that our algorithm can be used in driving non-cohesive or passive agents toward the target place and also coordinating multiple shepherds without a specific predefined cooperation controller.

2 Methods

2.1 'Sheep'-Flock Model

We consider a simple motion control method for the 'sheep'-flock with short-ranged repulsion between nearby 'sheep' to avoid collisions, and longer-ranged attraction to generate cohesive flocks. Figure 1 (a) shows the detailed interaction force for one focal

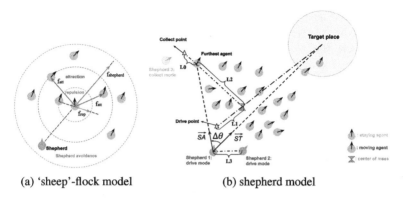

(a) 'sheep'-flock model (b) shepherd model

Fig. 1. Illustration of 'sheep'-flock model and shepherd model. (a) All the sheep agents are marked in green and the shepherds are marked in red. The focal sheep agent is in the center. Its local interaction forces includes mutual repulsion f_{rep}, attraction f_{att}, and shepherd avoidance $f_{shepherd}$. (b) Shepherds 1 and 2 are in the driving mode, while shepherd 3 in the collecting mode. $L0$ is the minimum distance that the shepherd will keep from the collected agent. $L1$ is the distance between the drive point and the center of the mass. $L2$ is the threshold that allows the maximum distance between the furthest agent and the center of mass: $d_{furthest}$. If $d_{furthest} > L2$, the shepherd will switch to collect mode. $L3$ is the equilibrium distance between shepherds. $\Delta\theta$ is the deviation angle between vector $\vec{SA}$ and $\vec{ST}$. The furthest agent has a maximum $\Delta\theta$ from the perspective of the shepherd.

sheep-like agent. $\vec{v}(t)$ and $\vec{\varphi}(t)$ are the sheep-like agent's linear and angular velocities that depend on the interaction forces $\vec{f}$ at time t. $\hat{\mu}_\perp$ is perpendicular to the unit vector $\hat{\mu}$ defining the sheep agent's heading direction φ (see Eq. 1). The position of the agent at time $v_i(t) = v_0 + \delta v_i(t)$, consists of a constant self-propulsion speed and a variable part $\delta v_i(t)$ determined by the social interactions. The equations of motions for agent i read:

$$\frac{d\delta v_i(t)}{dt} = \vec{f}_i(t) \cdot \hat{\mu}_i, \quad \frac{d\varphi(t)}{dt} = \vec{f}_i(t) \cdot \hat{\mu}_{\perp,i} \qquad (1)$$

with $\hat{\mu}_i = (\cos\varphi_i, \sin\varphi_i)^T$ and $\hat{\mu}_{\perp,i} = (-\sin\varphi_i, \cos\varphi_i)^T$. $\vec{f}_i(t)$ is the total interaction force of agent i at the time t (Eq. 2). Repulsion $\vec{f}_{rep}(t)$ and attraction $\vec{f}_{att}(t)$ act both along the unit vector $\hat{r}_{ij} = (\vec{r}_j - \vec{r}_i)/r_{ij}$ ($r_{ij} = |r_j - r_i|$), but with opposite sign: $\vec{f}_{rep/att,i}(t) = \mp \sum_j \hat{r}_{ij}(t)$. The index j runs over the N_r and N_a, respectively. N_r is the number of agents inside the repulsion zone ($r_{ij} < d_{rep}$), and N_a is the number of agents within the attraction zone ($d_{rep} \leq |r_j - r_i| < d_{att}$).

$$\vec{f}(t) = \begin{cases} k_{rep}\vec{f}_{rep}(t), & \text{if } N_r > 0 \\ k_{att}\vec{f}_{att}(t) + k_{shepherd}\vec{f}_{shepherd}(t), & \text{if } N_r = 0 \end{cases} \qquad (2)$$

The repulsion from the shepherd is $\vec{f}_{shepherd}(t) = -\sum_1^{N_{sh}} \frac{\vec{r}_{sa}(t)}{|\vec{r}_{sa}(t)|}$, with N_{sh} being the number of shepherds within the perception range $d_{shepherd}$. $\vec{r}_{sa}$ is the relative position vector of the shepherd with respect to the focal sheep agent. Parameters: k_{rep}, k_{att}, and

$k_{shepherd}$ are the interaction strengths for the different forces. Detailed values for all the parameters are listed in Table 1.

Table 1. Model parameters

Variables	Description	Selected values
N_{sheep}	the number of 'sheep' agents moving outside of the target area	100–300
R	radius of agent body	5
d_{rep}	repulsion distance between 'sheep' agents	10
d_{att}	attraction distance between 'sheep' agents	25
$d_{shepherd}$	safe distance from the shepherd	65
k_{rep}	repulsion strength from the 'sheep' agents	2
k_{att}	attraction strength from the 'sheep' agents	0.8
$k_{shepherd}$	repulsion strength from the shepherd	1.5
τ	numerical time step	0.01
D_θ	amplitude of noise	0.1
$L0$	safe distance between shepherd and agent	10
$L1$	distance between driving position and the centre of mass	$2R\lambda_1\sqrt{N_{sheep}}$
$L2$	maximum deviation of the 'sheep' agent from the center of mass	$2R\lambda_2\sqrt{N_{sheep}}$
$L3$	repulsion distance between shepherd	0, 20, 30
v_0	constant speed per tick for both shepherd and 'sheep' agent	1
α	acceleration rate for the shepherd	1
β	turning rate for the shepherd	0.1
$\Delta\theta$	the differential angle between the vector $\vec{ST}$ and the vector $\vec{SA}$	$[-\pi, \pi]$
λ_1	scaling parameter controlling the drive position to the center of mass	0.67
λ_2	scaling parameter controlling the time in driving mode	0.6, 0.67, 0.75
T	maximum running steps of one single simulation	$2*10^5$

2.2 'Shepherd'-Herding Model

Fig. 1 (b) demonstrates all the detailed information for the shepherd model. The shepherd constantly switches between drive mode and collect mode triggered by whether the distance between the furthest agent and the center of the mass $d_{furthest}$ is larger than $L2$, which is the threshold parameter related to the number of sheep-like agents. $L2 = 2\lambda_2 R\sqrt{N_{sheep}}$ (λ_2 : linear control parameter, R : the radius of the agent). In the collect mode, the shepherd attempts to collect the furthest agent toward the center of mass and thus keep the herds of sheep agents in a cohesive state (e.g. shepherd 3 in fig. 1 (b)). In the drive mode, the shepherd drives the 'sheep'-flock toward the target with previous knowledge of where the target place is (see shepherds 1 and 2 in fig. 1 (b)). Considering that the shepherds can not overlap in real-world systems, we introduce the repulsion distance $L3$ between the shepherds to avoid collision.

The speed of the shepherd evolves according to

$$\dot{x} = v_0 + \alpha(\vec{F} \cdot \hat{n}_i)\hat{n}_i, \quad \dot{\theta} = \omega[\beta\left(\vec{F} \cdot \hat{n}_i^\perp\right) + D_\theta \xi_\theta], \tag{3}$$

where α and β represent acceleration along heading direction (along unit vector $\hat{n}$) and angular direction (along unit vector $\hat{n}_i^\perp$), respectively. Furthermore, v_0 is the constant speed of the shepherd, $\omega = v_0^{-1}$ is the angular inertia, and D_θ is the strength of Gaussian white noise. The total force acting on a shepherd is a combination of an attraction towards either the collect point or drive point, and the repulsion from other shepherds: $\vec{F} = \vec{F}_{att} + \vec{F}_{rep}$. Here, $\vec{F}_{rep}(i) = \sum_{j \in S_i}(|\vec{r}_{ij}| - L3)a_{ij}\hat{r}_{ij}$ with $a_{ij} = 1$ if $r_{ij} < L3$ and $a_{ij} = 0$ otherwise, and $\vec{F}_{att}(i) = \vec{r}_{point}(i) - \vec{r}_i$. S_i is the collection of shepherds in the long-term perception range, and $\vec{r}_{point}(i)$ is the vector of the collect point or drive point position (see fig. 1 (b)). It depends on which mode the shepherd i is.

In this model, the angle information $\Delta \theta$ plays an important role for the shepherd to choose the furthest agent and switches its behavioral mode accordingly. In contrast to most models where position information relative to the center of mass is used, here the shepherd uses the angle information to select the furthest agent, who has the largest deviation angle $\Delta \theta_{max}$ from the shepherd to target (see $\vec{ST}$ in fig. 1 (b)) and from the shepherd to the agent (see $\vec{SA}$ in fig. 1 (b)).

Our approach does not rely on explicit (global) position information. The motivation is two-fold: First, biological agents use primary vision, where relative angles can be estimated more precisely than distances. Second, this information can also easily be extracted from simple camera sensors in artificial systems, which would allow robust operation when global positions are not available.

In fig. 1 (b), when the 'sheep' agent is outside of the target place, it is indicated in green. In this state, it will be moved due to interactions with the shepherd and other sheep agents. When it enters the target place, it switches to a 'staying' state indicated by a blue color. In this staying state, it exhibits an additional attraction towards the center of the target place to enable the accumulation of sheep agents in the target area. Furthermore, all sheep agents interact only with other agents of the same state. Mobile agents outside of the target place will ignore those inside. This ensures that all mobile agents are driven by the shepherds and other mobile agents, and prevents the "pulling-in" of sheep agents into the target arena by other agents already inside (staying mode) as well as the "pulling-out" of sheep agents within the target by others outside.

3 Results

In this section, we discuss the results of our agent-based simulations. We demonstrate how the algorithm proposed in this paper enables multiple shepherds to coordinate and perform collective shepherding of 'sheep'-flock of different sizes and types. Figure 2 showcases snapshots of a representative simulation, where three shepherds succeed in herding $N_{sheep} = 300$ 'sheep'-agents. All the agents are randomly initialized inside a certain area and without boundaries limitation. Before the 'shepherd' process starts, the 'sheep'-agents are allowed to relax into a self-organized stationary flocking state with equilibrated inter-individual distances.

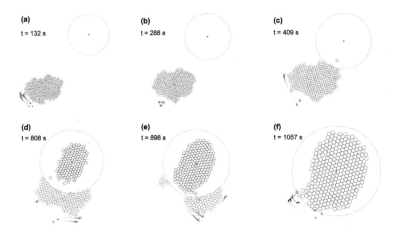

Fig. 2. Snapshots of three shepherds herding 300 agents. (a) All shepherds start to herd the 'sheep'-flock in collecting mode (blue trajectory). (b) The shepherds drive the 'sheep'-flock and maintain repulsion distance (red trajectory). (c) Two shepherds switch into the collecting mode while one shepherd remains in the driving mode. One 'sheep'-agent is already inside the target place and automatically switches into a staying state. (d) Near half of the 'sheep'-agents are in the target place. (e) The shepherds coordinate the rest of the moving agents with two in driving mode and one in collecting mode. (f) The experiment finishes with all 'sheep'-agents inside the target place.

In fig. 2, shepherds in 'drive' mode are marked in blue while the one in 'collect' mode are marked in red. For each shepherd, we also indicate the trajectory over a few past time steps to visualize the movement state. In the shown snapshots, the shepherd-shepherd repulsion distance is set to $L3 = 20$, which means that when the distance between two shepherds is closer than 20, they will avoid each other. The 'sheep'-agents in a moving state are shown in green and those in a staying state are shown in blue.

To investigate the performance of the shepherds, we quantified the time required to herd the flock of sheep agents, averaged over multiple runs: $T_{herding} = \sum_i T_i / n_{runs}$. Please note that, lower values indicate better performance and that the average times are bounded from above by the maximal simulation time of 2000 time steps. As a complementary measure, we also defined the average success rate $SR = \sum_i T_i^{-1} / n_{runs}$. Initially, we consider the case of no repulsion distance between shepherds ($L3 = 0$). Figure 3 shows the result of the finishing time and success rate as a function of the number of shepherds (up to 5). It indicates that with the same group size of 'sheep'-agents, increasing the number of shepherds reduces the herding time (fig. 3 (a)) and increases the success rate (fig. 3 (b)). In general, the larger the size of 'sheep'-flock is, the longer it takes to drive them.

3.1 Trade-Off Between Collect and Drive Mode for Shepherds

In the model, the shepherd has to switch between collect mode and drive mode to keep the 'sheep'-agent cohesive and moving toward the target. The distance of the furthest

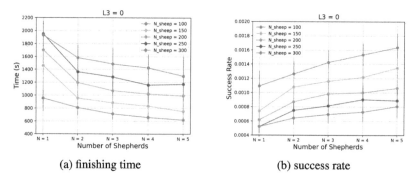

Fig. 3. Finishing time and success rate as a function of the number of shepherds. The success rate is calculated as $SR = \frac{1}{n}\sum_{i=1}^{n}\frac{1}{T_i}$, T_i is the finishing time of the repeated experiment i and n is the number of repetitions ($n = 25$). Due to the limited calculation resources, each experiment has maximum running time $T_{max} = 2 \times 10^3$ s. $L3$ is the repulsion distance between shepherds.

agent to the center of the mass serves as a threshold $L2 = 2R\lambda_2\sqrt{N_{agents}}$. Here λ_2 is a linear control parameter for $L2$, see Table 1), which decides when the shepherd should switch back and forth between the two behavior modes.

To analyze the correlation between the time the shepherd spends in drive mode and the final success rate, we compare the simulation results with only one single shepherd and increasing numbers of sheep (N_s varies from $[100, 300]$) (fig. 4). Figure 4 (a) shows the drive ratio as a function of the number of sheep for three different values of L_2 (controlled by the linear parameter $\lambda_2 = 0.6, 0.67, 0.75$). For the largest value of $\lambda_2 = 0.75$), and thus L2, the shepherd tends to spend the majority of time in the drive mode when the number of sheep ranges from $N_{sheep} = 100$ to $N_{sheep} = 300$. In addition, for the same parameter $\lambda_2 = 0.75$ we observe the highest success rate according to fig. 4 (b). It is noteworthy that for the larger group of sheep ($N_{sheep} = 250, 300$), although the drive ratio is higher than 0.6 when $\lambda_2 = 0.75$ (fig. 4 (a)), there is no distinguishable difference in success rate with respect to $\lambda_2 = 0.6$ (fig. 4 (b)). This is because the larger group has a larger dispersion, resulting in more difficult aggregation.

3.2 Coordination Among Multiple Shepherds

Our previous results in fig. 3 (a) and (b) show that increasing the number of shepherds can improve the success rate in our simulations. In general, the more shepherds are present, the higher the probability that different shepherds are at time t in different behavior modes. However, without any repulsion between shepherds ($L3 = 0$), the shepherds can in principle also overlap with each other most of the time. In these cases, they receive (almost) the same sensory information and in consequence, may exhibit the same switching between driving and collecting. The probability of shepherds in different modes is affected by either noise or initial spatial distribution.

To further understand the role of coordination among shepherds, we quantitatively compare the success rate and coordination ratio with the simulation results using different values of repulsion distance. For the multi-shepherd system ($N_{sh} \geq 2$), we speak

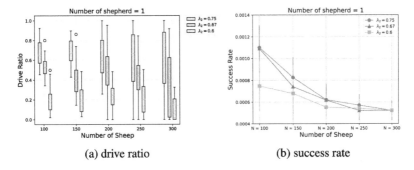

(a) drive ratio   (b) success rate

Fig. 4. Simulation results of one shepherd and different values of λ_2. (a) Drive ratio of one shepherd as a function of the number of sheep. The higher the value λ_2 is, the higher the drive ratio that shepherd takes in one experiment. (b) Success rate of one shepherd as a function of the number of sheep. λ_2 is the scaling parameter for the threshold distance $L2$. When the distance between the furthers agent and the center of the mass ($d_{furtherstagent} \geq L2$), the shepherd should switch to drive mode.

of coordination when not all of the shepherds are in the same mode. Lets define, that the state value for shepherd i at time equals 0 ($s_i(t) = 0$) for the collect mode, and 1 for the drive mode ($s_i(t) = 1$). The coordination state value at time t equals 1 when $\left(\sum_{i=1}^{i=N_{sh}} s_i(t)\right) \neq 0$, N_{sh}, otherwise it equals 0. The coordination ratio CR is is then the average coordination state value over the complete simulation time T. The maximum possible value $CR = 1$ indicates that at all times at least one shepherd was in a different behavioral mode than the others, while $CR = 0$ indicates that at all times all agents have been in the same behavioral mode.

For vanishing shepherd-shepherd repulsion distance ($L3 = 0$), where shepherds can overlap with each other, we observe only low levels of self-organized coordination emerging from random initial conditions. Moreover, we find that for the repulsion distance studied $L3 = 20, 30$, the success rate is not affected by shepherd-shepherd repulsion (fig. 5 (a)). In comparison, for the same sheep flock sizes and the number of shepherds, the coordination ratio increases strongly with increasing repulsion distance (fig. 5 (b)), showcasing how heterogeneity in spatial locations enforced by the repulsive interaction promotes a 'division of labour' through different agents performing different tasks at the same time.

3.3 Non-cohesive and Passive 'sheep'-Flock

Here, we test the adaptability of the shepherd strategies in two additional more challenging scenarios: the first one is the herding of non-cohesive sheep agents, which lack any attraction to other sheep agents; The second scenario is the herding of passive sheep agents which do not exhibit any self-propelled motion, and thus have to be actively pushed by shepherds through interactions with other driven sheep agents.

Given that the attraction force in the 'sheep' model enforces cohesive flocks, it requires less effort by the shepherds to keep the sheep agents aggregated. For non-cohesive sheep flocks, lacking attractive interactions, the shepherd agents must very

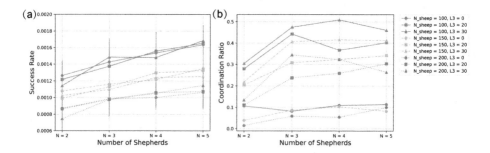

Fig. 5. Success rate (a) and coordination ratio (b) as a function of multiple shepherds. To control the variables, all the experiments have the same value of $\lambda_2 = 0.67$. $L3$ is the repulsion distance among shepherds.

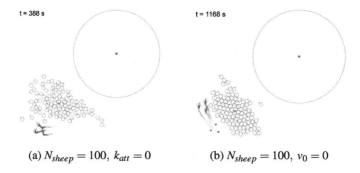

Fig. 6. Snapshots of herding non-cohesive and passive agents. (a) Non-cohesive agents without attraction. (b) Passive agents without constant speed. Other parameters are the same: $N_{sh} = 3$, $N_{sheep} = 100$, $L3 = 20$.

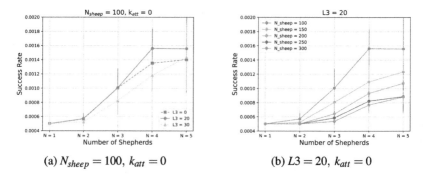

Fig. 7. Simulation result with non-cohesive 'sheep'-agents. (a) The success rate for different values of $L3$ while keeping the number of sheep $N_{sheep} = 100$. (b) The success rate for non-cohesive flocks with non-attraction force: $k_{att} = 0$.

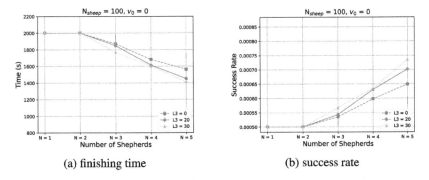

Fig. 8. Simulation result with passive 'sheep'-agents ($v_0 = 0$). (a) Time of finishing experiment as a function of the number of shepherds. (b) Success rate as a function of the number of shepherds. $L3$ is the repulsion distance between shepherds. The number of 'sheep'-agent: $N_{sheep} = 100$.

frequently switch to the collect mode to prevent dispersion of the flock. A snapshot of the herding non-cohesive 'sheep'-agents experiment is shown in fig. 6 (a), where three shepherds coordinate herding 100 sheep-like agents.

Figure 7 shows the result for the success rate SR of herding non-cohesive 'sheep'-agents. First, we compared the success rate SR with different values of $L3$ for a fixed number of 'sheep'-agents $N_{sheep} = 100$, (fig. 7a). Here, we observe a monotonous increase of the SR with the number of shepherds for all values of the shepherd-shepherd repulsion range $L3$, and the best performance for large N_{sh} for the intermediate value of $L3 = 20$. Second, we compared the success rate for different sheep flock sizes for a fixed value of $L3 = 20$ with N_{sheep} varying from 100 to 300 (fig. 7b). We observe a significant increase in the success rate SR as the number of shepherds increases, for all 'sheep'-flock sizes, with the largest flocks being most difficult to herd.

For the second scenario, investigating the herding of passive agents, we set the self-propulsion speed of the sheep agents to zero ($v_0 = 0$). The interaction of short-ranged repulsion and long-ranged attraction between neighbouring sheep can lead to the emergence of effective alignment at the group level. Here, flocks of self-propelled sheep agents, once they are aligned towards the target, can rather easily be driven by the shepherd. Therefore, in general, for passive agents without self-propulsion, it takes more effort and thus a longer time for the shepherd to drive the 'sheep'-flock.

A representative snapshot of herding passive agents is plotted in fig. 6 (b), while the corresponding results for the shepherding time T and success rate SR are summarized in fig. 8. If we consider the number of sheep agents $N_{sheep} = 100$, we observe the minimal average finishing time T of around $1400\,s$ (fig. 8 (a)). While in the original scenario, the time to shepherd a flock of active agents is significantly shorter $\langle T \rangle = 1000\,s$ (fig. 4 (c)). At the same time, the minimum success rate SR exceeds 0.0010 (fig. 4 (d)) and is higher compared to the passive agents' scenario when using five shepherds ($SR < 0.0080$, fig. 8 (b)).

Overall, despite possible decreases in performance, our results show that the proposed algorithm enables successful herding even in the more challenging scenarios of non-cohesive and passive flocks.

4 Discussion and Conclusion

In this work, we investigated an agent-based model of herding a flock of sheep-like agents by multiple shepherds. The sheep respond to the shepherds via a repulsive interaction, which allows the shepherds to control and drive the sheep flock through corresponding positioning. To ensure cohesion of the sheep flock, as well as the arrival at the target location, the shepherds switch in a self-organized manner between two different behavior modes: driving and collecting, which were initially proposed in [19] for the adaptive behavior of a single shepherd. We show that multiple shepherds can cooperate and move a large group of sheep agents toward the target location (fig. 2). We also observed that the coordination among shepherds, i.e. different shepherds performing different tasks during the herding process, emerges in a self-organized fashion, and can be further improved by adding repulsion interactions between shepherds (fig. 5). Finally, we could also demonstrate that the multiple shepherds are also able to perform well in more challenging scenarios, like herding non-cohesive or passive sheep agents, which suggests that it may apply to different real-world scenarios like herding non-cohesive self-propelled robots or collecting inert passive agents like blocks.

An important aspect of our work, different from the closely related single shepherd model [19], is the reduction of the amount of global information that shepherd agents rely on. Inspired by visual interactions, we assume that shepherds only use relative angle information to identify the furthest agents, that need to be collected to maintain flock cohesion. In contrast to distances, relative angles can be easily acquired through vision by biological agents. Individual sheep agents interact only with local neighbors within metric zones. We expect that the general herding ability will not be fundamentally affected by other local methods of choosing neighbors such as k-nearest neighbors [16], Voronoi [12], or visual interaction networks [15].

However, in our model, shepherds still use the information about the sheep flock center of mass location, when comparing the distance between the furthest agent and the center of mass with the threshold $L2$. This implicitly assumes the ability of shepherd agents to reliably estimate the center of mass of the sheep flock, which is a non-trivial task by vision alone. Here, alternative behavioral heuristics that rely only on limited visual information may be of interest in the future.

In conclusion, collective shepherding is a complex dynamical behavior with a broad range of applications. Its complexity arises from the interaction of two different collectives, namely the sheep and the shepherds. While our results extend our understanding of the self-organized dynamics of the coupled shepherd-sheep system, many questions remain open. Here, the presented modelling approach provides a starting point for future research for example on behavioral heuristics for shepherds with only limited, local (visual) information, on mechanisms for enhanced coordination of shepherds, e.g. via explicit communication, or on highly heterogeneous shepherd groups with heterogeneous movement or perception capabilities.

References

1. Bartashevich, P., et al.: Collective anti-predator escape manoeuvres through optimal attack and avoidance strategies. bioRxiv, pp. 2024–03 (2024)
2. Bennett, B., Trafankowski, M.: A comparative investigation of herding algorithms. In: Proc. Symp. on Understanding and Modelling Collective Phenomena, pp. 33–38 (2012)
3. Coppinger, L., Coppinger, R.: Dogs for herding and guarding livestock. In: Livestock Handling and Transport, pp. 245–260. CABI Wallingford UK (2014)
4. Dosieah, G.Y., Özdemir, A., Gauci, M., Groß, R.: Moving mixtures of active and passive elements with robots that do not compute. In: International Conference on Swarm Intelligence, pp. 183–195. Springer (2022).https://doi.org/10.1007/978-3-031-20176-9
5. Farivarnejad, H., Berman, S.: Multirobot control strategies for collective transport. Ann. Rev. Control, Robot. Auton. Syst. **5**, 205–219 (2022)
6. Hansen, M.J., Domenici, P., Bartashevich, P., Burns, A., Krause, J.: Mechanisms of group-hunting in vertebrates. Biol. Rev. **98**(5), 1687–1711 (2023)
7. Hughes, R.L.: The flow of human crowds. Ann. Rev. Fluid Mechanics **35**(1), 169–182 (2003)
8. Kachroo, P., Shedied, S.A., Bay, J.S., Vanlandingham, H.: Dynamic programming solution for a class of pursuit evasion problems: the herding problem. IEEE Transactions on Systems, Man, and Cybernetics, Part C (Applications and Reviews) **31**(1), 35–41 (2001)
9. King, A.J., et al.: Biologically inspired herding of animal groups by robots. Methods Ecol. Evol. **14**(2), 478–486 (2023)
10. King, A.J., et al.: Selfish-herd behaviour of sheep under threat. Curr. Biol. **22**(14), R561–R562 (2012)
11. Klamser, P.P., Gómez-Nava, L., Landgraf, T., Jolles, J.W., Bierbach, D., Romanczuk, P.: Impact of variable speed on collective movement of animal groups. Frontiers Phys. **9**, 715996 (2021)
12. Klamser, P.P., Romanczuk, P.: Collective predator evasion: putting the criticality hypothesis to the test. PLoS Comput. Biol. **17**(3), e1008832 (2021)
13. Lien, J.M., Bayazit, O.B., Sowell, R.T., Rodriguez, S., Amato, N.M.: Shepherding behaviors. In: 2004 IEEE International Conference on Robotics and Automation (ICRA). vol. 4, pp. 4159–4164. IEEE (2004)
14. Pierson, A., Schwager, M.: Bio-inspired non-cooperative multi-robot herding. In: 2015 IEEE International Conference on Robotics and Automation (ICRA), pp. 1843–1849. IEEE (2015)
15. Poel, W., Winklmayr, C., Romanczuk, P.: Spatial structure and information transfer in visual networks. Frontiers Phys. **9**, 716576 (2021)
16. Rahmani, P., Peruani, F., Romanczuk, P.: Flocking in complex environments-attention trade-offs in collective information processing. PLoS Comput. Biol. **16**(4), e1007697 (2020)
17. Ramos, D.G., Birattari, M.: Automatically designing robot swarms in environments populated by other robots: an experiment in robot shepherding. 2024 IEEE International Conference on Robotics and Automation (ICRA) (2024)
18. Rojas, J.: Plastic waste is exponentially filling our oceans, but where are the robots? In: 2018 IEEE Region 10 Humanitarian Technology Conference (R10-HTC), pp. 1–6. IEEE (2018)
19. Strömbom, D., et al.: Solving the shepherding problem: heuristics for herding autonomous, interacting agents. J. R. Soc. Interface **11**(100), 20140719 (2014)
20. Wilson, S., Pavlic, T.P., Kumar, G.P., Buffin, A., Pratt, S.C., Berman, S.: Design of ant-inspired stochastic control policies for collective transport by robotic swarms. Swarm Intell. **8**, 303–327 (2014)

DaNCES: A Framework for Data-inspired Agent-Based Models of Collective Escape

Marina Papadopoulou[1,2](✉)[iD], Hanno Hildenbrandt[2][iD], and Charlotte K. Hemelrijk[2][iD]

[1] Biosciences, School of Biosciences, Geography and Physics, Swansea University, Swansea, UK
[2] Groningen Institute for Evolutionary Life Sciences, University of Groningen, Groningen, The Netherlands
m.papadopoulou.rug@gmail.com

Abstract. The collective escape of predators by prey is a classic example of adaptive behavior in animal groups. Across species, prey has evolved a large repertoire of individual evasive maneuvers they can use to evade predators. With recent technological advances, more empirical data of collective escape is becoming available, and a large variation in the collective dynamics of different species is apparent. However, given the complexity of patterns of collective escape, we are still lacking the tools to understand their emergence. Computational models that can link rules of individual behavior to patterns of collective escape are needed, but species-specific motion and escape characteristics that will allow the link between behavior and eco-evolutionary dynamics of a given species are rarely included in agent-based models of collective behavior. Here, to tackle this challenge, we introduce a framework that uses individual-based state machines to model spatio-temporal dynamics of collective escape. A synthetic agent in our framework can switch its behavior between 'flocking' with different coordination specifics (e.g., quicker interactions when vigilant) and 'escape' with various maneuvers through a dynamic Markov-chain, depending on its local information (e.g., its relative position to the predator). A user can compose a new agent-based model adjusted to empirical data by choosing a set of states (which includes rules of motion, interaction, and escape), their temporal order, and a detailed parameterization. The flexibility and structure of our software allows substantial changes in a model with very minimal code alterations, showing great potential for future use to identify the underlying mechanisms of collective escape across species and ecological contexts.

Keywords: Collective escape · Agent-based modeling · Predator-prey interactions

---

Supported by the Netherlands Organization for Scientific Research (NWO), project 14723 awarded to CKH: "Preventing bird strikes: Developing RoboFalcons to deter bird flocks."

1 Introduction

Since the first model that reproduced patterns of collective escape in fish schools was published more than 20 years ago [19], few studies have attempted to model patterns of collective escape across taxa (e.g., [5,14,15,38]), leaving many gaps in our understanding of collective escape. Even though species differ a lot in the specifics of their escape reactions, that can also be predator-specific [6,7], models of collective escape have been usually generic (e.g., [1,2,15,38]). With technological advantages (e.g., unmanned aerial vehicles and underwater cameras) enabling the collection of empirical data of collective escape across many systems [12,32], a better link between data and theory is the next step forwards. However, species-specific adjustments in a model [11,18,27] can be time-consuming and challenging in an existing model, given the increasing model complexity as species-specific characteristics are added. Striving for a balance between mechanistic simplicity and data complexity is also of high importance, since overfitting a model to data may compromise the identification of self-organized mechanisms and make a model 'single-use', delaying advances in the field and limiting model comparisons [10].

To adjust a model of collective escape to an empirical system, the specifics of reaction of individual prey to the predator should be carefully considered. Most models include a single rule based on which prey avoids a predator. This rule is either 'continuous', balanced with interaction rules that make group members coordinate (for instance a tendency to turn away from the predator's position [19,38]), or 'discrete' (also referred to as 'fixed'), instantaneous turning that 'interrupts' the regular coordinated motion of the group (usually applied to study wave propagation, e.g., [14,15]). The escape reactions we see in nature may be represented by both continuous and discrete cognitive rules; which type of escape reaction individuals perform when their group is under attack is still hard to distinguish in empirical data. Additionally, more complex escape maneuvers (e.g. protean motion [17,21,22]) are rarely modeled in the context of collective behavior. Thus, in order to improve our modeling of collective escape, both continuous and discrete escape reactions should be tested at the individual level. Nevertheless, we first need to model the collective motion of a given species [27].

Most agent-based models of collective animal behavior are based on the rules of attraction, alignment and avoidance [5,30], even though species differ in their dynamics of collective motion (their specifics of order, shape and density) [23]. Mere changes in a model's parameters are often not enough to realistically simulate a group of a specific species, adjustments in the interaction rules are needed. For instance, in a model aiming to simulate groups of pigeons, an extra rule of speed adjustment was needed [27], with individuals deviating from their preferred speed to stay with the flock, in accordance to empirical findings in homing pigeons [29,31]. How exactly a single coordination rule of flocking is modeled may also vary; for instance, an individual may avoid collisions by turning away from the position of its neighbor, by aligning with the heading of its neighbor, or by decelerating. How these modeling decisions affect the emerging collective motion has been rarely studied. Additionally, most computational models of collective behavior assume that individuals are identical, even though individual variation in specific traits within a group has been identified in many species [20,31].

Here, we recognize the necessity for a new modeling structure that can counteract the aforementioned issues and model the diversity of collective escape we see in nature (Fig. 1). We want to create a model that is flexible and easily adjustable to different species and ecological contexts. Therefore, it should accommodate several (and perhaps diverse) types of individual motion (especially escape), from continuous coordinated motion to sudden reactions to the predator. To achieve these aims, we develop a modeling framework (named DANCES, DAta-iNspired Collective EScape) that uses individual-based state-machines to model collective escape. By combining internal states at the individual level with a composable software structure, we further highlight the potential of such conceptualization [36] to study collective escape. While developing our framework, we created three agent-based models of bird flocks [25,26,28], and present some of their technical details here as examples for demonstrating the full structure and functionality that our framework offers.

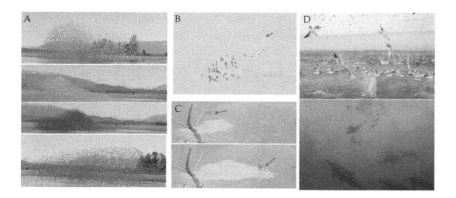

Fig. 1. Collective escape in animal groups **A.** Aerial displays of flock of dunlins (*Calidris alpina*) reacting to the attack of a peregrine falcon (*Falco peregrinus*). Screenshots taken from the video 'Dance of the Dunlins' (2013) by Ray Hamlyn. **B.** A flock of corvids turning away from a robotic falcon [33]. **C.** Propagation of a 'diving' escape wave in schools of sulphur mollies (*Poecilia sulphuraria*) as a response to the attack of a great kiskadee (*Pitangus sulphuratus*), screenshots taken from video by [8]. **D.** A group of Cape gannets (up, Mathieu Simonet/Galatée Films 2009) attacking a fish school surrounding by more predator (bottom, Galatée Films 2009), taken from [35].

2 Modeling Framework

Our framework is implemented in C++, using OpenGL for the visualization and DearImGui [4] for a user interface, and can be accessed in our GitHub repo [24].

2.1 Internal States and Social Interactions

An agent in our framework is represented by its 'internal state'. In the context of collective escape, this internal state consists primarily of the agent's position and velocity in a global coordinate system. Since agents in computational models of collective behavior move in space based on their interactions with one another, we conceptualize an individual-level 'rule' in our model that can alter the internal state of an agent (and thus control its motion), as a stand-alone 'internal-state control unit' (referred to as an *ISC-unit*). Examples of such units are an alignment interaction (that makes the individual align its heading with its neighbors), a roosting behavior (that makes it turn towards its roost), or a noisy motion (that adds a random error to its heading). Each ISC-unit owns a set of parameters (e.g. the number of interacting neighbors or the position of the roost). We treat each internal-state control unit as a building block; the modeler combines the pieces of interest according to their study system. Thus, all available ISC-units can be simply used to build a new model or examine alternative interaction rules in existing models, while modeling new units is facilitated (by copying the main structure of an existing unit) and does not risk interfering with other functionalities of the software.

All social ISC-units need information about the social environment of the agent at each timepoint. To provide that, a set of neighbors' information (`neighbor_info`) is defined and can be accessed by an agent (currently the id, state and stress of another agent, as well as the distance between them). An ISC-unit can access this information and indicate a set of neighbors for the agent to 'sense' (`while_topo.hpp`). For instance, at its current state, the framework supports metric and topological interactions along with the definition of a simple visual field: 'neighbors' are defined by a given number of agents (parameter `topo`), positioned within a given radius (parameter `maxView`), and within a field of view (an angle, parameter `foV`, e.g. prohibiting the sensing of individuals in the back). By parameterizing `topo` to be equal to the total population size and restricting the `maxView` one can have a 'traditional' metric model, and by setting `maxView` to a large value and restricting `topo`, a topological. These settings are ISC-unit specific, meaning that an agent can use different social information per behavioral rule (e.g., avoid collision with only a single neighbor but copy an escape maneuver from any surrounding neighbor).

2.2 Individual-Based State Machines

A set of ISC-units defines a 'state', that controls an agent's behavior (similar to a Markov-Chain [37]). For instance, a state that represents a 'flocking' behavior may consist of four ISC-units that affect the agent's behavior: alignment, centroid-attraction, avoidance and noise. Thus, a state is designed around the nature of its effect on individual motion. At its current state, our framework implements social interactions and individual motion through a 'social' force that is applied to the internal state of each individual. Thus, each ISC-unit adds a steering vector (glm::vec3 type) to the main social force that ultimately controls the motion of each agent (see the OODD Protocol of the HoPE model in [26]). The use of such steering vector is however not enforced by our framework; one can choose a different variable to be changed and control the agent's motion.

Apart from the parameters of each ISC-unit of a state, a state also has its own parameterization. This can include, for instance, the frequency with which the steering vector is updated (recalculating the effect of each ISC-unit, also known as 'reaction frequency' [2,18]). A description of example states in existing models based on our framework is given in *Section* 3. Overall, a model with a single state in our framework is the equivalent of a classic model of collective behavior, in which a constant set of rules at the individual level affects the motion of all agents identically.

To model discrete reactions and switches between different behaviors (related to the predator or the ecological context), our framework enables several states to be included in a model. An agent is thus defined by not only its internal state, but also a finite-state machine that controls its behavior. Each state can comprise a different combination of ISC-units or the same ISC-units with different parameterization. For instance, a state of 'flocking' can be followed by an 'escape' state, during which an agent flocks while also avoiding a predator. An example of the definition of a state machine of a prey agent is given in Listing 1.1.

Listing 1.1. Example of a definition of a state machine of a 'prey' agent with several states. A description of all ISC-units currently available in the framework is given in our repository [24]. The ones listed here represent alignment, cohesion, repulsion (by turning away from the position of its neighbors), roosting (attraction to a global point in space), random noise in the direction of motion (`wiggle`), and escape by a level turn or diving. To change the behaviour of an agent, the user needs to change the functions included in the defined 'package'(e.g., remove `roost_attraction` or replace `avoid_n_position` with `avoid_n_direction`: avoid collision points by turning parallel to the neighbour), and the parameters relating to each ISC-unit in the *.json* parameter file.

```
// States package of a PreyType agent
using AP = states::package<
    states::transient<iscus::package< // 1.regular flocking
        PreyType,
        iscus::align_n<PreyType>, // alignment
        iscus::cohere_centroid_distance<PreyType>, // cohesion
        iscus::avoid_n_position<PreyType>, // separation
        iscus::roost_attraction<PreyType>,
        iscus::wiggle<PreyType> >>,
    flee_state, // 2. escape multi-state
    states::persistent<iscus::package< // 3. refraction (persistent flocking)
        PreyType,
        iscus::align_n<PreyType>,
        iscus::cohere_centroid_distance<PreyType>,
        iscus::avoid_n_position<PreyType>,
        iscus::roost_attraction<PreyType>,
        iscus::wiggle<PreyType> >> >;

using flee_state = states::multi_state<PreyType, // multistate definition
    states::persistent<iscus::package<PreyType,
        iscus::random_t_turn_gamma_pred<PreyType>,
        iscus::wiggle<PreyType> >>, // 1.escape turn
    states::persistent<iscus::package<PreyType,
        iscus::dive<PreyType> >> >; // 2. dive
```

Transitions between states are controlled by a transition machine, owned by each agent. Each state has a duration that ranges from a single reaction step to any parameterized duration. For conceptual differentiation, a state is defined as 'transient' (effective duration of a single update cycle, an individual can switch state at any point) or 'persistent', with a user-define duration (only after a given period is passed the agent may switch state). At the exit of a state, the transition machine assigns the next state as a function of the current state and a set of probabilities, i.e. a transition matrix. This matrix can be dynamic or constant, and is predefined by parameterization. For instance, deterministic transitions are used in our existing models for the predator so its behavior follows a closed loop [25–27]: the predator has a probability of 1 to transition between its 'pursuit', 'attack' and 'retreat' states that all have a pre-defined duration.

Dynamic transition matrices are modeled for transitions that may depend on specific conditions during a simulation. For instance, an individual may switch to an escape state only if the predator is approaching. To account for this, each transition machine also owns a variable that can alter the probability of switching to a specific state. In the existing state of the framework, such transition-altering mechanism is modeled through a 'stress' variable that is affected by a set of 'stress source' functions, namely neighbors (*neighbors_stress*) or predator (*predator_distance*): the closer the predator or the higher the stress of the neighbors of an agent, the more stress is accumulated in each agent, and the higher its probability to perform an escape maneuver is. A decay rate parameter is used to stabilize this accumulating value, so that in the absence of a predator, the probability of escaping returns to 0. Thus, the exact transition matrix at a given update time step is calculated by the transition machine through a piecewise linear interpolator that uses the stress value, and a number of user-defined transition matrices and interpolation edges.

Overall, depending on the characteristics of the empirical systems, the transition probabilities between states for each agent can vary through time or be constant. Based on the probabilities of each current state, an agent selects its next state. Algorithm 1 presents a simplified version of the above mentioned processes.

In a real system, an individual may choose to react with an escape maneuver to a predator, but the exact maneuver may depend on the relative position of the prey to the predator [6]. Translated in our framework, to account for cases where the state selection should be performed by a specific ISC-unit and not the transition machine, we further included 'multi-states' in our framework. A multi-state is a mere collection of states. The states of a multi-state (sub-states) have a conceptual connection that affects their selection during a simulation. The transition machine may decide that an individual should perform an escape maneuver, but the probability of selecting each one can be altered by each ISC-unit during the simulation (for instance if the predator attacks from above, a diving maneuver may be evaluated as more effective than a turning one), or from the parameterization of the multi-state (for instance, empirical data may show that the frequency of turning is higher than the frequency of diving). An

Algorithm 1. Pseudo-code for the main structure of a simulation, showing the timing of processes and the handling of state transitions.

```
 1: for t_i : 1,..., T_max do                  //With T_max the total simulation time
 2:     integrate_motion(ψ)          //Calculate new position and velocity of agent
 3:     s_i ← s_{i−1} − r_decay                                        //Stress decay
 4:     if t_i = t_update then
 5:         ψ ← 0                                       //Start new steering vector (ψ)
 6:         ψ ← chain_ISC_units(S)    //Accumulate steering vector from the state
 7:         t_update ← t_i + t_react                                 //Next update step
 8:     end if
 9:     if t_i = t_exit_state then
10:         s_i ← accumulate_stress()
11:         TM ← get_transition_matrix(s_i)
12:         S ← sample_new_state(TM)                              //Get new state
13:         t_exit_state ← t_i + Δt_S   //Next switch from the duration of the state (Δt_S)
14:     end if
15: end for
```

example of the definition of a multi-state is given in Listing 1.1 ('flee_state'). This allows a model to be improved as new empirical data on individual escape reactions are becoming available, facilitating the link between observations and theory.

2.3 Multi-level Parameterization

Parameters in a model on our framework can be at the level of a state (for instance a state-specific speed), at the level of an ISC-unit within a specific state (such as the number of topological neighbors for alignment), and at the level of the agent (such as its mass). To control all these, a configuration file with all the necessary parameters should be given as input in the model (at run time). We implemented this as a JSON file given its nested nature that allows for parameterization of very complex state-machines. We can thus control every detail of a model, making conscious assumptions and allowing simulations of long sequences of collective escape seen in nature. This is particularly helpful when running a large set of simulations with different parameterization: ISC-units can be activated or deactivated and thus completely change the model between runs.

Our implementation of the parameterization of the transition machine is inspired by behavioral chains found in bird flocks [34], so that transition probabilities can be directly informed by empirical data. We define a Markov chain through a transition matrix in our configuration file. Given that these probabilities may need to be adjusted during a simulation, depending on external conditions such as the proximity of an individual to the predator, we implemented an interpolator between several transition matrices. In detail, by giving the model several matrices (3 by default, for low, medium and high threat) for some extreme conditions, the exact transition matrix is created at every state-switching step based on the parameter of the transition machine (*stress*). For instance, the interpolator could be used when the probability of an individual to perform an extreme escape maneuver should be 0 when the predator is not present and only increase when the predator gets very close.

It is important to note that ISC-units are not specific to the type of the agent in a model. They can be used by any agent as long as all its necessary variables (parameters) are there. For instance, predator and prey can use the same ISC units to move around, but to use the ISC-unit that allows an agent to select a target from a group (conceptually a predatory behavior), the agent needs to own a 'target' variable. This flexibility allows for all of the ISC-units in our framework to be reusable, even when introducing a completely new type of agent.

2.4 Customizable Locomotion Type

The type of motion of grouping individuals (e.g. flying, swimming, or walking) is crucial to the emerging collective behavior [11,13]. Currently, our framework has been used to model bird flocks, and thus the locomotion module resembles flying. For our framework to be extendable to non-flying species, the way that a steering vector is affecting the motion of an agent is controlled by a stand-alone motion integrator. This integrator functions independently of the specifics of an agent (its variables and states). This allows the adjustment of the core of a model to other study systems, without changes being necessary on the structure of the agents, the ISC-units or the states.

2.5 Real-Time Visualization and Data Analysis

To incorporate analysis and visualization in our simulations we use the concept of 'observers' (a software design pattern mainly used in event-driven computing [9]). Observers access variables of a simulation in real time without interfering with the model. Based on a chain of specialized observers, we can perform data analysis on the simulated trajectories in real time (for instance calculating the average nearest neighbor distance or the neighbor stability). With a specific observer per analysis type, our simulations can output analyzed data along with raw trajectories in '.csv' format. All exported files are stored within a unique folder, along with a copy of the configuration file of the simulation. The folder is created at the beginning of each simulation, within a user-defined folder (parameter `data_folder`) in the *"bin/sim_data"* directory. This is extremely valuable for pairwise measurements, e.g., neighbor stability, that are computationally very demanding [25] .

The observers enable the real-time visualization of a simulation (for visual debugging and calibration). By changing the representation of an agent, effects of body coloration that affect the observed patterns of collective escape can also be studied (e.g. Fig. 1A). A 'head-less' version (without visualization) can also be turned-on in order to speed up the simulation, for instance when running a large number of simulations during data collection. Finally, a graphic user interface

Fig. 2. Screenshot of the GUI (created in *Dear ImGui*) and visualization (in OpenGL) of a 3D model of bird flocks based on *DaNCES*. Information from the simulation is collected through *Observers* and thus the visualization and data analysis do not interfere with the model itself, ensuring that the model can be run without them (headless version) and software complexity does not increase. The color of the trails of each agent shows its current state.

(GUI) allows the real-time plotting and visualization of several variables in the model, such as the current speed or state of each agent (Fig. 2).

An overview of a state machine and its connection with other elements of our framework is given in Fig. 3.

3 Use Cases

Our framework has so far been used for four models of bird flocks [25–28]. *HoPE*, a model adjusted to the collective escape of pigeons, first consisted of a single state with 5 ISC-units: alignment, centroid-attraction, speeding attraction, noise, and predator avoidance [27]. Since predator avoidance is modeled as a continuous tendency to turn away from the predator while coordinating, similar to many models of collective escape [19,38], an extra state to model escape was not needed. Given that this model couldn't capture all patterns of collective escape seen in pigeons, an extension of the model was created to study the propagation of escape maneuvers [26]. Patterns of collective escape were conceptualized to being initialized by one individual that stops coordinating with its neighbors and maneuvers to escape the predator for a specific time period. Additionally, a refractory time period is added after an escape maneuver, during which an individual does not coordinate with its neighbors to ensure that the emergence of splitting and collective turning is the effect of the group's 'decision'. Thus, the extended *HoPE* model consists of three states: a flocking state (the one from its previous version [27]), an escaping state (with two ISC-units, an escape maneuver and noise), and a refraction state (with a single ISC-unit that adds noise to the straight heading of the initiator).

A model of higher complexity based on our framework is the 3-dimensional model of starlings (*Sturnus vulgaris*), *StarEscape* [28]. Aiming to reproduce

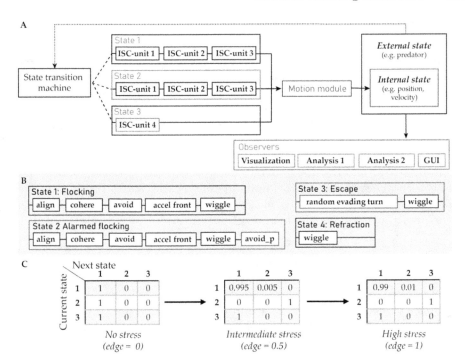

Fig. 3. A. Schematic representation of a state machine and its connection with the other elements of our framework. The transition machine selects a state depending on the agent's internal (e.g., its position, the previously selected state), and external state (e.g., its distance to the predator). States can be composed of a different combination of ICS-units and have their own parameterization. The different border colors represent different parameter values. State 1 and 2 are identical in their ISC-units but differ in their state-level parameter values (for instance in their duration or the agent's reaction frequency). The output of the ISC-units changes the internal state of an agent depending on the specifics of the motion module. Observers have access to an agent's external and internal state. Several observers with different functions (e.g., to export, analyze or visualize the simulated data) can be added in a model. **B.** An example of the states and ISC-units similar to the ones used for the extended HoPE model [26]. Red borders relate to escape-related blocks. **C.** An example of 3 user-defined transition matrices for a model with 3 states (1:flocking, 2:escape, 3:refraction). As stress increases, the probability to escape (transition from state 1 to 2) should also increase. The transition machine will interpolate across the given values depending on the value of stress and the input edges of the interpolation.

their aerial displays [3], the model includes 4 states: flocking, alarmed flocking, escape (which as mentioned earlier is a multi-state), and refraction. Flocking in *StarEscape* includes the following 6 ISC-units: alignment, centroid-attraction (distance biased, ensuring sharp edges of the flock similar to real starlings [18]), avoidance, noise, roost attraction, and altitude attraction. The last two ISC-units constrain the movement of the flock within the 3-dimensional 'infinite' space: they ensure that individuals will stay within a given radius and flying at a preferred altitude. The alarmed flocking state is structurally identical with the flocking state, apart from being parameterized with a higher reaction frequency (imitating more vigilant individuals when the predator is pursuing the flock [2,16]) and an extra ISC-unit that affects the transition machine: a 'copy-escape' behavior. This ISC-unit checks whether any neighbor is in the escape state. If so, it sets the future state in the transition machine to be the same as its neighbors. The escape multi-state has two states that include the coordination-related ISC-units of the flocking states, and an additional maneuvering ISC-unit, one for a diving maneuver and one for a turning maneuver. Finally, the refraction state, has the coordination-related ISC-units of the flocking states but a predefined duration. Separating these states ensures that individuals will not keep reacting to the predator in an unrealistic manner.

4 Summary and Potential

In our new framework, a model is composable, flexible and reusable. By the user changing a few lines of code in the definition of an agent's state machine and the configuration file, a new model can be created. A large set of already available ISC-units can be found in our repo [24]. Furthermore, given the stand-alone nature of ISC-units, adding a new one requires minimal coding and does not interfere with the chain of the model itself. Apart from adding new rules, the user can also quickly remove elements of a model for comparing simulations of decreasing complexity. This is extremely valuable when searching for a simple self-organized process to explain the emergence of a collective pattern.

Following the example of the 'Medawar' zone of pattern-oriented models [10], we should emphasize that the increasing complexity of models that our framework supports should be used with caution: a very large parameter space can limit the explanatory power of the model and constrain our understanding of the underlying mechanisms of the complex system. Finding the balance between minimal models and models derived from complex empirical data is challenging, and should be reached while keeping the study's research question in mind. We hope that DaNCES supports modelers while seeking this balance, trains students in the modeling of animal behavior, and promotes clear (and conscious) assumptions and modeling decisions.

Apart from the flexibility of our models, our framework's structure enables the modeling of different behaviors that are crucial to collective escape, for instance the interplay between continuous and instantaneous reactions. Because of its 'building-blocks', we can develop complex models that reproduce long sequences of collective escape similar to the ones seen in nature without increasing the complexity of the code itself. This reduced code complexity also includes the avoidance of long series of nested conditional blocks (if-statements) which, as a model grows, become error prone, difficult to work with, and introduce optimization issues (code that is rarely used increases the computation time). Every block in our framework (being an ISC-unit, a state, a transition machine or a motion interpolator) is a pure function that owns a set of parameters and effectively alters the state of an agent. Because of their simplicity, they are reusable, can be more easily optimized, while providing code clarity and debugging efficiency. Thus, the extension and growth of a model based on our framework is less error prone and less time consuming.

To conclude, our simulation framework provides a skeleton to build a new agent-based model rather than being just a function library. Its building-block structure and detailed parameterization allows for a model to be informed with as many empirical data are available, increasing its biological relevance. Programming-wise, it does not impose our implementation, allowing a more experienced user to work in their own modeling style. Overall, if one can conceptualize a set of rules and actions that agents should follow, along with the interconnection (time sequence) of these actions, it can be modeled in our framework. We hope that it will enable the modeling of many systems for which our theoretical understanding is currently lacking, such as collective escape from multiple predators [12] or with highly complex displays [34] (Fig. 1A,1D). Despite our focus on collective escape, our framework can be used for the spatially explicit modeling of collectives across contexts.

References

1. Angelani, L.: Collective predation and escape strategies. Phys. Rev. Lett. **109**(11), 1–5 (2012). https://doi.org/10.1103/PhysRevLett.109.118104
2. Bode, N.W., Faria, J.J., Franks, D.W., Krause, J., Wood, A.J.: How perceived threat increases synchronization in collectively moving animal groups. Proc. R. Soc. B Biol. Sci. **277**(1697), 3065–3070 (2010)
3. Carere, C., Montanino, S., Moreschini, F., Zoratto, F., Chiarotti, F., Santucci, D., Alleva, E.: Aerial flocking patterns of wintering starlings, Sturnus vulgaris, under different predation risk. Anim. Behav. **77**(1), 101–107 (2009)
4. Cornut, O.: Dear imgui: Bloat-free graphical user interface for c++ with minimal dependencies (2024). https://github.com/ocornut/imgui
5. Couzin, I.D., Krause, J.: Self-organization and collective behavior in vertebrates. Adv. Study Behav. **32**, 1–75 (2003)
6. Domenici, P.: Context-dependent variability in the components of fish escape response: integrating locomotor performance and behavior. J. Exp. Zool. Part A Ecol. Genet. Physiol. **313**(2), 59–79 (2010)
7. Domenici, P., Blagburn, J.M., Bacon, J.P.: Animal escapology I: theoretical issues and emerging trends in escape trajectories. J. Exp. Biol. **214**(15), 2463–2473 (2011). https://doi.org/10.1242/jeb.029652
8. Doran, C., et al.: Fish waves as emergent collective antipredator behavior. Curr. Biol. **32**(3), 708–714 (2022). https://doi.org/10.1016/j.cub.2021.11.068
9. Gamma, E., Helm, R., Johnson, R., Johnson, R., Vlissides, J.: Design patterns : elements of reusable object-oriented software. Pearson Deutschland GmbH (1995)
10. Grimm, V., et al.: Pattern-oriented modeling of agent-based complex systems: lessons from ecology. Science **310**(5750), 987–991 (2005). https://doi.org/10.1126/science.1116681

11. Gyllingberg, L., Szorkovszky, A., Sumpter, D.J.: Using neuronal models to capture burst-and-glide motion and leadership in fish. J. R. Soc. Interface **20**(204), 20230212 (2023)
12. Hansen, M.J., Domenici, P., Bartashevich, P., Burns, A., Krause, J.: Mechanisms of group-hunting in vertebrates. Biol. Rev. **98**(5), 1687–1711 (2023)
13. Hemelrijk, C.K., Hildenbrandt, H.: Schools of fish and flocks of birds: their shape and internal structure by self-organization. Interface Focus **2**(6), 726–737 (2012)
14. Hemelrijk, C.K., van Zuidam, L., Hildenbrandt, H.: What underlies waves of agitation in starling flocks. Behav. Ecol. Sociobiol. **69**(5), 755–764 (2015)
15. Herbert-Read, J.E., Buhl, J., Hu, F., Ward, A.J.W., Sumpter, D.J.: Initiation and spread of escape waves within animal groups. Royal Soc. Open Sci. **2**(4), 140355–140355 (2015). https://doi.org/10.1098/rsos.140355
16. Herbert-Read, J.E., et al.: How predation shapes the social interaction rules of shoaling fish. Proc. Roy. Soc. B Biol. Sci. **284**, 1861 (2017)
17. Herbert-Read, J.E., Ward, A.J.W., Sumpter, D.J., Mann, R.P.: Escape path complexity and its context dependency in Pacific blue-eyes (Pseudomugil signifer). J. Exp. Biol. **220**(11), 2076–2081 (2017)
18. Hildenbrandt, H., Carere, C., Hemelrijk, C.K.: Self-organized aerial displays of thousands of starlings: a model. Behav. Ecol. **21**(6), 1349–1359 (2010)
19. Inada, Y., Kawachi, K.: Order and flexibility in the motion of fish schools. J. Theor. Biol. **214**(3), 371–387 (2002)
20. Jolles, J.W., King, A.J., Killen, S.S.: The role of individual heterogeneity in collective animal behaviour. Trends Ecol. Evol. **35**(3), 278–291 (2020). https://doi.org/10.1016/j.tree.2019.11.001
21. Jones, K.A., Jackson, A.L., Ruxton, G.D.: Prey jitters; protean behaviour in grouped prey. Behav. Ecol. **22**, 831–836 (2011)
22. Mills, R., Hildenbrandt, H., Taylor, G.K., Hemelrijk, C.K.: Physics-based simulations of aerial attacks by peregrine falcons reveal that stooping at high speed maximizes catch success against agile prey. PLoS Comput. Biol. **14**(4), 1–38 (2018). https://doi.org/10.1371/journal.pcbi.1006044
23. Papadopoulou, M., et al.: Dynamics of collective motion across time and species. Philos. Trans. R. Soc. B **378**(1874), 20220068 (2023)
24. Papadopoulou, M., Hildenbrandt, H.: DaNCES framework (2024). https://github.com/marinapapa/DaNCES_framework
25. Papadopoulou, M., Hildenbrandt, H., Hemelrijk, C.K.: Diffusion during collective turns in bird flocks under predation. Front. Ecol. Evol. **11**, 1198248 (2023)
26. Papadopoulou, M., Hildenbrandt, H., Sankey, D.W.E., Portugal, S.J., Hemelrijk, C.K.: Emergence of splits and collective turns in pigeon flocks under predation. Roy. Soc. Open Sci. **9**, 211898 (2022)
27. Papadopoulou, M., Hildenbrandt, H., Sankey, D.W.E., Portugal, S.J., Hemelrijk, C.K.: Self-organization of collective escape in pigeon flocks. PLoS Comput. Biol. **18**(1), e1009772 (2022). https://doi.org/10.1371/journal.pcbi.1009772
28. Papadopoulou, M., Hildenbrandt, H., Storms, R., Hemelrijk, C.K.: Starling murmurations under predation. In prep. (2024)
29. Pettit, B., Perna, A., Biro, D., Sumpter, D.J.: Interaction rules underlying group decisions in homing pigeons. J. Roy. Soc. Interface **10**(89), 20130529 (2013)
30. Reynolds, C.W.: Flocks, herds and schools: a distributed behavioral model. ACM Comput. Graph. **21**(4), 25–34 (1987). https://doi.org/10.1145/37402.37406
31. Sankey, D.W., Shepard, E.L., Biro, D., Portugal, S.J.: Speed consensus and the 'Goldilocks principle' in flocking birds (Columba livia). Anim. Behav. **157**, 105–119 (2019). https://doi.org/10.1016/j.anbehav.2019.09.001

32. Sankey, D.W., Storms, R.F., Musters, R.J., Russell, T.W., Hemelrijk, C.K., Portugal, S.J.: Absence of "selfish herd" dynamics in bird flocks under threat. Curr. Biol. **31**(14), 3192-3198.e7 (2021). https://doi.org/10.1016/j.cub.2021.05.009
33. Storms, R.F., Carere, C., Musters, R., Van Gasteren, H., Verhulst, S., Hemelrijk, C.K.: Deterrence of birds with an artificial predator, the robotfalcon. J. R. Soc. Interface **19**(195), 20220497 (2022)
34. Storms, R.F., Carere, C., Zoratto, F., Hemelrijk, C.K.: Complex collective motion: collective escape patterns of starling flocks under predation. Behav. Ecol. Sociobiol. **73**, 10 (2019). https://doi.org/10.1007/s00265-018-2609-0
35. Thiebault, A., Semeria, M., Lett, C., Tremblay, Y.: How to capture fish in a school? Effect of successive predator attacks on seabird feeding success. J. Anim. Ecol. **85**(1), 157–167 (2016). https://doi.org/10.1111/1365-2656.12455
36. Tunstrøm, K., Katz, Y., Ioannou, C.C., Huepe, C., Lutz, M.J., Couzin, I.D.: Collective states, multistability and transitional behavior in schooling fish. PLoS Comput. Biol. **9**(2), e1002915 (2013). https://doi.org/10.1371/journal.pcbi.1002915
37. Wilson, A.D., et al.: Dynamic social networks in guppies (poecilia reticulata). Behav. Ecol. Sociobiol. **68**(6), 915–925 (2014)
38. Zheng, M., Kashimori, Y., Hoshino, O., Fujita, K., Kambara, T.: Behavior pattern (innate action) of individuals in fish schools generating efficient collective evasion from predation. J. Theor. Biol. **235**(2), 153–167 (2005)

Evolutionary Approaches to Adaptive Behavior

The Role of Energy Constraints on the Evolution of Predictive Behavior

William Kang, Christopher Anand, and Yoonsuck Choe[✉]

Department of Computer Science and Engineering, Texas A&M University, College Station, TX 77843-3112, USA
{rkdvlfah1018,chrisanand,choe}@tamu.edu

Abstract. Prediction is an important foundation of cognitive and intelligent behavior. Recent advances in deep learning heavily depend on prediction, in the form of self-supervised learning based on prediction and reinforcement learning (reward prediction). However, how such predictive capabilities emerged from simple organisms has not been investigated fully. Prior works have shown the relationship between input delay and predictive function to compensate for such delay. In this paper, we investigate other key factors that may contribute to the emergence of predictive behavior in evolving neural networks. We set up a delayed reaching task with a two-segment articulated arm. The arm is controlled to reach a moving target, where the target's coordinate information is received with a delay. Following our previous work, we introduced a tool to extend the reach, when the target is beyond the arm's reach. In this task, without predicting the trajectory of the moving target, the controller cannot reach the target. For the controller, we used the NeuroEvolution of Augmenting Topologies (NEAT) algorithm. Our results indicate that an important (auxiliary) fitness criterion for the emergence of predictive behavior is that of reduced energy usage (in the form of economy of motion). Further analysis shows that the number of recurrent loops correlates with target reaching performance, but more strongly so with the energy constraint. We expect our findings to lead to further investigations on the role of energy constraints on the evolution of predictive behavior.

Keywords: Prediction · Evolution · Neuroevolution · Fitness

1 Introduction

Prediction forms an important foundation of cognitive and intelligent behavior [1,17,18]. Recent advances in deep learning heavily depend on prediction, in the form of self-supervised learning based on prediction and reinforcement learning (e.g., reward prediction) [13,14].

However, how such predictive capabilities emerged from simple organisms through evolution has not been investigated sufficiently [7,12]. Furthermore, what kind of external/internal factors and constraints could have influenced the development of prediction is unclear. Prior works have suggested the relationship between input delay and predictive function for its compensation [10,11].

In this paper, we investigate other key factors that may contribute to the emergence of predictive behavior in evolving neural networks. We set up a reaching task with a two-segment articulated arm, with added input delay [9]. (Note that predicting an unknown true location like this is different from predicting the sensory consequence of action, which requires the true sensory input to compute the prediction error, as in [17].) The arm is controlled to reach a moving target, where the target's coordinate information is received with a fixed delay. Following our previous work [9], we introduced a tool to extend the reach, when the target is beyond the arm's reach. This task is not solvable without predicting the trajectory of the moving object, since reaching for the coordinate location based on the immediate input would lead to a location previously occupied by the moving object. For the controller, we used the NeuroEvolution of Augmenting Topologies (NEAT) algorithm [15] (cf. CTRNN [17] and FORCE [6,16]).

Our results indicate that an important (auxiliary) fitness criterion for the emergence of predictive behavior is that of reduced energy usage (in the form of economy of motion). Further analysis shows that the number of recurrent loops correlate with target reaching performance, but more strongly so with the energy constraint. We expect our findings to lead to further investigations on the role of energy constraints on the evolution of predictive behavior.

2 Background

In this section, we will briefly review existing works relating to the evolution of prediction, and provide an overview of the NEAT algorithm, which we will use.

2.1 Evolution of Prediction

In our previous works, we show that predictive dynamics emerge in evolved neural network controllers when environmental conditions change [8,21]. Aside from these, the few papers that discuss prediction in an evolutionary context include [7] and [12]. [7] mentioned the lack of computational studies on evolution of prediction, and proceed to propose how utility and predictive capabilities can co-evolve in a simulated agent. The work is based on the "Gap Theory" in evolutionary economics which exactly states the co-evolutionary nature of utility and prediction. [12], on the other hand, observed that curiosity (a form of intrinsic motivation) helps in the improvement of prediction, and these developmental structures can constrain evolution in return. Thus, the work is more about the emergent predictive ability shaping evolution, not about how evolution gives rise to predictive abilities.

2.2 NeuroEvolution of Augmenting Topologies (NEAT)

Topological neuroevolution methods evolve both topology and weights of neural networks. Our perspective is that natural evolution includes changes in the network topology in the brain, thus they mimic the natural evolution better than traditional weight-only neuroevolution methods. Moreover because the functionality of a neural network can be constrained by its topology, allowing the topology to evolve will set free the structural constraints and result in new capabilities such as recurrent dynamics. Amongst many variations of such an approach [2,19,20], we will use NeuroEvolution of Augmenting Topologies (NEAT) because of its advantages over other topological evolution methods [15]. Figure 1 illustrates the main concepts of NEAT.

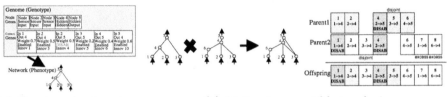

(a) Genotype to phenotype (b) Mating incompatible topologies

Fig. 1. NEAT. (a) Nodes and connections are separately encoded. (b) Connections with the same innovation number can be crossed over. Adapted from [15].

Historical marking is a major feature in the NEAT algorithm. By enumerating each innovation, NEAT solves the competing conventions problem, which is one of the main problems in neuroevolution [15,19]. The crossover operation in NEAT happens between two genomes with identical historical marking (also called "innovation number"), regardless of their locations and size in the network (Fig. 1). NEAT encodes the genome in two arrays, node genes and connection genes. Innovation number is assigned to each connection gene according to the order of its appearance throughout the evolutionary stages. The connections can also be enabled or disabled through mutation. Since connections can be generated arbitrarily between neurons, recurrent connections can also be generated. There are several other important facilities such as speciation, where a subpopulation of individuals are isolated from other subpopulations, forming a species.

3 Methods

In this section, we will discuss the details of the task and how NEAT is hooked up to the environment. We will also discuss the various fitness factors we used, and the performance metrics we employed for the evaluation.

3.1 Delayed Reaching Task

The delayed reaching task is illustrated in Fig. 2. A two-segment arm with two joints can be controlled by its two joint angles (θ_1, θ_2) to reach a moving target (black square). To test the predictive capability, the task is modified so that the coordinate of the target object is fed to the controller with a delay (red square).

If the controller tries to reach the delayed target, based on the delayed input, it will not be able to reach the true target. An additional obstacle is that the movement of the target can take it beyond the reach of the arm. The controller can decide to pick up a stick (green) to extend its reach. The stick automatically snaps onto the hand (circle) and extends the arm's reach once the stick handle is touched by the hand.

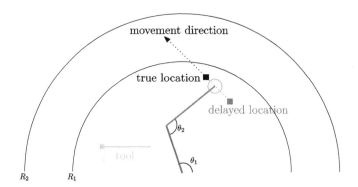

Fig. 2. Delayed Reaching Task. The task is to control the limb angles θ_1 and θ_2 to reach the moving target (black square). The location of the target is received by the controller with a delay (red square). If the target is beyond the arm's reach (R_1), the controller needs to pick up the stick (green) to extend the reach.

The target was initially placed randomly within one of four regions (upper left, lower left, lower right, upper right), and moved in one of four directions (0°, 45°, 135°, and 180°, respectively). The target was moved for 500 simulations steps in the beginning (which was the amount of delay), before the controller can start movement.

3.2 NEAT Controller

For control of the arm, we used NEAT. Figure 3 shows the initial topology of the controller network. The 9 inputs and 2 outputs of the network were as follows.

Input: (1) θ_1: joint angle 1, (2) θ_2: joint angle 2, (3) *target_dist*: distance between hand and delayed target, (4) *target_angle* angle from hand to delayed target, (5) *tool_dist* distance between hand and tool handle (not delayed, since the tool is static), (6) *tool_angle* angle from hand to tool handle, (7) θ_{1limit}: triggered when joint 1 limit is reached, (8) θ_{2limit}: triggered when joint 2 limit is reached, (9) *target_touch*: tactile sensor triggered when hand touches the true target.

Output: (1) θ_1: joint angle 1 adjustment, (2) θ_2: joint angle 2 adjustment. The adjustments were limited to $-1.5° \leq \theta_i \leq 1.5°$.

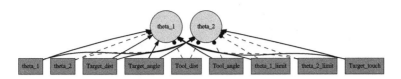

Fig. 3. Initial NEAT Topology. The initial NEAT controller topology is shown. Red: input, Green: output. Note that there are no hidden neurons, since this is the initial topology. See text for details.

3.3 Fitness

We tested a combination of different fitness factors, each scaled between 0.0 and 1.0, and the factors multiplied, as shown below (see the Appendix for details):

$$F = \alpha \times \prod_i F_i, \qquad (1)$$

where α is a product of common factors used in all experimental conditions, and $\prod_i F_i$ a product of the fitness factors that are to be compared. The common factor α is defined as $\alpha = D \times N \times C$, where D is based on the mean square distance between the true target and the hand location (1 - this quantity), N is the average number of abrupt change of direction to discourage oscillating behavior (1 - this quantity), and C is the total hit count in each behavioral attempt. The main factors we compared were $F_i \in \{E, R, T\}$, where E is the energy factor (1 - normalized total hand travel distance), R the predictive reach factor (number of 5 or more consecutive hits), and T the tool factor (tool pick up). In our case, we tested these factor combinations: $R, RT, ER,$ and ERT.

3.4 Performance Metrics

Other than the fitness, each individual was tested in multiple random behavioral attempts ($N_{max} = 100$) to measure the performance.

The first metric is the total number of "hits" (touching the target) in each behavioral attempt. (Note that all attempts were 5,000 steps. Also, the maximum value of total hits in each attempt depends on the relative position of the randomly initialized hand, tool, and target positions.) Although this is a good general metric, it does not measure the predictive capability.

$$total\_hits = \sum_{n=1}^{N_{max}} \text{total number of all hits in attempt } n \qquad (2)$$

Metric 1_hit measures the number of attempts in which the agent had reached the target at least once. The 1_hit is computed with Eq. 3. While this metric is standardized and can show the effectiveness of a network, it cannot distinguish between an accidental hit vs. predictive reach.

$$1\_hit = \sum_{n=1}^{N_{max}} h_n, \text{ where } h_n = \begin{cases} 1 & \text{if target hit at least once in attempt } n \\ 0 & \text{else} \end{cases} \quad (3)$$

Hence, we also introduce the 5_hit metric (Eq. 4) which counts the number of attempts in which the agent had reached the target at least 5 consecutive time steps (a different value may also be used, e.g., 10). This is to ensure that the reach of a target was intentional and therefore is used as our main metric to measure the predictive capabilities of the agents.

$$5\_hit = \sum_{n=1}^{N_{max}} h_n, \text{ where } h_n = \begin{cases} 1 & \text{if target hit 5+ consec. steps in attempt } n \\ 0 & \text{else} \end{cases}$$
(4)

All the metrics mentioned above apart from the fitness scores, act as a standard metric to compare the successes of different agents. They share common extremes ([0, 500,000] for the number of reaches and [0, 100] for 1_hit or 5_hit) and are computed using the same equations resulting in standardized scores.

4 Experiments and Results

Each of the four fitness types (R, RT, ER, ERT) was evolved for 150 generations.

4.1 Population Average over Generations

Figure 4 shows the population average of the fitness and the number of reaches over the generations. We find that the ER factor combination significantly outperforms the others (note that in this case, the evolutionary "trial" ended early, since it exceeded the preset fitness threshold).

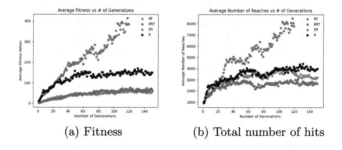

(a) Fitness   (b) Total number of hits

Fig. 4. Population Average of Fitness and Total Number of Hits.

4.2 Performance of Best Individuals

Next, we tested the best individuals for each fitness type, using the 1_hit and 5_hit metric (note that E is not a good metric due to trival cases: e.g., when the agent did not move). For each individual tested, we ran the individual in a task 100 times ("attempts"), and counted the number of times it was able to hit the target once or five consecutive times, respectively. We repeated n (= 31) such "runs" to measure the performance (Fig. 5). Note that 1_hit and 5_hit can be at the most 100 ($N_{max} = 100$). For 1_hit, the results are mixed (Fig. 5(a)). ERT shows the best performance, followed by R, ER, then RT. Thus, there is no clear difference between (R, RT) vs. (ER, ERT). However, Fig. 5(b) shows that for 5_hit, (ER, ERT) clearly outperforms (R, RT). Furthermore, Fig. 5(c) shows that most of the 1_hit events are also 5_hit events for (ER, ERT), showing that most reaching behavior is prediction based. However, this is not the case for the (R, RT) condition, where most reaching behavior is blind waving.

Considering that 5_hit is an indicator of predictive behavior (continuously and intentionally tracking the true target location, rather than randomly waving to hit the target by luck), we can conclude that the fitness types that include the Energy factor facilitates the emergence of predictive behavior.

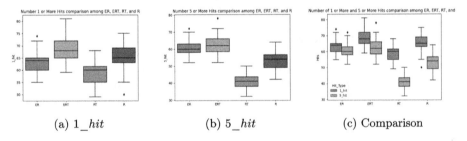

(a) 1_hit   (b) 5_hit   (c) Comparison

Fig. 5. Performance of Best Individuals. The (a) 1_hit and (b) 5_hit results are shown for the four fitness types. In (b): Mann-Whitney test: $ERT > RT$ [$n = 31, p = 1.36e - 11$], $ER > R$ [$n = 31, p = 9.71e - 07$]. (c) Compares (a) and (b) in a single plot.

4.3 Behavior

Observing the behavior can provide some insights on the relevance of our 1_hit and 5_hit metrics. Figure 6 show some typical behaviors by fitness type. We plotted the behavior in time lapse (vivid color = most recent frame). We can also see the target's moving direction (black = true target, red = delayed target): For example, in fig. 6(b), the target is moving from lower left to upper right. For each fitness type, representative behavior that exhibit predictive property (a through d) and those that do not (e through h) are shown. For the top row (predictive), we see that the hand dwells close to the true target location (black square), ahead of the delayed target (red square). This kind of behavior may not be possible without some form of prediction, and may score high on the 5_hit criterion. For the bottom row (non-predictive), the hand makes broad sweeping gestures. This could lead to a high 1_hit score, but a low 5_hit score. With this, we can view the results in fig 5(c) in a new light: The fitness types that involve the Energy factor may be exhibiting predictive behavior, while those that do not are merely successful in reaching the target through undirected broad sweeping behavior.

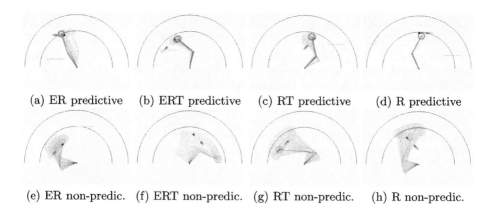

(a) ER predictive (b) ERT predictive (c) RT predictive (d) R predictive

(e) ER non-predic. (f) ERT non-predic. (g) RT non-predic. (h) R non-predic.

Fig. 6. Representative Behaviors. Time lapse of representative behaviors are shown (vivid color = most recent frame). Black square: true target. Red square: delayed input. Top row (a~d): predictive behavior. Bottom row (e~h): non-predictive behavior. Note that in (g) and (h), the tool is picked up, so the reach is extended. See text for details.

4.4 Network Topology

The evolved network topology also gave us further insights on how different fitness types shaped the controller's behavior (Fig. 7). At first glance, there are no distinguishable differences in appearance, other than some having more hidden neurons than others. Further analysis reveals an interesting property. Our previous work on analyzing evolved network topology showed a positive correlation between the number of loops and the performance [9]. We conducted the same kind of analysis, by counting the number of simple cycles in the evolved network (we used NetworkX for this [3]). The results are shown in Fig. 8. Interestingly, for the fitness types that include the Energy factor (ER, ERT), the number of loops are strongly correlated with the success rate ($r = 0.59$ and 0.51, respectively). However, for those without the Energy factor (RT, R), the correlation is weak ($r = 0.17$ and 0.38, respectively). These results suggest that the recurrent loops in the networks evolved with the energy constraints may be supporting predictive function better than those without. How these loops contribute to prediction need to be investigated (for initial attempts, see lesion studies in [5]).

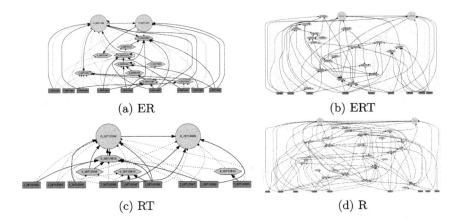

Fig. 7. Typical Evolved Topologies. Typical evolved topologies are shown for the different fitness types. Neurons: Red = input, Orange = hidden, Green = output. Connections: Solid lines with arrows = excitatory, Dashed lines with discs = inhibitory.

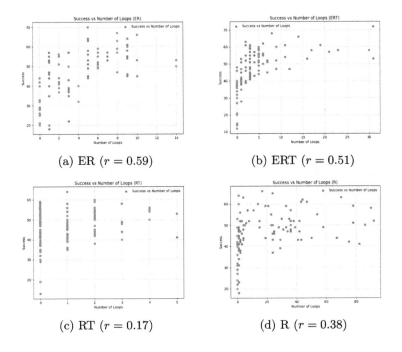

Fig. 8. Number of Loops vs. Success Rate. The number of loops (simple cycles in graph theory) in an individual vs. its success rate is plotted. The r values are the correlation coefficients. Each point corresponds to one individual in the population. We can see that the correlation is higher for the fitness types that includes the Energy factor (ER and ERT), compared to those that do not include this factor (RT and R).

5 Discussion

The main contribution of this paper is the identification of constraint on energy as a major factor in the emergence of predictive behavior in the context of evolution. With naive performance metrics such as the number of times the target has been reached, evolution found equally effective strategies to reach the target such as broad sweeping, but the behavior was not indicative of prediction. It is also interesting to note that even with the addition of a fitness factor that explicitly rewards predictive behavior (the R factor, which basically uses 5_hit), that alone was not able to give rise to predictive behavior, i.e., E was needed.

What could be some other factors that contributed to prediction? As mentioned in the introduction, delay in the input could be one such factor [10,11]. Due to delay in the sensory input, the need to compensate for this arises, which may be resolved by prediction: prediction of the present from the past. We have also found another curious factor that may be involved, which is environmental change, as briefly mentioned in the background. In a simple 2D pole balancing task, we found that individuals with predictive internal dynamics can be more robust in a changing environment (in this case, the change in the initial pole angle from $\theta < 5°$ to $\theta < 10°$) [8,21]. With our research reported here, we can now summarize the seemingly unrelated factors that contribute to the emergence of prediction as (1) energy constraint, (2) delay, (3) environmental change.

6 Conclusion

In this paper, we investigated factors contributing to the emergence of predictive behavior in an evolutionary context. We used a delayed reaching task, where reaching the true target requires some form of prediction based on the delayed input from the past. We found that the energy factor plays an important role, allowing the evolved controllers to be more successful in the reaching task, and to exhibit more focused predictive reaching behavior. The controllers evolved without the energy factor were moderately successful, but the strategy were mostly based on undirected, systematic sweeping, not indicative of any prediction of the target's trajectory. Furthermore, we found that structural innovations like recurrent loops in the evolved controllers play a more cohesive role in support of the predictive function, when the energy factor was included. These results suggest a key role of energy constraints in the evolution of predictive behavior.

Acknowledgments. Based on the undergraduate honor thesis [5]. Environment simulation and NEAT based on Qinbo Li's code [9], using ANJI [4]. We would like to thank the anonymous reviewers for helping us clarify our points.

Appendix

The fitness factors were computed as follows. All factors were computed from N_{max} behavioral attempts (= 100) using the same individual chromosome, with each attempt running for a maximum of M_{max} simulation steps. W is the width of the arena (= 512 pixels), and L is the initial lead up steps which corresponds to the delay (= 500 steps). The factors were D: distance, N: turn, C: total hit count, E: energy, R: predictive reaches, T: tool.

$$D = 1 - \sum_{n=1}^{N_{max}} \sum_{m=1}^{M_{max}} \frac{d_{nm}^2}{(WN_{max}M_{max})^2} \quad N = 1 - \sum_{n=1}^{N_{max}} \frac{n_n}{N_{max}M_{max}}$$

$$C = \sum_{n=1}^{N_{max}} \frac{c_n}{N_{max}M_{max}} \quad E = 1 - \sum_{n=1}^{N_{max}} \frac{e_n}{N_{max}(M_{max}-L)}$$

$$R = \sum_{n=1}^{N_{max}} \frac{r_n}{N_{max}} \quad T = \sum_{n=1}^{N_{max}} \frac{t_n}{N_{max}}$$

where, in each attempt, d_{nm}=distance between hand and true target in attempt n at time step m, n_n = number of sharp turns (= reversal of hand movement direction > 90°), c_n = total number of hits, e_n = consumed energy (= cumulated hand travel distance), $r_n = 5\_hit$, $t_n = 1$ if tool is held and 0 otherwise.

Table 1. NEAT (ANJI) Hyper-parameters

Evolution	
num.generations = 150	popul.size = 100
topology.mutation.classic = false	add.connection.mutation.rate = 0.02
remove.connection.mutation.rate = 0.01	remove.connection.max.weight = 100
add.neuron.mutation.rate = 0.01	prune.mutation.rate = 1.0
weight.mutation.rate = 0.75	weight.mutation.std.dev = 1.5
weight.max = 10.0	weight.min = -10.0
survival.rate = 0.2	selector.elitism = true
selector.roulette = false	selector.elitism.min.specie.size = 1
Speciation	
chrom.compat.excess.coeff = 1.0	chrom.compat.disjoint.coeff = 1.0
chrom.compat.common.coeff = 0.4	speciation.threshold = 0.25
Fitness Function	
stimulus.size = 9	response.size = 2
fitness.func.adjust.for.netw.size.factor = 0	fitness.threshold = 1000
fitness.target = 1200	
Activation Function	
initial.topology.activation = sigmoid	
Network Architecture	
initial.topology.fully.connected = true	init.topology.num.hidden.neurons = 0
initial.topology.activation.input = linear	recurrent = best_guess
recurrent.cycles = 1	ann.type = anji

References

1. Bubic, A., Von Cramon, D.Y., Schubotz, R.I.: Prediction, cognition and the brain. Front. Hum. Neurosci. **4**, 1094 (2010)

2. Gruau, F., Whitley, D., Pyeatt, L.: A comparison between cellular encoding and direct encoding for genetic neural networks. In: Koza, J.R., Goldberg, D.E., Fogel, D.B., Riolo, R.L. (eds.) Genetic Programming 1996: Proceedings of the First Annual Conference, pp. 81–89. MIT Press, Cambridge (1996)
3. Hagberg, A., Conway, D.: Networkx: network analysis with python (2020). https://networkxgithub.io
4. James, D., Tucker, P.: ANJI: another NEAT java implementation (2004). https://anji.sourceforge.net/index.html
5. Kang, W., Anand, C.: Emergence of prediction in delayed reaching task through neuroevolution. In: Engineering Honors in Computer Science and Engineering Thesis, Texas A&M University (2023)
6. Kashyap, H.J., Detorakis, G., Dutt, N., Krichmar, J.L., Neftci, E.: A recurrent neural network based model of predictive smooth pursuit eye movement in primates. In: 2018 International Joint Conference on Neural Networks (IJCNN), pp. 1–8. IEEE (2018)
7. Korb, K.B., Brumley, L., Kopp, C.: An empirical study of the co-evolution of utility and predictive ability. In: 2016 IEEE Congress on Evolutionary Computation (CEC), pp. 703–710. IEEE (2016)
8. Kwon, J., Choe, Y.: Internal state predictability as an evolutionary precursor of self-awareness and agency. In: Proceedings of the Seventh International Conference on Development and Learning, pp. 109–114. IEEE (2008). http://faculty.cs.tamu.edu/choe/ftp/publications/kwon.icdl08.pdf
9. Li, Q., Yoo, J., Choe, Y.: Emergence of tool use in an articulated limb controlled by evolved neural circuits. In: Proceedings of the International Joint Conference on Neural Networks (2015). http://faculty.cs.tamu.edu/choe/ftp/publications/li-ijcnn15.pdf. https://doi.org/10.1109/IJCNN.2015.7280564
10. Lim, H., Choe, Y.: Compensating for neural transmission delay using extrapolatory neural activation in evolutionary neural networks. Neural Inf. Process. Lett. Rev. **10**, 147–161 (2006). http://faculty.cs.tamu.edu/choe/ftp/publications/lim.niplr06-reprint.pdf
11. Nijhawan, R.: Motion extrapolation in catching. Nature **370**, 256–257 (1994)
12. Oudeyer, P.Y., Smith, L.B.: How evolution may work through curiosity-driven developmental process. Top. Cogn. Sci. **8**(2), 492–502 (2016)
13. Shwartz Ziv, R., LeCun, Y.: To compress or not to compress-self-supervised learning and information theory: a review. Entropy **26**(3), 252 (2024)
14. Stachenfeld, K.L., Botvinick, M.M., Gershman, S.J.: The hippocampus as a predictive map. Nat. Neurosci. **20**(11), 1643–1653 (2017)
15. Stanley, K.O., Miikkulainen, R.: Evolving neural networks through augmenting topologies. Evol. Comput. **10**, 99–127 (2002)
16. Sussillo, D., Abbott, L.F.: Generating coherent patterns of activity from chaotic neural networks. Neuron **63**(4), 544–557 (2009)
17. Tani, J.: Exploring Robotic Minds: Actions, Symbols, and Consciousness as Self-Organizing Dynamic Phenomena. Oxford University Press, Oxford (2016)
18. Tjøstheim, T.A., Stephens, A.: Intelligence as accurate prediction. Rev. Phil. Psychol. **13**(2), 475–499 (2022)
19. Whitley, D., Dominic, S., Das, R., Anderson, C.W.: Genetic reinforcement learning for neurocontrol problems. Mach. Learn. **13**, 259–284 (1993)
20. Yao, X.: Evolving artificial neural networks. Proc. IEEE **87**(9), 1423–1447 (1999)
21. Yoo, J., Kwon, J., Choe, Y.: Predictable internal brain dynamics in EEG and its relation to conscious states. Front. Neurorob. **8**, 00018 (2014). http://journal.frontiersin.org/article/10.3389/fnbot.2014.00018/full

Influence of the Costs of Acquisition of Private and Social Information on Animal Dispersal

Antoine Sion[1](✉), Matteo Marcantonio[2], and Elio Tuci[1]

[1] Faculty of Computer Science, University of Namur, Grandgagnage 21, 5000 Namur, Belgium
{antoine.sion,elio.tuci}@unamur.be
[2] Landlife Ecospatial Labs, Via del Giontech 1, 38016 Mezzocorona, Italy

Abstract. Habitat fragmentation is currently speeding up due to the impact of human activities. This leads to higher variations in habitat qualities for different species, influencing their behaviours. One key behaviour directly linked to species survival is dispersal (displacement from one habitat patch to another). In order to take successful dispersal decisions, animals rely on different sources of information. Private information is derived from the physical environment and is directly linked to its quality, whereas social information is derived from the behaviour of conspecifics. Few modelling studies include both information types and their associated acquisition costs in their modelling frameworks. We fill this gap by adding genetic factors influencing the evolution of information acquisition to an existing agent-based model. By varying total and relative acquisition costs for different environmental conditions and perceptual ranges, we show that dispersal strategies and information usage are heavily influenced by the type of environment and information acquisition costs. As the total cost of information rises, the use of information progressively disappears under all environmental conditions. In stable environments with a low cost of information, the acquisition of the cheapest type of information results in an increase in fitness. In environments where patch quality varies greatly, the type of information used also depends on the perceptual range of the agents: agents with a restricted perceptual range often select both types of information while, agents with a larger perceptual range almost exclusively use private information.

Keywords: Dispersal · Acquisition costs · Information · Evolutionary modelling

1 Introduction

Human-Induced Rapid Environmental Changes (HIREC) are altering the living conditions of many species worldwide [23]. These additional environmental stresses compel species to adjust their behaviours for survival. Understanding how animals respond to these changes from an evolutionary point of view is an expanding area of research [20]. Specifically, understanding the interplay between individual behaviours and their repercussions on global population dynamics remains a subject of ongoing investigation [4]. Dispersal, defined as the movement from one habitat patch to another [19], has consequences at both individual and population levels and is especially impacted by habitat fragmentation. This behaviour is resource-driven, non-cyclical and short-ranged as opposed to migration [2]. Individuals may choose to disperse for various reasons, such as resource competition, environmental shifts, kin competition, or to avoid inbreeding, thus enhancing their individual fitness [4]. At the population level, dispersal can also help to mitigate extinction risks by colonising empty habitats or moving towards declining populations.

Animals rely on diverse information sources to make successful dispersal decisions, for example they can gather private or social information to assess environmental quality [6,18]. Private information is directly obtained through sensory observation of the physical world while social information is obtained indirectly by monitoring others' interactions with the environment. Conspecific density is one of the most widespread social information and it is known to influence species dispersal [7]. The acquisition of information comes with costs, often in the form of energy expenditure, potentially leading to reduced fecundity. Additionally, the perceptual ranges of animals can differ between species or in the same species across time. A perceptual range is defined as "the distance from which a particular landscape element can be perceived as such (or detected)" [10]. Conducting experimental research to quantify how information acquisition affects dispersal is thus fundamental, but it poses practical challenges due to the involvement of multiple information sources and environmental stochasticity [12]. Evolutionary models based on multi-agent simulations may offer a way forward to study the interactions between costs of acquisition of information, dispersal tactics and habitat fragmentation in a controlled setting [3].

Despite a widespread application of modelling frameworks to study animal dispersal, they often focus solely on conspecific density as the basis for dispersal decision-making processes [1,9,13,17]. Recent studies have begun to integrate both socially derived and individually gathered information to explore how multiple knowledge sources affect dispersal decisions [8,15,16]. Yet, these models often overlook the reproductive costs linked to the acquisition of both private and social information. Our study aims to fill this gap by integrating private and social information in dispersal decisions as well as their potential reproductive costs. Our model builds upon and extends the framework introduced by [8], integrating genetic factors that influence the evolution of both social and private information gathering strategies. We systematically vary both the overall cost of information acquisition and the relative cost of social and private information across varying environmental conditions. To this end, we explore two scenarios with different quantities of available information: 1) a scenario where agents have less available information due to a restricted perceptual range and 2) another with more available information linked to an extended perceptual range. This approach enables us to investigate a wide spectrum of scenarios while emphasising the population dynamics induced by these associated costs.

2 Material and Methods

We simulate a scenario in which a population of 40,000 female/male butterflies (ratio of 1:1) are located in a 2D toroidal square environment, made of 400 patches of different quality. The quality of each cell bears upon the fitness (i.e., the number of offspring) of the hosted female butterflies. To increase their fitness, butterflies have to choose to reproduce either on the patch where they have been initialised, or on a randomly chosen proximal patch, reached by dispersal. Generations are non-overlapping and only female agents can disperse to an adjacent patch once in their lifetime (males do not disperse for simplicity, see [8]). The decision on whether to disperse or not is influenced by genetically-controlled traits that determine whether and the extent to which private and/or social information contribute to the agent decisions.

Information that can be acquired privately by agents is represented by the overall habitat quality in a patch, where information that can be acquired socially is represented by the number of conspecifics present in a patch. We consider two scenarios for our agent's sensing range. Agents with a restricted perceptual range can only acquire information from the patch where they were born (local patch) while agents equipped with an expanded perceptual range can sense both their local patch and a randomly chosen adjacent patch. Two binary genes are used by the agents to determine whether they use social (s_1) or private (s_2) information. If these two genes are activated, additional genes are used for the contribution of each information type in the dispersal decision. With a restricted perceptual range, two real-valued genes (i.e., genes with values expressed as real numbers) are associated to private (a_1) and social (b_1) information from the natal patch. With an expanded range, four genes intervene for the acquisition of information from both the natal and the adjacent patch (a_1, a_2, b_1, b_2). Agents reproduce asexually at each generation and genes mutate at each generation (see Table 1 for an overview of all the model state variables and their possible values). For each generation, our model is organised in four different consecutive steps: 1) reproduction, 2) setting the quality of the environment, 3) mating and 4) dispersal.

Reproduction: new agents are born in a patch as a function of the number of female agents in the previous generation. The number of new agents in patch i is thus computed as:

$$N_{next_i} = min(2N_{f_i} exp(r(1 - \frac{2N_{f_i}}{K_i})), N_{max}) \quad (1)$$

where N_{f_i} is the number of female agents in the previous generation in patch i, $r = 1$ a fixed population growth rate parameter, K_i the quality of the patch and

Table 1. State variables used in the model

Parameter	Values/bounds
Probability of dying while dispersing (μ)	0.1
Gene enabling the use of social information (s_1)	$\{0;1\}$
Gene enabling the use of private information (s_2)	$\{0;1\}$
Gene weighting social information in the natal patch (b_1)	$[-0.2, 0.2]$
Gene weighting social information in the adjacent patch (b_2)	$[-0.2, 0.2]$
Gene weighting private information in the natal patch (a_1)	$[-0.2, 0.2]$
Gene weighting private information in the adjacent patch (a_2)	$[-0.2, 0.2]$
Lower bound for the reproductive success of a female (r_{min})	$[0.75, 1.0]$
Upper bound for the reproductive success of a female (r_{max})	1.0
Relative cost for acquiring social information (c_1)	$[0.0, 1.0]$
Relative cost for acquiring private information (c_2)	$[0.0, 1.0]$
Probability of mutation of an allele (P_{mut})	0.00005
Initial quality of a patch (K_0)	100
Minimum quality of a patch (K_{min})	5
Standard deviation of the quality of a patch (σ)	$\{1; 10; 100\}$
Upper limit for the local population on a patch (N_{max})	500

N_{max} the maximum possible number of new agents in a patch [8]. Total population size is thus dependent of the number of females on each patch as well as environmental variation. A high total population size is obtained if females disperse themselves homogeneously on high quality patches, translating an implicit fitness improvement following successful dispersal decisions. Each new agent inherits the genes of a mother present in the patch at the previous generation. Each mother has a reproductive weight that is function of the amount of information that she uses for the dispersal decision. Each new agent is assigned a mother from the local pool of female agents, selected randomly without replacement based on a weighted distribution. The reproductive weight is computed using:

$$w = r_{max} + (c_1 s_1 + c_2 s_2)(r_{min} - r_{max}) \qquad (2)$$

with r_{min} and r_{max} the possible bounds of the weight, c_1 the relative cost for acquiring social information and $c_2 = 1 - c_1$ the relative cost for acquiring private information. We study scenarios with an increasing cost of information by varying r_{min} while keeping r_{max} fixed. A fully informed female has the lowest possible reproductive weight (r_{min}) and thus has a lower probability to pass its genes to the next generation, but has an increased probability to disperse to a suitable patch. Conversely, an uninformed female agent will have the highest reproductive weight (r_{max}). We also vary the relative costs of information acquisition to study scenarios where one type of information is more costly than the other, as well as cases where one type of information is completely free. Each real-valued gene is composed of two alleles, their sum being the value of the gene. Each allele can mutate during reproduction with a probability P_{mut}, whereas binary genes have the same probability to mutate (i.e., their value is flipped) at each generation. When mutating, the value of the allele is incremented by a value drawn from a Laplacian distribution with mean zero and a standard deviation corresponding to the total interval of the bounds of the gene divided by 40. At the end of the reproduction phase, all agents from the previous generation die.

Varying quality: each patch has an associated quality corresponding to its carrying capacity. This quality varies at each generation following a Gaussian distribution:

$$K_i(t) = max(K_0 + \phi, K_{min}) \qquad (3)$$

with ϕ an independent normal variable with a mean of zero and a standard deviation σ. Quality values are not correlated over time. By varying σ, we increase or decrease the amount of environmental variation experienced by agents. This allows us to test challenging environments (i.e., larger σ), where quality varies greatly over time as well as more stable environments, where quality is approximately uniform on the entire grid.

Mating: if a male is present on the patch, females mate and can pass their genes during reproduction. In the rare event of a patch containing no male, females are assumed to stay unmated and will not be taken into account for reproduction.

Dispersal: females choose to disperse following the probability:

$$P = \frac{1}{1 + exp(-a_1 Q_{1s} - a_2 Q_{2s} - b_1 N_{1s} - b_2 N_{2s})} \qquad (4)$$

with Q_{1s} the sensed quality of the natal patch, Q_{2s} the sensed quality of the adjacent patch, N_{1s} the sensed number of conspecifics on the natal patch and N_{2s} the sensed number of conspecifics on the adjacent patch. Information is enabled or not following:

$$N_{1s} = s_1 N_1 \quad \text{and} \quad N_{2s} = s_1 N_2 \qquad (5)$$

$$Q_{1s} = s_2 Q_1 \quad \text{and} \quad Q_{2s} = s_2 Q_2 \qquad (6)$$

If one of the two binary genes is disabled, the associated quantity is equal to zero and it is not used in Eq. 4. Females have a probability $\mu = 0.1$ of dying in the dispersal process [8].

To assess the impact of information usage and its corresponding fitness outcomes, our study considers two specific sensing ranges for agent perception: a restricted range and an expanded range. These ranges are subjected to varying degrees of environmental variability, quantified by $\sigma = \{1; 10; 100\}$, which creates a spectrum of dispersal challenges for the population. Within this framework, we explore six levels of information acquisition costs, represented by $r_{min} = \{0.75; 0.80; 0.85; 0.90; 0.95; 1.0\}$. Furthermore, we explore the cost-benefit dynamics of private versus social information through the relative information cost parameter $c_1 = \{0.0; 0.25; 0.5; 0.75; 1.0\}$. Findings presented in the next section are obtained after 20,000 generational cycles and ten independent iterations of the simulation, ensuring that the population reaches an evolutionary steady state.

3 Results

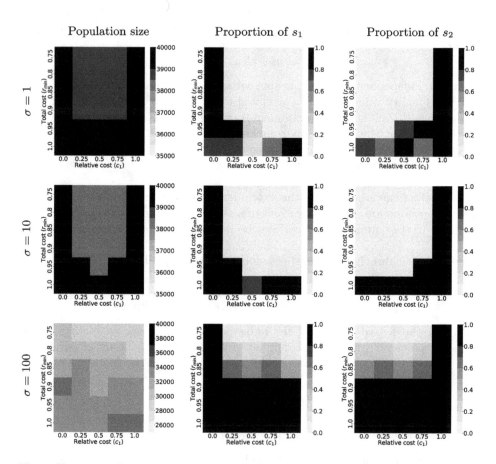

Fig. 1. Heatmaps showing the median values (blue palette) for the population size (first column) and the proportions of s_1 and s_2 (second and third columns) in the total agent population after 20,000 generations for scenarios with a restricted perceptual range. Environmental variation increases from heatmaps in the top row to the heatmaps in the bottom row. (Color figure online)

Heatmaps showing the median values over 10 iterations of the population size as well as the proportions of s_1 (social information) and s_2 (private information) in the agent populations are displayed in Figs. 1 and 3. For completeness,

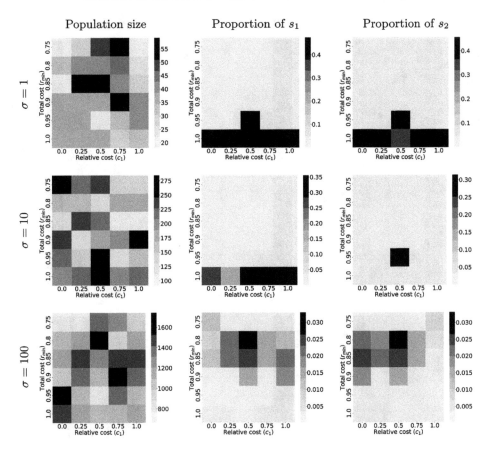

Fig. 2. Heatmaps showing the standard deviation values for the population size and the proportions of s_1 and s_2 in the total agent population after 20,000 generations for scenarios with a restricted perceptual range.

standard deviation values are also shown in Figs. 2 and 4. High standard deviation values for the proportions of s_1 and s_2 indicate the selection of various dispersal strategies in some cases. Population size can be considered a direct measure of the fitness of the populations since the offspring of agents making optimal dispersal decisions will have a higher chance to survive and contribute to the population size in the next generation. In all environmental conditions for the restricted perceptual range (see Fig. 1), population sizes decrease with increasing cost of acquisition of information (r_{min} diminishing). However, population sizes do not decrease with the total cost of information rising when one of the two sources of information is free for low environmental variations ($\sigma = 1$ and $\sigma = 10$ with a relative cost $c_1 = 1$ or $c_1 = 0$). With a high environmental stochasticity ($\sigma = 100$), this effect is not appearing clearly but this could be due to the higher standard deviations for the population sizes between the runs (see Fig. 2). It can also be noted that highly variable environmental conditions result

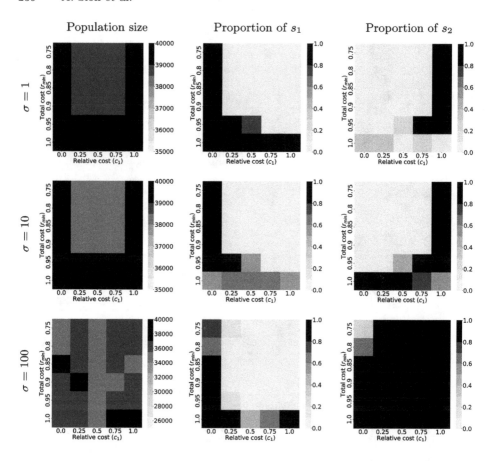

Fig. 3. Heatmaps showing the median values for the population size and the proportions of s_1 and s_2 in the total agent population after 20,000 generations for scenarios with an expanded perceptual range.

overall in lower population sizes than stable conditions. The numbers of agents using any type of information decreases when the total cost rises, except for cases where the acquisition of private or social information is free. We also see that with a high environmental stochasticity, at equal total costs, more agents still acquire information: when $r_{min} = 0.85$, half of the agents acquire information for $\sigma = 100$ but no information is collected for $\sigma = 10$ or $\sigma = 1$. Standard deviations values for s_1 and s_2 proportions are included in the following intervals: $[0, 0.45]$ for $\sigma = 1$ and $[0, 0.35]$ for $\sigma = 10$ but only when $r_{min} = 1.0$, close to zero otherwise; $[0, 0.03]$ for $\sigma = 100$ (see Fig. 2). Different dispersal strategies leading to the same fitness outcomes are thus selected at low environmental variations when information is free.

With an extended perceptual range (see Fig. 3), the same general trend can be observed for population sizes except at high environmental variations. For $\sigma = 1$

and $\sigma = 10$, population size decreases when the total cost rises but stay constant when one information source is free. For $\sigma = 100$, no clear trend can be seen and the population size remains approximately constant for all conditions. However, the population size is overall higher than with a restricted perceptual range. For the use of information, agents keep the same strategy as with the restricted range for low environmental variations ($\sigma = 1$ and $\sigma = 10$). The difference arises at $\sigma = 100$, where agents predominantly rely on the use of private information in almost all costs conditions. Social information is only used when free with either $r_{min} = 1.0$ or $c_1 = 0$. As before, agents tend to invest more into information acquisition when the environment is challenging at equal acquisition costs but only for private information. Standard deviations values for s_1 and s_2 proportions are included in the following intervals: $[0, 0.5]$ for $\sigma = 1$ and $[0, 0.25]$ for $\sigma = 10$ but only when $r_{min} = 1.0$, close to zero otherwise; with $\sigma = 100$, $[0, 0.03]$ for s_2 and $[0, 0.35]$ for s_1 when social information is free, otherwise close to zero (see Fig. 4). Different dispersal strategies are selected for the same data point only when the total or relative cost is free.

4 Discussion

Our results highlight that the cost of information acquisition has a major impact on the evolution of dispersal strategies. In conditions where agents must pay a fecundity cost to acquire social or private information, they only do so if the total cost of information acquisition is low enough compared to the magnitude of the environmental fluctuations (similarly to the fertility-information trade-off found in [14]). In stable environments, they quickly abandon the use of information when the cost rises and prefer to disperse blindly. This is most likely due to the low induced environmental pressure: agents dispersing in an adjacent patch will often find themselves in a configuration similar to the natal patch. However, if environmental fluctuations are high, dispersal strategies that rely on the use of information are more likely to offer evolutionary advantages and are thus selected at higher total costs, as seen in [15]. This stronger reliance on information acquisition for dispersal movements in variable environments aligns with the increased explorative movements of butterflies from fragmented habitats reported in several empirical studies [5, 11, 22]. Explorative movements that ensure the acquisition of accurate information in variable habitats may be a key behavioural strategy for making successful dispersal decisions.

We also studied edge cases where information acquisition is completely free. With a restricted perceptual range, the use of free information is always selected in challenging environments. Nevertheless, the proportion of agents relying on both information types in stable environments varies greatly between model iterations, resulting in different dispersal strategies with the same population fitness overall. For the extended perceptual range, a similar pattern appears with some key differences. At very low environmental variability, the variety of information acquisition choices remains but with a stronger reliance on the use of social information. With moderate environmental stochasticity, there is

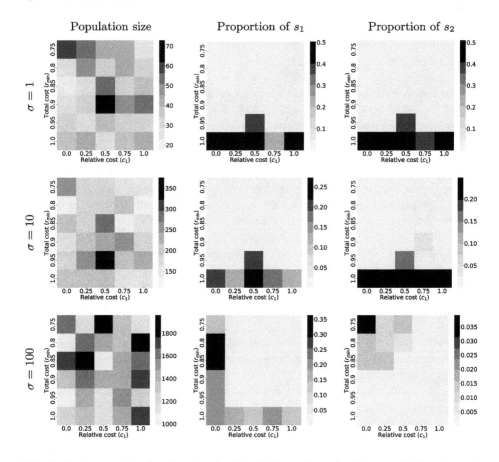

Fig. 4. Heatmaps showing the standard deviation values for the population size and the proportions of s_1 and s_2 in the total agent population after 20,000 generations for scenarios with an expanded perceptual range.

a slight shift towards preference for the use of private information. Finally, in the most challenging environments, there is a clear preference for employing private information, while social information is used only in rare cases. The observed difference between the two perceptual ranges stems from the amount of information sources that are available to the agents. Having only access to information from the natal patch selects for a dispersal behaviour relying equally on private and social information. However, to take successful dispersal decisions agents may need to diversify their information sources, thus if agents can sample information from additional patches, they can rely more heavily on one of the two information sources according to context. In very stable environments, social information is selected because environmental quality is homogeneous. Dispersing towards patches with a lower number of conspecifics provides a better evolutionary advantage than sampling the patches qualities, due to resource competition. On the contrary, when environmental quality fluctuates widely, the best dispersal strategies heavily rely on the acquisition of private information. By using updated private information, individuals can functionally shrink environmental variability, turning variable environments into more predictable and homogeneous habitats [21]. Information that can be derived from conspecifics, such as population density, becomes rapidly "outdated" in conditions of fast environmental fluctuations [12]. Thus, relying on social information may often represent an "ecological trap" (sensu [23]), for example indicating that an environment has been of high quality when the current quality is instead low (for example see [11]). Strategies that rely on the acquisition of more updated private information are positively selected in rapidly changing environments.

In conclusion, our findings demonstrate that the type of information, social or private, selected by a species is influenced by the relative fitness costs of acquiring and using this information. When the total information cost remains sufficiently low, it provides an evolutionary benefit, leading to the predominant selection of the cheapest type of information in stable environments. Social and private information are two important modes of acquiring knowledge on the local and surrounding habitats that can be selectively exploited by animals under different environmental contexts. Animals with an extended perceptual range often show improved fitness in challenging environments due to better-informed dispersal decisions. However, as the overall cost of information increases, population fitness declines. This trend holds unless environmental stochasticity justifies high information costs due to the critical need for accurate decision-making.

Acknowledgments. This work was supported by Service Public de Wallonie Recherche under grant n° 2010235 - ARIAC by DIGITALWALLONIA4.AI. Computational resources have been provided by the Consortium des Équipements de Calcul Intensif (CÉCI), funded by the Fonds de la Recherche Scientifique de Belgique (F.R.S.-FNRS) under Grant No. 2.5020.11 and by the Walloon Region. M.M. was partially supported by the Fonds National de Recherche Scientifique (F.R.S.-FNRS) grant number T.0169.21.

References

1. Bach, L.: On the evolution of conditional dispersal under environmental and demographic stochasticity. Evol. Ecol. Res. **9**(4), 663–673 (2007)
2. Bhaumik, V., Kunte, K.: Dispersal and migration have contrasting effects on butterfly flight morphology and reproduction. Biol. Let. **16**(8), 20200393 (2020). https://doi.org/10.1098/rsbl.2020.0393
3. Bonte, D., et al.: Costs of dispersal. Biol. Rev. **87**(2), 290–312 (2012). https://doi.org/10.1111/j.1469-185X.2011.00201.x
4. Bowler, D.E., Benton, T.G.: Causes and consequences of animal dispersal strategies: relating individual behaviour to spatial dynamics. Biol. Rev. **80**(2), 205–225 (2005). https://doi.org/10.1017/S1464793104006645
5. Braem, S., Turlure, C., Nieberding, C., Van Dyck, H.: Oviposition site selection and learning in a butterfly under niche expansion: an experimental test. Anim. Behav. **180**, 101–110 (2021). https://doi.org/10.1016/j.anbehav.2021.08.011

6. Dall, S.R.X.: Defining the concept of public information. Science **308**(5720), 353–356 (2005). https://doi.org/10.1126/science.308.5720.353c
7. Enfjäll, K., Leimar, O.: Density-dependent dispersal in the glanville fritillary, melitaea cinxia. Oikos **108**(3), 465–472 (2005). https://doi.org/10.1111/j.0030-1299.2005.13261.x
8. Enfjäll, K., Leimar, O.: The evolution of dispersal - the importance of information about population density and habitat characteristics. Oikos **118**(2), 291–299 (2009). https://doi.org/10.1111/j.1600-0706.2008.16863.x
9. Kun, Á., Scheuring, I.: The evolution of density-dependent dispersal in a noisy spatial population model. Oikos **115**(2), 308–320 (2006). https://doi.org/10.1111/j.2006.0030-1299.15061.x
10. Lima, S.L., Zollner, P.A.: Towards a behavioral ecology of ecological landscapes. Trends Ecol. Evol. **11**(3), 131–135 (1996). https://doi.org/10.1016/0169-5347(96)81094-9
11. Marcantonio, M., Levier, M.L., Kourtidis, A., Masier, S.: Social cues and habitat structure affect the behaviour of a non-social insect (2024). https://doi.org/10.22541/au.169686036.60259743/v1
12. Nieberding, C.M., Marcantonio, M., Voda, R., Enriquez, T., Visser, B.: The evolutionary relevance of social learning and transmission in non-social arthropods with a focus on oviposition-related behaviors. Genes **12**(10), 1466 (2021). https://doi.org/10.3390/genes12101466
13. Poethke, H.J., Hovestadt, T.: Evolution of density-and patch-size-dependent dispersal rates. Proc. Roy. Soc. Lond. Series B: Biol. Sci. **269**(1491), 637–645 (2002). https://doi.org/10.1098/rspb.2001.1936
14. Poethke, H.J., Kubisch, A., Mitesser, O., Hovestadt, T.: The evolution of density-dependent dispersal under limited information. Ecol. Model. **338**, 1–10 (2016). https://doi.org/10.1016/j.ecolmodel.2016.07.020
15. Ponchon, A., Garnier, R., Grémillet, D., Boulinier, T.: Predicting population responses to environmental change: the importance of considering informed dispersal strategies in spatially structured population models. Divers. Distrib. **21**(1), 88–100 (2015). https://doi.org/10.1111/ddi.12273
16. Ponchon, A., Travis, J.M.J.: Informed dispersal based on prospecting impacts the rate and shape of range expansions. Ecography **2022**(5), e06190 (2022). https://doi.org/10.1111/ecog.06190
17. Ruxton, G.D., Rohani, P.: Fitness-dependent dispersal in metapopulations and its consequences for persistence and synchrony. J. Anim. Ecol. **68**(3), 530–539 (1999). https://doi.org/10.1046/j.1365-2656.1999.00300.x
18. Schmidt, K.A., Dall, S.R.X., Van Gils, J.A.: The ecology of information: an overview on the ecological significance of making informed decisions. Oikos **119**(2), 304–316 (2010). https://doi.org/10.1111/j.1600-0706.2009.17573.x
19. Schtickzelle, N., Joiris, A., Van Dyck, H., Baguette, M.: Quantitative analysis of changes in movement behaviour within and outside habitat in a specialist butterfly. BMC Evol. Biol. **7**(1), 4 (2007). https://doi.org/10.1186/1471-2148-7-4
20. Snell-Rood, E.C., Kobiela, M.E., Sikkink, K.L., Shephard, A.M.: Mechanisms of plastic rescue in novel environments. Annu. Rev. Ecol. Evol. Syst. **49**(1), 331–354 (2018). https://doi.org/10.1146/annurev-ecolsys-110617-062622
21. Snell-Rood, E.C., Steck, M.K.: Behaviour shapes environmental variation and selection on learning and plasticity: review of mechanisms and implications. Anim. Behav. **147**, 147–156 (2019). https://doi.org/10.1016/j.anbehav.2018.08.007

22. Van Dyck, H., Baguette, M.: Dispersal behaviour in fragmented landscapes: routine or special movements? Basic Appl. Ecol. **6**(6), 535–545 (2005). https://doi.org/10.1016/j.baae.2005.03.005
23. Wong, B.B., Candolin, U.: Behavioral responses to changing environments. Behav. Ecol. **26**(3), 665–673 (2015). https://doi.org/10.1093/beheco/aru183

Integrated Information in Genetically Evolved Braitenberg Vehicles

Hongju Pae[1](✉) and Jeffrey L. Krichmar[1,2]

[1] Department of Cognitive Sciences, University of California, Irvine,
Irvine, CA 92697, USA
hjpae@uci.edu, jkrichma@uci.edu
[2] Department of Computer Science, University of California, Irvine,
Irvine, CA 92697, USA

Abstract. Integrated information, denoted as Φ, quantifies the intrinsic information within causal systems. Despite its profound theoretical implications, applications of Φ have mostly taken place in simulations of arbitrary systems, particularly in terms of biological realism. This study applies Φ calculations to biologically inspired robotic agents that adapt to environmental conditions, thus providing a novel context for observing changes in information integration. The agents' neural network is evolved to demonstrate behavior similar to Braitenberg's Vehicles. The neuro-mechanical design of these evolved agents are then suitable for Φ analysis. Interestingly, early generations had higher Φ values. In later generations the diversity of connection weights and the Φ values decreased, leading to simpler and more reactive neural activations.

Keywords: Integrated information · Neurorobotics · Artificial neurobiological systems · Genetic evolutionary algorithm

1 Introduction

Integrated information, Φ, quantifies the intrinsic causal information of a network system, by measuring the information above and beyond the sum of its parts [1,13,14]. Inspired by subjective consciousness, Φ aims to measure information intrinsic to the system. Despite its intriguing motivation, practical applications in neurobiological systems are rare due to the complexity of calculation. Previous simulation studies primarily focused on verifying its theoretical aspects, using logic gates to represent causal relationships rather than actual neuronal wiring [2,5,11]. However, the emergence of conscious processes that Integrated Information Theory aims to explain through Φ fundamentally occurs within biological systems. Thus, our study shifts the focus towards biological plausibility by calculating Φ in biologically inspired systems. We aim to compare information integration across behavioral scenarios using robotic agents that show adaptive behaviors.

In this study, the neurobiological robots are modeled on the Braitenberg vehicle, a biological agent introduced by Valentino Braitenberg in his book *Vehicles: Experiments in Synthetic Psychology* [3]. Although these vehicles feature a relatively simple structure with two light sensors connected to two motor wheels, they exhibit distinct behavioral patterns based on the neuron's firing and wiring mechanisms, which aligns well with the study's objectives. The complexity within these vehicles is sufficient for Φ calculations, making them suitable for simulations that examine the relationship between biological functionality and Φ.

Evolutionary computation will be used to optimize the behavioral patterns of these robots by altering the connection weights. Based on the genetic representation and fitness function, starting from randomly generated genes and selection towards higher fitness, this process is also can be considered as supporting the biological basis of the simulation [9]. This study will utilize artificial neural networks to build robots, using the connectivity weights of neural networks as the genetic representation to simulate the evolution toward the behavior of Braitenberg vehicles. This lets us explore the comparison of connectivity weights and integrated information in a neural network system.

The key contributions of this study are as follows: 1) Identify the relationship among the functional behavior of biologically inspired agents and Φ. 2) Identify the relationship between the connectivity weights selected by the evolutionary algorithm and the Φ values.

2 Methods

2.1 Design of the Neural Network

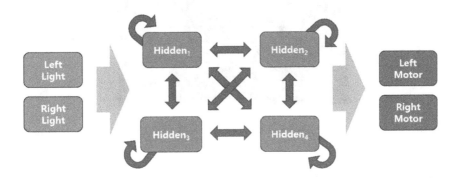

Fig. 1. An overview of the neural network architecture of simulated robots.

The robot's neural network architecture comprises 8 neurons: 2 neurons in the input layer, 4 neurons in the hidden layer, and 2 neurons in the output layer. Each input neuron is connected to all 4 neurons in the hidden layer, and each output neuron is also connected to these 4 hidden layer neurons. The hidden layer neurons are fully-connected, including self-connections. The input neurons function as sensors detecting light, with their activation directly proportional to the intensity of light detected from the left side and right side. The output neurons control the motors of the robot, corresponding to the motors on either side of a Braitenberg vehicle. The neurons in the hidden layer receive activation signals from the input layer and transmit them to the output layer. Here, the hyperbolic tangent (tanh) activation function is used for all neurons. All connections are weighted connections that are free to evolve to excitatory (positive) or inhibitory (negative) values. These weights correspond to the genome in the evolving process, and are determined through the following evolutionary computation process to emulate the functions of the Braitenberg vehicle (Fig. 1).

2.2 Evolving the Vehicles

The robots were evolved to mimic Braitenberg vehicles for fear and aggression (2A and 2B) and for lover and explorer (3A and 3B) [3]. A trial consisted of 200 movements in the Webots environment [16] during which the robot's position was collected. A trial began with the robot 0.75 m from the light source and from a position either 0.2 m to the left or right of the arena's center. Figure 2 shows the starting position of the trial on the Webots environment.

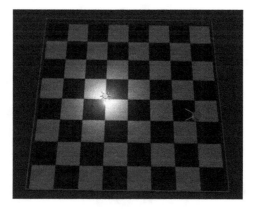

Fig. 2. Rendering of the simulated vehicle and arena under Webots environment.

Evolutionary computation was used to shape vehicle behavior [9]. The *genome* consisted of the neural network connection weights and was initialized with a uniform random distribution between -1 to 1. A population contained 100 individuals. After all individuals completed the two trials, the *fitness* was calculated for each individual. Fitness was judged by how closely the trajectory of an evolved agent matched an ideal agent by calculating the overall mean squared error (MSE) between trajectories.

The ideal agent was based on the vehicle simulations described in Chap. 1 of [6]. It has the input layer (light sensor) directly connected to the output layer (motor) without any hidden layers. The core part of the underlying code is as follows.

```
# Vehicle 2A                          # Vehicle 2B
leftSpeed = leftLight                 leftSpeed = rightLight
rightSpeed = rightLight               rightSpeed = leftLight


# Vehicle 3A                          # Vehicle 3B
leftSpeed = maxLight - leftLight      leftSpeed = maxLight - rightLight
rightSpeed = maxLight - rightLight    rightSpeed = maxLight - leftLight
```

From the given pseudocode, Speed refers to the motor's speed, and Light refers to the activation of the light sensor. Vehicle 2 had excitatory connections and Vehicle 3 had inhibitory connections from the light sensors to the robot's motors. As a result, 2A and 3B will trace a path heading away from the light source, and 2B and 3A will trace a path heading toward the light source. [7] further provides simulation code of Braitenberg vehicles.

If an individual was successful, that is, its fitness had a lower MSE than the previous generation, it survived into the next generation. Otherwise, it was deleted and replaced with the prior individual. A full evolutionary run or *set* consisted of 1000 *generations* at which point asymptotic population performance was observed. The parameters for the evolutionary algorithm were chosen based on a series of pilot runs and guidance from [9].

After a generation, agents were subjected to *crossover* and *mutation*. Multi-point crossover was carried out by selecting the top 25% as parents, which would then generate 10 children from the remaining 75% of the population. Both parents and children were drawn from a uniform random distribution. A child contained 50% of each parent's genome as drawn from a uniform random distribution. After the crossover operation, mutations were carried out on 10% of the genes for all individuals in the population. The gene to be mutated was drawn from a random distribution. A mutation consists of taking the existing gene, a connection weight, and altering it by drawing the new gene from a normal distribution with a mean of the old gene and a standard deviation. To promote exploration, the standard deviation varied. On the first generation and every 200 generations, the standard deviation was set to 1. For each subsequent generation, the standard deviation was either increased by 0.01 if the number of successful individuals was greater than 20 (5% of the population), or decreased otherwise. This had the effect of initially exploring the evolutionary landscape and later refining when the population as close to a solution [9]. The standard deviation was kept between 0.1 and 1.0.

A total of 5 sets for each vehicle type (2A, 2B, 3A, 3B) were simulated. For each set, individuals of the top 25% in a generation, that is, 25 individuals in order of high fitness, were selected as the *representative individuals* for that generation.

2.3 Calculating Φ

The calculations and descriptions of Φ in this study are based on IIT v3.0. Φ is derived from the informational difference between a system's transition probabilities when a given connection exists versus when it does not [13,14]. Since a network's connection shows whether a particular node would contribute to another node's state change, the information from each connection can be retrieved by selecting only a subset of nodes out of the system and dropping out the rest. This results in *partitioning* the network, cutting the connection between selected subsets and dropped nodes.

In IIT, the nodes of interest are called *mechanism*, and the subset of the system as *purview*. Among all possible purviews, the smallest informational distance is defined as the irreducible intrinsic information of the mechanism.

$$intrinsic\_information = min(D(p(system_{unpartitioned})||p(system_{partitioned})\,))$$

Here, iteration of the same calculation over every possible subset selection inside the mechanism and purview is performed (i.e. partitioning the mechanism and purview), and the specific partitioning that results in the smallest informational distance is the Minimum Information Partition (MIP). Then, the largest informational distance between the purview under MIP and the mechanism is defined as ϕ (small phi) of the specific mechanism.

$$MIP_{mechanism} \sim min(D(p(mechanism_{partitioned})||p(purview_{partitioned})))$$

$$\phi_{mechanism} := max(D(p(mechanism_{unpartitioned})||p(purview_{MIP})))$$

Now the same iteration is done for the original system. The partitioning over the original system, not just over mechanisms and purviews, is performed to discover the MIP of the system level. The difference is that the sum of $\phi_{mechanism}$ contributes to calculating informational distance among system-level connection cuts. Finally, Φ of the system is defined by the largest informational distance among candidate mechanisms.

$$\Phi_{system} := max(D(p(system_{mechanism})||p(system_{MIP})))$$

To optimize iteration, this study limited the hidden layer to four neurons. Computation of Φ is achieved with the Python library PyPhi [12]. The theoretical basis of the calculation of integrated information is further detailed in [12] and [14]. This section will mostly focus on detailing how the activation data was specifically transformed in order to calculate Φ.

The activation of the neural network over time captures the state transitions for the robot vehicles during behavior. *Testing runs* were conducted using genes from representative individuals to capture the activity of the neural network. Each testing run was performed under the same conditions as during vehicle evolution. While testing runs, the neural network activity and the trajectory of

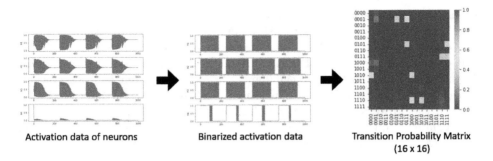

Fig. 3. Schematic process of obtaining TPM from activation data.

individuals are recorded over 1000 timepoints. Hence, the functional behavior of each individual robot and its corresponding Φ during those movements are measured.

The informational difference of state changing from one state to another can be captured as the probabilistic difference of such transition. Thus, the *transition probability matrix (TPM)* plays the key role in the computation of Φ [14]. From the calculation of Φ, the dropped-out nodes do not affect state transitions, thus calculating informational distance over different mechanisms and purviews is done by reducing the dimension of the TPM, by adding up the transition probabilities related to those nodes.

To obtain the TPM of an individual neural network, the system's activity must be expressed as either 0 or 1. Therefore, each neuron's activation data is normalized and then binarized using the median as the threshold to determine if it is considered active. The transition probability for each state was then extracted by counting how many times a transition from a particular state to a particular state has occurred over the 1000 timepoints, and dividing this count by 1000. Only the activity of the hidden layer from the neural network was used in the calculation of Φ, and since the hidden layer contains 4 nodes, the total number of possible states for the system is $4^2 = 16$. Specifically, at timepoint t_i, the state of the neural network could be one of 0000, 0001, 0010, 0011, ..., 1111, and at the next timepoint t_j, it will change to one of these states. The state transition probability from t_i to t_j is thus represented by a 16-by-16 matrix. Each entry (i, j) in this matrix gives the probability that the current state i will transition into the next state j. Following the TPM convention in PyPhi, this is referred to as a state-by-state TPM [12]. This process is illustrated in Fig. 3.

Importantly, the interpretation concerns cases where a state transition never occurs within the provided activity data, leading to the sum of the probability not equal to 1. For a state-by-state TPM, the sum of each row of any column must always equal 1. Since integrated information fundamentally represents causal information, it is impossible to calculate Φ in nondeterministic systems. Hence, the system must be assumed deterministic if Φ is to be calculated, which in turn, any row of a state-by-state TPM that does not sum to 1 would indicate that the activity data lacks sufficient information to fully represent transitions for that case. That is, unobserved transitions in the data should be deemed due to transitions to states external to the dataset. Under such interpretation, if the sum of a specific row in the TPM is 0, it indicates insufficient observations of the current state; if the sum is between 0 and 1, it indicates insufficient observations of the next state. In this study, instances occurred where the sum of certain rows in the TPM is 0. To interpret these cases, a *context node* is introduced. This context node is assumed to activate under such unobserved transitions, and it is marginalized during the calculation process to obtain only the information derived from the activation of the actual hidden layer neurons.

The Φ values are non-negative, with a lower bound of zero and no defined upper bound [1,14]. Since Φ is not an absolute metric, it is unsuitable for comparing information integration across systems under different structures directly. It should be used to explore relative differences under different conditions within the same-structured system.

2.4 Analysis of Results

To examine changes in Φ across evolution, testing runs were conducted for the representative individuals of the 0th, 100th, 200th, 500th, and 1000th generations. To determine whether the differences in Φ values between generations are statistically significant, repeated measures ANOVA is used. Additionally, differences across vehicle types are assessed using the Kruskal-Wallis test.

To explore the relationship between Φ and the connectivity weights (genes), t-distributed stochastic neighbor embedding (t-SNE) analysis [10] was conducted on evolved weights from both early (100th) and late (1000th) generations. Vehicle type, set, and the magnitude of Φ were labeled for t-SNE clustering and visualized accordingly. For Φ values, agglomerative hierarchical clustering, which is less affected by outliers [8], was used to categorize them into three clusters: high, medium, and low based on their magnitude.

3 Results

3.1 Evolution of Vehicles

For all four vehicle types (2A, 2B, 3A, 3B), fitness approached an asymptote around the 500th generation. As shown in Fig. 4, the best-so-far fitness individual in the final generation exhibited functional behavior characteristics of the expected Braitenberg vehicle.

Furthermore, Φ from each vehicle type appeared to be statistically different according to the Kruskal-Wallis test. As in Table 1a, the probability that Φ calculated from each vehicle type would share the same distribution decreases significantly with increasing generation. Therefore, it can be safely concluded that the population of integrated information would differ depending on the vehicle type. Based on this result, each vehicle type was considered independent for the remaining analyses.

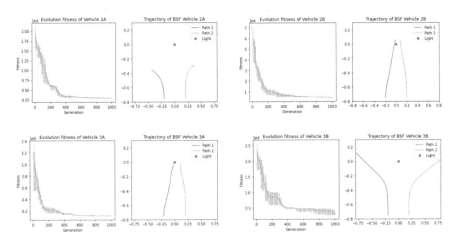

Fig. 4. Fitness change over generation and movement trajectory of evolved best-so-far (BSF) individuals of each vehicle type.

Table 1. Statistical test results.

(a) Test statistic and corresponding p-value of Kruskal-Wallis test of Φ over different vehicle types from each generation.

Generation	test statistic	p-value
0	5.3053	0.1508
100	25.4399	< .0001
200	57.9434	< .0001
500	59.1832	< .0001
1000	110.0884	< .0001

(b) F-statistic and corresponding p-value of repeated measures ANOVA of Φ over generation from each vehicle type.

Vehicle type	F-statistic	p-value
2A	18.7849	< .0001
2B	22.9872	< .0001
3A	7.8776	< .0001
3B	24.0576	< .0001

3.2 Changes in Φ over Generation

For all four vehicle types, Φ decreased as generations progressed. The trend and its statistics can be found in Fig. 5. Additionally, the difference in Φ over generations was found to be statistically significant for each vehicle type, as confirmed through repeated measures ANOVA conducted for each type in Table 1b.

The average value of Φ mostly falls below 1. While Φ cannot be directly compared in a strict sense across different studies due to its non-absolute scale, however, other studies with systems of similar node numbers also report Φ values mostly ranging from 0 to 2, suggesting there is no significant anomaly in the case of our calculation [2,4,5,11].

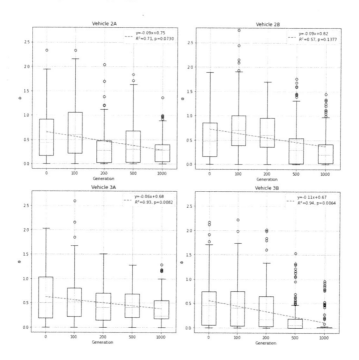

Fig. 5. Changes of Φ under increasing generation among four different types of vehicles. Regression line illustrated as dotted red line. (Color figure online)

One notable point is that, although the average value of Φ generally decreases as generations progress, there are still a considerable number of outliers exceeding the average throughout all generations. Consequently, in the subsequent t-SNE analysis, vehicle type, set, and the magnitude of Φ were labeled to facilitate tracking of individuals with high Φ values through visualization.

3.3 Changes in Connection Weights over Generation

The t-SNE analysis in Fig. 6 demonstrates how evolution has influenced connection weights. In this simulation setting, evolution appears to have led individuals to converge on specific genes (weights). Interestingly, this convergence of weights resulted in decreased Φ values. The bottom right plot in Fig. 6 shows that only one cluster maintains a high Φ after the evolutionary process, predominantly distributed among the set 4 group that evolved the functional behavior of vehicle 2B. In contrast, most late-generation genes measured lower Φ values, especially when compared to the highly dispersed early-generation genes, suggesting that earlier generations could be more prone to maintain higher Φ values due to more integration between hidden layer nodes. Interestingly, as behavior becomes more stereotyped and accurate as Φ decreases. This transition has been observed in animals as they progress from planned behavior to habits when learning a task [15].

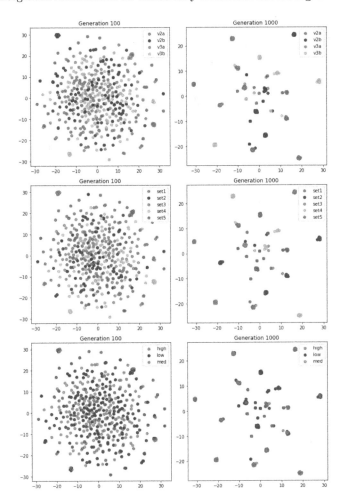

Fig. 6. t-SNE analysis. The first column shows the weights at the 100th generation, and the second column shows the 1000th generation. The first row is labeled with vehicle types, the second row is labeled with different runs (set1 - set5), and the third row is labeled with the Φ magnitude.

3.4 Tracking the Sample Individuals

In Fig. 7, individuals with distinct characteristics are selected to specifically examine their evolved behavior, connection weights, and Φ. It is notable that individuals of earlier generations, despite less accurate trajectories, can have higher Φ values. Note that the weights should not be similar since all individuals are selected from different t-SNE clusters. As seen in Fig. 6, there are multiple types of weights the individual can converge even for the same behavior. Figure 7 also shows that the TPM also can be different for the same behavior, though the TPM alone cannot directly tell the range of Φ value.

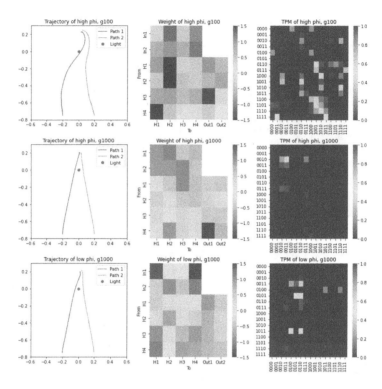

Fig. 7. Trajectory, neural network weight, and TPM of selected individuals. All individuals perform vehicle 2B behavior. **First row.** Early-generation high-Φ individual ($\Phi = 2.08529$). **Second row.** Late-generation high-Φ individual ($\Phi = 1.4463$). **Third row.** Late-generation low-Φ individual ($\Phi = 0$).

4 Conclusion

The present paper applied IIT, which is used to investigate conscious systems, to a classic thought experiment from synthetic psychology. Rather than applying supervised or reinforcement learning methods that would limit the neural architecture, we used an evolutionary approach that explored different weight configurations. As the evolutionary process proceeded, there was a reduction in weight diversity, and at the same time, a decrease in the Φ values. In the early stages of evolution where the robots were "exploring" to find the optimal path, the weights were diverse, but as weights became relatively consistent later on, the neurons' activation was also optimized, similar to behavior observed in rats [15]. Although the results were unexpected, they are consistent with the initial transition from random behavior to stereotyped, goal-driven behavior. However, to determine whether high Φ can be consistently observed in various instances of random behavior and low Φ in goal-driven behavior beyond the scope of this simulation, further studies in more complex systems including real-world experiments are necessary.

References

1. Albantakis, L., et al.: Integrated information theory (iit) 4.0: formulating the properties of phenomenal existence in physical terms. PLOS Comput. Biol. **19**(10), 1–45 (2023). https://doi.org/10.1371/journal.pcbi.1011465
2. Albantakis, L., Hintze, A., Koch, C., Adami, C., Tononi, G.: Evolution of integrated causal structures in animats exposed to environments of increasing complexity. PLOS Comput. Biol. **10**(12), 1–19 (2014). https://doi.org/10.1371/journal.pcbi.1003966
3. Braitenberg, V.: Vehicles: Experiments in Synthetic Psychology. MIT Press, Cambridge (1986)
4. Gomez, J.D., Mayner, W.G.P., Beheler-Amass, M., Tononi, G., Albantakis, L.: Computing integrated information (ϕ) in discrete dynamical systems with multi-valued elements. Entropy **23**(1), 6 (2021). https://doi.org/10.3390/e23010006
5. Hoel, E., Albantakis, L., Marshall, W., Tononi, G.: Can the macro beat the micro? integrated information across spatiotemporal scales. Neurosci. Consciousness (2016). https://doi.org/10.1093/nc/niw012
6. Hwu, T., Krichmar, J.: Neurorobotics: Connecting the Brain, Body and Environment. MIT Press, Cambridge (2022)
7. Hwu, T., Krichmar, J.: Neurorobotics: Connecting the brain, body and environment. (2022). https://github.com/jkrichma/NeurorobotExamples/tree/main
8. Jain, A., Dubes, R.: Algorithms for Clustering Data. Prentice Hall advanced reference series, Prentice Hall, Upper Saddle (1988). https://books.google.co.kr/books?id=7eBQAAAAMAAJ
9. Jong, K.A.D.: Evolutionary Computation: A Unified Approach. MIT Press, Cambridge (2016)
10. van der Maaten, L., Hinton, G.: Visualizing data using t-sne. J. Mach. Learn. Res. **9**(86), 2579–2605 (2008). http://jmlr.org/papers/v9/vandermaaten08a.html
11. Marshall, W., Albantakis, L., Tononi, G.: Black-boxing and cause-effect power. PLOS Comput. Biol. **14**(4), 1–21 (2018). https://doi.org/10.1371/journal.pcbi.1006114
12. Mayner, W.G.P., Marshall, W., Albantakis, L., Findlay, G., Marchman, R., Tononi, G.: Pyphi: a toolbox for integrated information theory. PLOS Comput. Biol. **14**(7), 1–21 (2018). https://doi.org/10.1371/journal.pcbi.1006343
13. Moon, K., Pae, H.: Making sense of consciousness as integrated information: evolution and issues of integrated information theory. J. Cogn. Sci. **20**(1), 1–52 (2019). https://doi.org/10.17791/jcs.2019.20.1.1
14. Oizumi, M., Albantakis, L., Tononi, G.: From the phenomenology to the mechanisms of consciousness: integrated information theory 3.0. PLOS Comput. Biol. **10**(5), e1003588 (2014). https://doi.org/10.1371/journal.pcbi.1003588
15. Redish, A.D.: Vicarious trial and error. Nat. Rev. Neurosci. **17**(3), 147–159 (2016). https://doi.org/10.1038/nrn.2015.30
16. Webots: http://www.cyberbotics.com. Open-source Mobile Robot Simulation Software

Motor Learning

Neural Chaotic Dynamics for Adaptive Exploration Control of an Autonomous Flying Robot

Vatsanai Jaiton and Poramate Manoonpong[✉]

Bio-inspired Robotics and Neural Engineering Laboratory,
School of Information Science and Technology, Vidyasirimedhi Institute of Science and Technology, Rayong, Thailand
poramate.m@vistec.ac.th

Abstract. Efficient search and exploration behaviors observed in animals are significant for their survival, encompassing foraging, mate-finding, and predator avoidance. Animal movement patterns underlying the search and exploration behaviors are typically modeled as Lévy flights. Despite their efficacy, current Lévy flight models rely on mathematical formulations and lack connection to biological neural systems. To address this, we propose an adaptive neural system which encodes Lévy distributions for generating and controlling efficient search and exploration behaviors. The neural system consists of three subnetworks: a neural chaotic network, a directional control network, and a step length control network. The neural chaotic network generates chaotic output signals, which are used as exploration seeds, further processed by the directional and step length control networks to control the search and exploration of a flying robot. By simply adjusting the scaling input of the step length control network, our neural control system can generate adaptive exploration behaviors with a combination of global and local search strategies. Our system is implemented on a simulated autonomous flying robot, evaluated across diverse environments, and compared with Gaussian random walk and traditional Lévy flight strategies.

Keywords: Exploration behavior · Navigation control · Chaotic dynamics · Neural dynamics · Neural control · Lévy flight

1 Introduction

Animals' efficient search and exploration in nature are crucial behaviors for their survival, used for activities such as food foraging, finding mates, and avoiding predators. Biological studies have analyzed movement trajectories of various animals, including albatrosses flying over the ocean [2], sharks and other marine predators swimming underwater [10], and Drosophila larvae crawling in experimental arenas [9] while performing foraging tasks. These studies indicate that the most common and effective method for such tasks is a Lévy flight (Lévy walk) model [12]. The model effectively represents the complex behaviors that arise from a combination of local search, which involves collecting nearby food, and global search, which entails moving across large areas to explore new food resources. Inspired by this, several robotic studies have been made to understand and replicate efficient search and exploration behaviors by integrating the Lévy flight model into robotic control systems for environmental exploration [4,13]. These efforts aim to reproduce exploration behavior based on Lévy flight patterns. However, current Lévy flight models, which are primarily based on mathematical formulas with Lévy distributions, are still difficult to relate to the biological neural systems of animals.

From this point of view, we propose an adaptive neural exploration control system (see Fig. 1(a)) composed of three sub-networks: a neural chaotic network, a directional control network, and a step length control network. The core mechanism relies on the neural chaotic network, a two-neuron recurrent network that generates output signals exhibiting chaotic neural dynamics. These signals are subsequently processed by the directional and step length control networks to finally generate control commands. The directional and step length control commands are applied together to control the exploration behavior of a flying robot. By adjusting the scaling input of the step length control network, we can generate efficient adaptive exploration behaviors exhibiting both global and local search patterns (see Figs. 1(b), 1(c), 1(d)). The proposed neural control system is implemented on an autonomous flying robot (drone) to evaluate its

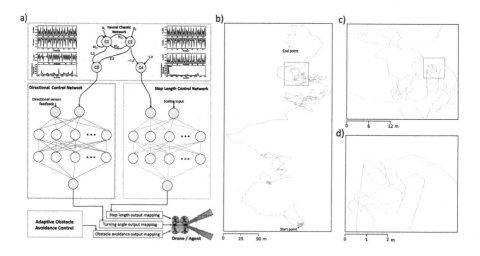

Fig. 1. a) An adaptive neural exploration control system consists of three sub-networks: a neural chaotic network (depicted in blue), a directional control network (depicted in orange), and a step length control network (depicted in yellow). b) A flying trajectory of a flying robot (global search) through an open area. c) Zoomed-in view of the robot's flying trajectory (local search) within the red box in (b). d) Zoomed-in view of the robot's flying trajectory (local search) within the red box in (c). (Color figure online)

performance and demonstrate adaptive exploration behaviors in both open and closed environments of various sizes. These behaviors are compared with two other well-known exploration strategies that use different step length-defined methodologies: Gaussian random walk and Lévy flight.

2 Materials and Methods

This study proposes a method for generating adaptive exploration behaviors using a neural control system, which enables a flying robot to execute both local and global search strategies. The proposed neural control system is based on a neural chaotic network, which exhibits chaotic neural dynamics output. The output is then utilized to produce control commands, which are used for generating adaptive exploration behaviors. The neural chaotic network acts as an internal pattern generator, with its output signals serving two primary functions: directional control via a post-processing directional control network and step length control through a post-processing step length control network. By integrating the outputs from both networks, the flying robot can effectively perform adaptive exploration in various environments.

2.1 Neural Chaotic Network

The neural chaotic network is developed based on a two-neuron recurrent network, as shown in Fig. 2. The neurons $C1$, $C2$, $C3$, and $C4$ in the network are modeled as discrete-time non-spiking neurons where their output is governed by the following equation:

$$o_i(t) = f\left(\sum_{j=1}^{n} W_{ij} \cdot o_j(t-1) + B_i\right) \quad i = 1, 2, 3, 4, \tag{1}$$

where n denotes the number of neurons, and B_i is the internal bias term to neuron i. W_{ij} represents the synaptic weight of the connection from neuron j to neuron i. $f()$ denotes the activation function, which uses the hyperbolic tangent (tanh) function (for the blue neurons in Fig. 2) and linear function (for the green neurons in Fig. 2).

The network parameters, including synaptic weights and bias inputs, are set according to the neural chaotic system detailed in [3,11]. Specifically, the synaptic weights are configured as follows: W_{11} is set to -5.5, W_{12} to -1.65, and W_{21} to 1.48. Bias inputs are determined as 5.73 for $B1$ and 0.25 for $B2$. However, initial values of $C1$ and $C2$ are independently generated using random processes. The outputs of $C1$ and $C2$ exhibit chaotic neural dynamics, as shown in Figs. 2(b) and 2(c), characterized by irregularity, aperiodicity, determinism, and high sensitivity to initial conditions [6]. Notably, these outputs are then processed by two linear neurons in distinct ways to produce the directional ($C3$) and step length ($C4$) control inputs (see Figs. 2(d) and 2(g)). $C3$ receives inputs from both $C1$ and $C2$ through synaptic weights of 0.5, while $C4$ solely receives

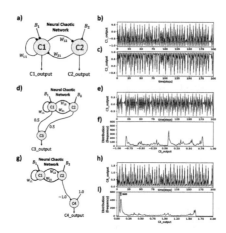

Fig. 2. a) Neural chaotic network. b) Output signal of $C1$. c) Output signal of $C2$. d) Chaotic processing neuron ($C3$) transmitting its output to the directional control network. e) $C3$ output serving as a directional control input. f) Distribution of the directional control input. g) Chaotic processing neuron ($C4$) transmitting its output to the step length control network. h) $C4$ output serving as a step length control input. i) Distribution of the step length control input.

an input from $C2$ through synaptic weight of -1.0. Additionally, $C4$ also receives a bias input of 1.0. The outputs of $C3$ and $C4$ exhibit different dynamics, as illustrated in the output signal graphs in Figs. 2(e) and 2(h), and in the output signal distribution graphs in Figs. 2(f) and 2(i), respectively. These output signals ($C3$ and $C4$ outputs) serve as directional and step length control inputs for the directional and step length control networks, described in the following section.

2.2 Directional and Step Length Control Networks

The directional and step length control networks utilized here are constructed based on multilayer feedforward neural networks. All neurons in the networks are modeled as discrete-time non-spiking neurons, as described in Eq. 1, using the linear (for green neurons, Figs. 3(a) and 4(a)) and ReLU (for yellow neurons, Figs. 3(a) and 4(a)) activation functions.

The directional control network, is employed to compute the new heading direction of the drone. This network receives two inputs: the directional control input generated by $C3$ and the current heading directional feedback. The network consists of two hidden layers, each containing 30 neurons, and has a single output representing the drone's new/target heading directional control command.

The synaptic weights in the network are trained using the supervised backpropagation method, with mean absolute error (MAE) as the loss function; see Eqs. 2 and 3. The training dataset is prepared using a state machine for directional computation as described in Eqs. 4 and 5. The training framework and

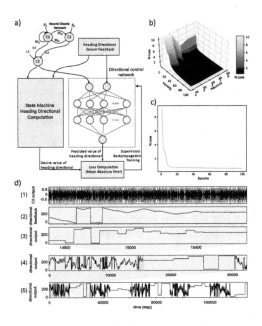

Fig. 3. a) Framework for training the directional control network. b) Surface graph (heatmap) depicting the landscape of loss percentage over the training period. c) Graph showing the loss percentage of the directional control network with 30 neurons for each hidden layer. d) Directional control network input and output signals. The green areas in 1–3 show zoomed-in views of $C3$ output, directional (input) feedback, and directional output from the green area in 4. The blue area in 4 shows a zoomed-in view of the directional output from the blue area in 5, which shows the overall directional output. (Color figure online)

output evaluation are illustrated in Fig. 3.

$$\Delta W_{ij} = -\mu \frac{\partial E}{\partial W_{ij}}, \tag{2}$$

$$E = \frac{1}{n} \sum_{i=1}^{n} |D_{target\_i} - D_{predict\_i}|, \tag{3}$$

where W_{ij} denotes the weight of the connection between neuron i and neuron j. μ is the learning rate. E is an error or loss calculated using the mean absolute error. D_{target_i} represents the target heading direction for sample i. $D_{predict\_i}$ represents the predicted heading direction for sample i.

$$D_{t\_cal}(t) = 90 \cdot o_3(t) + D_c(t), \tag{4}$$

$$D_{target}(t) = \begin{cases} D_{t\_cal}(t) + 360 & D_{t\_cal}(t) < 0, \\ D_{t\_cal}(t) & 0 \leq D_{t\_cal}(t) \leq 360, \\ D_{t\_cal}(t) - 360 & D_{t\_cal}(t) > 360, \end{cases} \tag{5}$$

where D_{t_cal} denotes the calculated target heading direction, o_3 represents the output of $C3$, D_c is the current heading direction value (heading directional feedback), and D_{target} is the final target heading direction.

The step length control network is responsible for computing the step length control command. Empirically, the synaptic weights in this network are set to 2.0. Similar to the directional control network, the step length control network receives two inputs: the step length control input from $C4$ (I_C) and a scaling input (I_S). It consists of two hidden layers, each containing 10 neurons, and has a single output representing the drone's step length control command. The primary function of this network is to scale the step length control input based on the scaling input, which is preprogrammed to cyclically alternate between the empirical values of -1.5 and 3.0. The network architecture and parameter settings allow for simply adjusting the step length according to the scaling input value, enhancing the drone's exploration capabilities. The output (scaling capability) of the network is shown in Fig. 4.

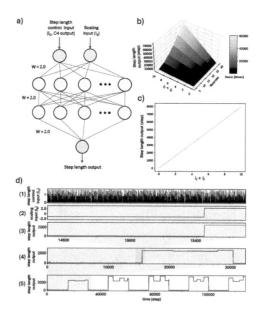

Fig. 4. a) Step length control network. b) Surface graph (heatmap) depicting the output landscape of the network. Note that $I_C + I_S$ represents the summation value of the network inputs. c) Graph showing the output value of the step length control network using ten neurons for each hidden layer. d) Step length control network input and output signals. The green areas in 1–3 show zoomed-in views of inputs I_C, I_S, and step length output from the green area in 4. The blue area in 4 shows a zoomed-in view of the step length output from the blue area in 5, which shows the overall step length output. Here, the step length is based on a computation time step where 100 steps correspond to approximately 0.5 m. (Color figure online)

2.3 Neural System Integration

To achieve adaptive exploration control, we integrate two control commands: directional and step length control commands. These commands are initially generated by the neural chaotic network (Fig. 1(a), blue neurons) and then undergo neural dynamics handling through the linear neurons (Fig. 1(a), green neurons) to ensure that the output signals are suitable for further processing. Subsequently, two multilayer feedforward neural networks (Fig. 1(a), orange and yellow boxes) are employed for post-processing. The overall neural system integration is shown in algorithm 1.

Algorithm 1. Neural System integration

1: Initialize all variables
2: $C1, C2 \sim \mathcal{U}(-1.0, 1.0)$
3: $W_{11} = -5.5, W_{12} = -1.65, W_{21} = 1.48, B_1 = -5.73, B_2 = 0.25$
4: Deploy all weights of directional control network as trained values
5: Set all weights of step length control network $= 2.0$
6: **while** $Terminate\ is\ not\ TRUE$ **do**
7:     Fly the drone forward
8:     Get current heading directional feedback   ▷ receive input information
9:     Update $C1$ and $C2$ ▷ generate chaotic output signals
10:     Update $C3$ ▷ compute directional control input
11:     Update $C4$ ▷ compute step length control input
12:     **if** $Step\_count > Step\ Length\ Command$ **then**
13:         Compute step length control network   ▷ get new step length command
14:         Compute directional control network   ▷ get new directional command
15:         $Step\_count = 0$
16:     **end if**
17:     **if** $Loop < 15000$ **then**   ▷ perform cyclically alternate the scaling input
18:         **if** $Temp\ \%\ 2 = 0$ **then**   ▷ perform local search
19:             $Scaling\_input = -1.5$ ▷ use low scaling input value
20:         **else**   ▷ perform global search
21:             $Scaling\_input = 3.0$ ▷ use high scaling input value
22:         **end if**
23:     **else**
24:         $Temp + +$
25:         $Step\_count = 0$
26:         $Loop = 0$
27:     **end if**
28:     Map the directional command to the turning direction
29:     Execute turning   ▷ turn until achieve the target heading direction
30:     $Step\_count + +$
31:     $Loop + +$
32: **end while**

3 Experiments and Results

The proposed adaptive neural exploration control system underwent performance evaluation and exploration behavior demonstration using a simulated drone within CoppeliaSim as the experimental platform. A robot operating system (ROS) was used to interface between the drone and the neural control system. To achieve this, two experiments were implemented under five varying sizes environments (see Fig. 5) in both open and closed areas. During the tests, the drone explored the given environment by maintaining a constant forward motion (pitch) at an approximate speed of 0.1 m/s. Concurrently, the heading direction (yaw) was controlled using the proposed adaptive neural exploration control system to facilitate adaptive exploration behaviors.

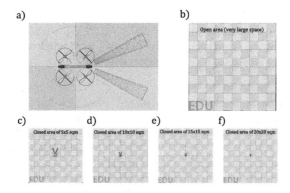

Fig. 5. a) A simulated drone used as the experimental platform. b)-f) Experimental environments with different sizes.

The first experiment aimed to investigate global search performance within an open environment. Three exploration strategies were compared: Gaussian random walk, Lévy flight, and adaptive neural exploration control utilizing neural chaotic dynamics. Each test was approximately 120 min. The results in Fig. 6(a) show three flying trajectories corresponding to the three exploration strategies, all shown within the same map. This visualization allows for the observation of different behaviors compared among each exploration strategy. Specifically, when employing the Gaussian random walk (blue) exploration strategy, the drone performed more looping flight patterns, leading to focused exploration within a specific small area.

In contrast, when utilizing the Lévy flight (orange) and adaptive neural exploration control (green) strategies, the drone exhibited prolonged forward flight patterns. This behavior results in traversing large areas, interspersed with periods of exploration in particular small areas. Figure 6(b) shows step length information compared among each exploration strategy. The results show that the step lengths of the Gaussian random walk strategy consist mostly of low-range values, although with the highest frequency. While for both the Lévy

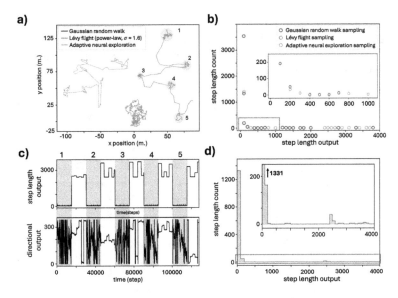

Fig. 6. a) Comparison of flying trajectories among three exploration control strategies: Gaussian random walk (depicted in blue), Lévy flight (depicted in orange), and adaptive neural exploration control (depicted in green). b) Comparison of step length information. The inset plot shows a zoomed-in view of the red box. c) Step length and directional control output signals. The shaded gray areas represent the output signals corresponding to the shaded flying patterns in (a). d) Step length histogram of the adaptive neural exploration control strategy. The inset plot shows a zoomed-in view of the red box. A video showing adaptive exploration flying behaviors can be seen at www.manoonpong.com/AEC/video.mp4. (Color figure online)

flight and adaptive neural exploration control strategies, their step lengths were predominantly low-range values at a moderate frequency, and the rest was distributed across high-range values. Additionally, Fig. 6(c) shows the step length and directional output signals. The shaded gray areas represent the output signals with low step length output values and high-frequency directional outputs, corresponding to the shaded flying trajectory in Fig. 6(a), which exhibits the local search pattern. Figure 6(d) shows the histogram of the step length using the adaptive neural exploration control strategy. Based on the step length distribution and histogram of the adaptive neural exploration control in Figs. 6(b) and 6(d), we conducted a Kolmogorov-Smirnov test to assess the distribution similarity between the adaptive neural exploration control and the Lévy flight (power-law model with exponent $\alpha = 1.6$). The result shows that there was no significant difference between the two distributions at $p > 0.05$.

The second experiment focused on assessing local search performance within four closed environments, with map sizes of 5×5, 10×10, 15×15, and 20×20 sqm. The experiments were conducted for approximately 10, 20, 30, and 40 min, respectively, according to the map sizes. Each map underwent testing from three initial angles (0, 120, and 240°), and the experiments were repeated ten times for each map. The results in Figs. 7(a), 7(b), and 7(c) provide a comparison of the flying trajectories for each exploration strategy in each map. The gray boxes in the graphs indicate the unexplored grid areas.

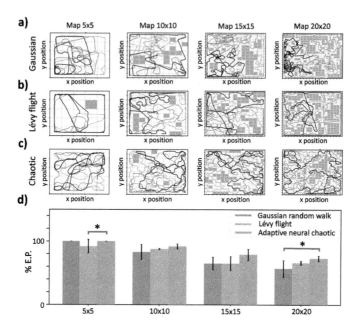

Fig. 7. Drone flying trajectories in closed areas of sizes 5×5, 10×10, 15×15, and 20×20 sqm using three exploration control techniques: a) Gaussian random walk, b) Lévy flight, and c) adaptive neural (chaotic) exploration control. The red, blue, and pink lines represent three initial angles (0, 120, and 240°). d) Comparison of exploration performance percentage (%E.P.) among each map and technique. * denote significant differences with the following p ≤ 0.05. (Color figure online)

Figure 7(d) illustrates a comparison of exploration performance percentage (%E.P.), calculated as below:

$$\% E.P. = (\frac{C_A}{A})(\frac{C_G}{A}) \times 100\%, \tag{6}$$

where A denotes the total exploration area (grid count). C_A is the largest covered area. C_G is the total covered grid or covered density. Note that grid resolution directly affects the exploration performance percentage. Here, we used a grid size of 1×1 sqm.

The overall result, as depicted in Fig. 7(d), indicates that the proposed adaptive neural exploration control consistently outperformed the other two exploration strategies across all maps. Interestingly, the performance of the Gaussian random walk exploration strategy significantly decreased in a map size of

20×20 sqm (a large closed environment), suggesting its efficiency for exploration in smaller environments. Conversely, the performance of the Lévy flight exploration strategy dropped notably in a map size of 5×5 sqm (a small closed environment), indicating its suitability for larger environment exploration. This observation emphasizes the versatility of our proposed adaptive neural exploration control, which surpasses both Gaussian random walk and Lévy flight strategies in both small and large environments. Hence, it emerges as the most efficient exploration strategy for environments of all sizes.

4 Discussion and Conclusion

In this study, we introduced an adaptive neural exploration control system designed to generate adaptive exploration behaviors for an autonomous drone. The proposed system leverages i) the neural chaotic dynamics of a small recurrent network to generate complex patterns, and ii) nonlinear transformations of multilayer feedforward networks to translate these patterns into appropriate directional and step length control commands for the drone. Additionally, at an abstract level, the proposed system is comparable to a biological neural system in two components: first, an internal pattern generator [1,5,9], which here is a neural (chaotic) pattern generator network, and second, premotor control networks [7], which are here realized as directional and step length premotor control networks.

This neural exploration control strategy enables the drone to perform both global search, by traversing large areas, and local search, by intermittently investigating specific regions. These flying behaviors mimic the efficient exploration patterns observed in nature [2,9,10], surpassing the performance of both Gaussian random walk and Lévy flight strategies across all closed environments.

Future work will focus on integrating adaptive neural obstacle avoidance control [8] with the exploration control system (see Fig. 1(a), green box). This integration will enable safe exploration missions by allowing the drone to avoid obstacles. We will then validate the combined system on a real drone. Additionally, we will investigate the use of sensor feedback for adaptive step-length control, potentially refining the balance between global and local search strategies.

Acknowledgments. This work was supported by the startup grant on Bio-inspired Robotics (P.M., the project PI) of Vidyasirimedhi Institute of Science and Technology.

References

1. Clement, L., Schwarz, S., Wystrach, A.: An intrinsic oscillator underlies visual navigation in ants. Curr. Biol. **33**(3), 411–422.e5 (2023). https://doi.org/10.1016/j.cub.2022.11.059. https://www.sciencedirect.com/science/article/pii/S0960982222018504

2. Humphries, N.E., Weimerskirch, H., Queiroz, N., Southall, E.J., Sims, D.W.: Foraging success of biological lévy flights recorded in situ. Proc. Natl. Acad. Sci. **109**(19), 7169–7174 (2012). https://doi.org/10.1073/pnas.1121201109. https://www.pnas.org/doi/abs/10.1073/pnas.1121201109
3. Jaiton, V., Manoonpong, P.: Chaotic neural oscillator for navigation and exploration of autonomous drones. In: The 9.5th International Symposium on Adaptive Motion of Animals and Machines (AMAM 2021) (2021). https://doi.org/10.18910/84868
4. Katada, Y., Hasegawa, S., Yamashita, K., Okazaki, N., Ohkura, K.: Swarm crawler robots using lévy flight for targets exploration in large environments. Robotics **11**(4) (2022).https://doi.org/10.3390/robotics11040076. https://www.mdpi.com/2218-6581/11/4/76
5. Maye, A., Hsieh, C.h., Sugihara, G., Brembs, B.: Order in spontaneous behavior. PLOS ONE **2**(5), 1–14 (2007).https://doi.org/10.1371/journal.pone.0000443
6. Muthu, J.S., Murali, P.: Review of chaos detection techniques performed on chaotic maps and systems in image encryption. SN Comput. Sci. **2**(5), 1–24 (2021). https://doi.org/10.1007/s42979-021-00778-3
7. Namiki, S., Dickinson, M.H., Wong, A.M., Korff, W., Card, G.M.: The functional organization of descending sensory-motor pathways in *Drosophila*. eLife **7**, e34272 (2018).https://doi.org/10.7554/eLife.34272
8. Pedersen, C.K., Manoonpong, P.: Neural control and synaptic plasticity for adaptive obstacle avoidance of autonomous drones. In: Manoonpong, P., Larsen, J.C., Xiong, X., Hallam, J., Triesch, J. (eds.) SAB 2018. LNCS (LNAI), vol. 10994, pp. 177–188. Springer, Cham (2018). https://doi.org/10.1007/978-3-319-97628-0_15
9. Sim, D.W., Humphries, N.E., Hu, N., Medan, V., Berni, J.: Optimal searching behaviour generated intrinsically by the central pattern generator for locomotion. eLife **8**, e50316 (2019).https://doi.org/10.7554/eLife.50316,
10. Sims, D.W., et al.: Scaling laws of marine predator search behaviour. Nature **451**, 1098–1102 (2008). https://doi.org/10.1038/nature06518
11. Steingrube, S., Timme, M., Wörgötter, F., Manoonpong, P.: Self-organized adaptation of a simple neural circuit enables complex robot behaviour. Nat. Phys. **6**, 224–230 (2010). https://doi.org/10.1038/nphys1508
12. Viswanathan, G.M., Buldyrev, S.V., Havlin, S., da Luz, M.G.E., Raposo, E.P., Stanley, H.E.: Optimizing the success of random searches. Nature **401**, 911–914 (1999). https://doi.org/10.1038/44831
13. Zedadra, O., Guerrieri, A., Seridi, H.: Lfa: A lévy walk and firefly-based search algorithm: application to multi-target search and multi-robot foraging. Big Data Cogn. Comput. **6**(1) (2022).https://doi.org/10.3390/bdcc6010022

Non-instructed Motor Skill Learning in Monkeys: Insights from Deep Reinforcement Learning Models

Laurène Carminatti[1,2], Lucio Condro[3], Alexa Riehle[3,4], Sonja Grün[4], Thomas Brochier[3], and Emmanuel Daucé[3,5(✉)]

[1] Department of Informatics, Bioengineering, Robotics, System Engineering, University of Genova, Genova, Italy
[2] Department of Robotics, Brain and Cognitive Science, Italian Institute of Technology (IIT), Genova, Italy
[3] Institut de Neurosciences de la Timone (INT), Aix-Marseille Univ, Marseille, France
emmanuel.dauce@univ-amu.fr
[4] Institute of Neuroscience and Medicine and Institute for Advanced Simulation and JARA Institut Brain Structure-Function Relationships, Jülich Research Centre, Jülich, Germany
[5] Ecole Centrale Méditerranée, Marseille, France

Abstract. We employ Reinforcement Learning (RL) models to unravel the mechanisms behind the learning behavior of two macaque monkeys engaged in a free-moving multi-target reaching task. The study was conducted using computer simulations reflecting the animal's learning conditions, and compared with the actual arm movements recorded on two macaque monkeys on a Kinarm apparatus. Our paper thus provides important insights for the design of motor control learning systems, combining end-effector control design with the learning of motor chunks associations. Our research is of interest for the modeling and understanding of natural motor learning systems, but also heads toward the design of more "brain-inspired" adaptive robotic manipulators.

Keywords: Animal Behavior and Cognition · Computational neuroscience · Reinforcement learning · Motor control

1 Introduction

The survival of animals in complex environments relies on their ability to continually learn and refine their skills to enhance both efficiency and competitiveness. [11,12]. Motor skill learning takes place over long time scales (from weeks to years), reflecting the slow acquisition of combination/coordination rules of the motor apparatus to control its many degrees of freedom [4,8,12,30]. Complex skills generally require a specific combination of elementary motor commands [16,23,24,26], and involves visual/perceptual monitoring at different steps of the process [27,31]. The shaping of composite actions, made of many elementary movements, has led experimentalists to formulate the *chunking hypothesis*, which

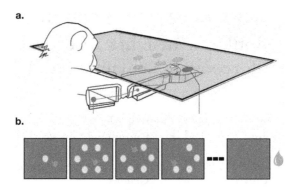

Fig. 1. Self-paced multi-target reaching task a. Experimental setup used to for the experimental task. Joint and arm kinematics were recorded at 1kHz (figure adapted from [17]) **b.** Visual display. Once the central target was reached, six peripheral targets appeared simultaneously on the screen. Each target was extinguished when reached by the hand feedback cursor. Once all the targets were successfully reached, the monkey was rewarded with a drop of water.

is the presence of a hierarchy in motor skill learning based on a composition of motor chunks [8,13,15,16].

Importantly, the many degrees of freedom of the motor apparatus allow to achieve a singular task objective through different patterns of joint coordination [4,28,34,37]. Classical motor learning models focus on optimizing kinematic criteria such as minimum jerk [21], variance control [19], energy cost [2], and achieving explicit goals through inverse models [37]. These principles of optimality readily combine with reinforcement learning models, enabling the interpretation of motor control as a trade-off between minimizing the costs of movement and achieving the task (e.g. reaching a target) [10,33].

In this study, our main objectives is to assess the capacity of a deep reinforcement learning agent, specifically Soft Actor Critic (SAC) [18], to reproduce animal's learning behavior, encompassing learning dynamics, motor pattern selection, and variability throughout training. We shed light on two fundamental issues in motor learning: (1) the relation between sequential motor patterns and the cumulative integration of reward and (2) the exploration of motor response variability across trials and individuals, and its significance in the optimization process.

2 Experiment

The motor task used in this study was designed to explore the neural basis of complex motor behavior [17] and can be related to the implementation of a (simplified) traveling salesman problem. It requires to reach multiple targets in any order, from a central starting position in a 2D visuo-motor space (Fig. 1). Since the targets are always presented at the same location, the task facilitates motor exploration as well as planning and anticipation, beyond the classical

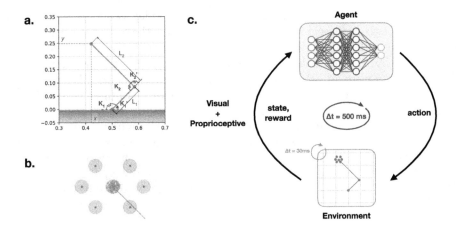

Fig. 2. Simulation setup **a.** Diagram of the 2-DOF mechanical model used in the simulations; (x, y) are the coordinates of the hand end position. **b.** The blue dots illustrate the distribution of the initial hand positions across trials. **c.** Schematic description of the closed-loop perception/action training setup, composed of an agent and an environment. The environment iterates the dynamics of the double-joint arm (with a 30 ms resolution), controls the disappearance of the 6 targets, and provides state observations and rewards at a 500 ms interval to the agent.

constraints imposed by single target reach tasks [5]. The hypothesis behind this task is that without any constraints on the order for target reach, the monkeys should develop integrated motor plans to optimize their hand trajectory along with the precision and speed of movement [7].

Two rhesus macaques (Macaca Mulatta) were trained to perform the task using positive reinforcement : Monkey E and Monkey J[1]. Behavioral data were recorded throughout the learning process [5,17]. Pacing of task execution and order of target extinction were fully determined by the monkey itself. The spatial location of the peripheral targets differed slightly between the two monkeys according to their ability to reach the farthest targets, resulting in "Setting 1" and "Setting 2" for monkey E and J respectively. During the entire learning period, Monkey E performed a total of 7,346 trials across 68 sessions lasting in total 23,872 s and Monkey J a total of 11,853 trials over 118 sessions for a total of 25,400 s.

3 Model

Mass-spring model. Our simulated environment uses a minimal mass-spring model [9], enabling the generation of a control in equilibrium point [21], directing

---

[1] Monkey E is a 6 years old female weighting 7 kg and Monkey J a 7 years old male of 9 kg. All procedures on macaque monkeys were approved by the local ethical committee (C2EA 71; authorization Apafis#13894-2018030217116218v4) and conformed to the European and French government regulations.

the arm towards a unique position in space. The model as shown in Fig 2a is composed of two rigid segments: the upper arm and the forearm, with a total of two degrees of freedom: the elbow joint and the shoulder joint. The dynamics is controlled by a set of ordinary differential equations. The weight and length of the segments were adjusted to match that of a macaque's arm. Each segment has two "springs" (anterior and posterior) for which the stiffness is modified to actuate a change in the angular position of the segment.

The state of the system is defined by the 2 joints' angular position θ_1, and θ_2, as well as their angular velocity $\dot{\theta}_1$ and $\dot{\theta}_2$. The length of the rigid segments is l_1 and l_2, and their mass is m_1 and m_2. The angular position of each joint $i = \{1, 2\}$ is controlled by two antagonist forces: $F_i(t)$ and $F'_i(t)$, generated by two opposite springs whose stiffness are $K_i(t)$ and $K'_i(t)$, plus a damping factor C. When the stiffness is constant, the dynamics converges to an equilibrium point in the planar space. The final movement equations are as follows:

$$\ddot{\theta}_1 = \frac{m_2 l_1 \dot{\theta}_1^2 \sin(\theta_2 - \theta_1)\cos(\theta_2 - \theta_1) + m_2 l_2 \dot{\theta}_2^2 \sin(\theta_2 - \theta_1)}{(m_1 + m_2)l_1 - m_2 l_1 \cos(\theta_2 - \theta_1)^2}$$
$$- C\dot{\theta}_1 - \theta_1 K_1(t) - (\theta_1 - 2\pi)K'_1(t) \quad (1)$$

$$\ddot{\theta}_2 = -\frac{l_1}{l_2}\frac{m_2 l_2 \dot{\theta}_2^2 \sin(\theta_2 - \theta_1)\cos(\theta_2 - \theta_1) + (m_1 + m_2)l_1 \dot{\theta}_1^2 \sin(\theta_2 - \theta_1)}{(m_1 + m_2)l_1 - m_2 l_1 \cos(\theta_2 - \theta_1)^2}$$
$$- C\dot{\theta}_2 - (\theta_2 - \theta_1 + \pi)K_2(t) - (\theta_2 - \theta_1 - \pi)K'_2(t) \quad (2)$$

The physical engine is a custom first-order ODE integrator[2], with a 30 ms temporal resolution. The practical values used in simulations are the following : l_1=12 cm, l_2=23 cm, m_1=200 g, m_2 = 500 g, C= 10 (damping factor). Two distinct settings (Settings 1 and 2) are employed to reproduce the target positions and coordinate values used in the monkey experiments for monkey E and J, respectively. Last, to increase the trial-to-trial variability, the hand initial position is normally distributed within a 1cm disc around the position of the central target, and a random initial angular velocity with a 0.1 rad s^{-1} standard deviation (see Fig 2b).

Equilibrium-point motor command. The upper arm and forearm are controlled by the agent through the stiffness of the springs (K_1, K'_1, K_2, K'_2). The motor command $\boldsymbol{a} = (a_1, a_2)$ defines a stiffness value for the four springs of the model, that is $K_1 = 12.5 + 2.5 \times a_1$, $K'_1 = 7.5 - 2.5 \times a_1$, $K_2 = 5 + 5 \times a_2$, and $K'_2 = 15 - 5 \times a_2$, with $(a_1, a_2) \in [-1, 1]^2$. Each command thus defines the value of two angular fixed points to which the arm should converge in the long run. In our setting, the motor command is updated every 500 ms, reflecting a putative response time interval. Each execution step of our environment interface thus lasts 500 ms of the simulated physical time. The environment also records the disposition of the visual targets in the 2D physical space. During an execution step, the targets disappear as soon as the extremity of the second segment (the "hand") reaches the target within a radius $R = 1$ cm, at any time during the

---

[2] adapted from https://matplotlib.org/2.0.2/examples/animation/double_pendulum_animated.html.

execution (the course of the hand is allowed to cross several targets during a single execution step). An example of the Kinarm hand surrounded by 6 targets is shown in Fig 2c.

Sensory input. At the end of an execution step, the environment interface returns a state vector s containing visual, proprioceptive, and motor information. The state of each target (lit or extinguished) is represented by a binary pair: (1,0) if the target is lit on, and (0,1) if the target is extinguished. For a total of $n = 6$ targets, the state vector is of size $2n+4$. It contains $2n$ binary values representing the state of the targets, plus the angular positions of the two joints, and a copy of the previous motor command ("efferent copy").

Controller. Reinforcement learning (RL) algorithms were originally developed to resolve non-linear/non-invertible control problems in robotics [36]. Over the past decade, significant advancements have been made in the physical environment simulators, and have been coupled with recent methods that use auto-differentiation-based learning and gradient backpropagation within deep neural networks. These have greatly improved the handling of continuous motor learning tasks [18,29,35], particularly self-paced motor tasks (e.g., learning of locomotor patterns [20]). Classically, RL agents learn to estimate the (cumulative) value of their actions from observing their final consequences, and retrospectively putting more weight on the actions for which to expect the highest total reward [3,36]. Formally, learning takes place under a closed-loop control setup, where the current action a_t elicits an observation s_{t+1}, and a reward r_{t+1}, which subsequently guides the selection of a new action a_{t+1}, and so forth. The Soft Actor-Critic (SAC) algorithm employed here encourages exploration by incorporating a maximum entropy criterion into its optimization formula [18]. It is noteworthy that, in the SAC procedure, the Q-value does not coincide with the Bellman optimum, but rather includes the action entropy term over the future pathway, i.e.

$$Q(s_t, a_t) \simeq r_{t+1} + \gamma(Q(s_{t+1}, a_{t+1}) - T \log \pi(a_{t+1}|s_{t+1}, \theta))$$

with γ being the discount factor, $T = 0.2$ the temperature, $Q(s, a)$ a parametric action value function (the critic), and $\pi(a|s, \theta)$ a parametric policy (or actor), viewed as a probability distribution over actions, with θ the policy parameters optimized by gradient descent (backpropagation).

In practice, the controller is structured as two multi-layered perceptrons. The first network, referred to as the "actor", takes the observation vector as input and computes the motor response. This network is coupled with a second neural network, known as the "critic", which assigns a value to the actions produced by the actor. Both networks are trained simultaneously. The algorithm version employed in this study is based on the implementation provided by OpenAI [1].

Reinforcement Signal. In the original task, monkeys are rewarded with a drop of water upon successfully extinguishing all six targets within the allocated time

(9 s). We thus introduce a positive extrinsic reward (+10) granted at the end of a trial, when the controller effectively accomplishes the task within the time limit. Unlike the monkeys, the model is completely new to the task and is therefore allowed a 70 s time limit. In addition to the final reward, we incorporate a penalty term (negative reward) that accounts for the costs associated with task execution. This penalty equates the metabolic cost associated with arm control in space. By disregarding dissipative effects from damping, this energy cost can be quantified as the mechanical energy spent controlling the arm. Specifically, the control of the springs stiffness constant induces a displacement that corresponds to a transfer of potential energy into kinetic energy. We describe this cost as the integrated kinetic energy of both segments of the arm over the time interval, defined as $dW = \frac{1}{2}(m_1 v_1^2 + m_2 v_2^2)dt$, where v_1 represents the velocity at the end of the first segment (arm), v_2 represents the velocity at the end of the second segment (forearm). To keep the penalty comparable across trials, it's scaled to give an average penalty of around 1 per second. This integrated cost is received at the end of each execution step and is scaled by a factor of 300 (so that the average penalty is $\simeq 1$ per second, depending on the velocity), i.e. $r_{\text{work}} = -300 \times \int_t^{t+0.5} dW$.

Learning. The learning process is organized into trials called "episodes" (a sequence of movements up to extinguishing all 6 targets), and structured into epochs, each encompassing 200 execution steps (equivalent to approximately 100 s of interaction with the environment). The overall training spans 400 epochs, corresponding to a total of 40,000 simulated seconds ($\sim 11h$) of interaction with the environment. Ten learning agents were trained for each of the different task-condition combination, differing by the seed used for the random draw of the neural networks initial parameters. The model's hyperparameters were optimized to both achieve the learning of the task and provide hand trajectories approaching those observed in primates. Specifically, the optimizer used for the gradient backpropagation is Adam [25], the learning rate is equal to 10^{-3}, and the hyperparameters are $\beta = 5$ (inverse temperature) and $\gamma = 0.95$ (time horizon discount factor).

4 Results

Figure 3-A presents the temporal evolution of the average cumulative reward (trial return) for the 10 agents of the 2 different settings. The values account for both the 6-targets reaching bonus (+10) and the cost associated with the condition (energy or duration spent). The trend of the curve reflects the degree to which the controllers improve over time (heading toward value 7–8 , with 10 being the upper limit).

For the purpose of analysis, the learning sessions were divided into three phases. The first phase is a warm-up phase characterized by rapid convergence, during which the agent learns to perform the task through random trial and error. The second phase is a consolidation phase (red shaded region), where it

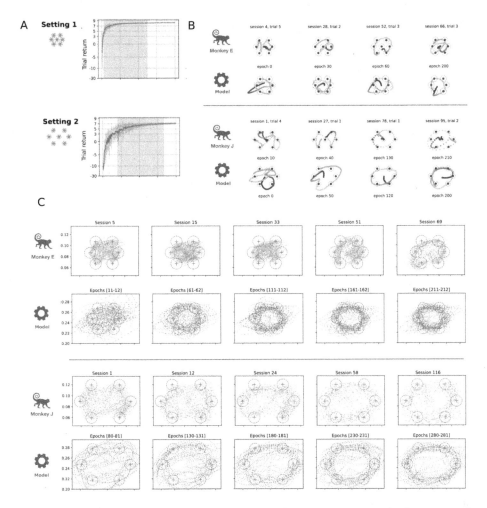

Fig. 3. Learning curves and comparisons. A. Average cumulative reward. (trial return) over the course of 400 epochs and across 10 agents, in the 2 settings. The blue shaded regions indicate the episode return standard deviation among the agents. The red shaded regions represent the consolidation phase of the training. A semi-logarithmic scale is used on the y-axis. Hyperparameters : $\beta = 5, \gamma = 0.95$. **B. Single trial comparisons.** Example of hand/end effector trajectories of monkeys and a chosen agent of each model condition, at different stages of learning, showing the motor control evolution along epochs. **C. Multi trials comparisons.** Superimposed end-effector trajectories recorded on 3000 time steps (picked over all the agents for the models on a given epoch interval) or 40 trials (picked in the same monkey session every 0.5 s), at different stages of learning.

optimizes the efficiency of its motor response. Finally, there is a plateau phase, corresponding to the stabilization of the motor response. For the comparison with the monkey's learning data, we consider the consolidation periods. These periods encompass the 240 epochs (24 000s) following the point at which the algorithm achieves the task in less than 6 movements on average (like monkeys at the start of their training). The initial learning period ("warm-up" period) is excluded from the comparison.

Figure 3-B provides illustrative examples of trajectories spanning the learning periods of both monkeys and models. These trajectories can be distinguished by their sequence (the order in which targets are reached) and by the overall trajectory shape. The monkeys start the experiment with a target-to-target reaching strategy. Over time, this behavior gradually evolves towards greater integration, wherein single movements are merged in a single trajectory targeting multiple consecutive targets. At the end of the experiment, a more reproducible behavior is obtained with a majority of circles being performed in a smooth manner. The models' trajectories closely resemble those of the monkeys. Notably, the progressive combination of short, discrete movements into larger ones (principle of binding motor "chunks" [32]), allowing to reach several targets at once, are reproduced in the models. Moreover, an intriguing and consistent behavior is observed both for the monkeys and in the model: when a target is missed, the primarily intended sequence is often *repeated entirely* with a slight inflexion (see last column of Fig. 3-B), enabling to reach previously missed target, rather than selecting a straight corrective trajectory towards the remaining target. This behavior, called a "double circle", reflects a certain lack of flexibility, as if the full motor sequence was "automatized". This strategy comes at a significant cost in terms of distance traveled. This behavior is characteristic of an "open-loop" motor response. This suggests that both monkeys and models seem to opt for a form of "open-loop" control, with limited contribution of the visual feedback. This is consistent with our initial assumption that motor learning may follow a form of model free learning, that tends to develop systematic action sequence patterns over time.

In a more comprehensive manner, Fig. 3-C provides a visual representation of the control law at various stages of learning, both in the monkey and in simulation. The control law can be characterized by its flow, which refers to the mapping that assigns a direction and movement speed to each point in space. This estimate is based on scattering the trajectories on 30 ms intervals, estimating the velocity in both directions, and representing the velocity with the direction of an arrow at each hand position, on 3000 observation points both in the monkeys and the models. By aligning 5 different flows at different stages of learning along a single row, we attain a comprehensive view of the behavioral changes attained throughout the learning process. This confirms the general trend observed in previous examples, that is to form and *stabilize movement sequences* that are then reproduced, with some variation, from trial to trial, reflecting the presence of an optimization process at play, resulting in a convergence towards a circular trajectory that approaches the mathematical optimum for this task. This convergence indicates that the energy cost adequately reflects the constraints of the task. Despite its simplicity, our model adequately adapts to the mechanical constraints posed by the Kinarm apparatus.

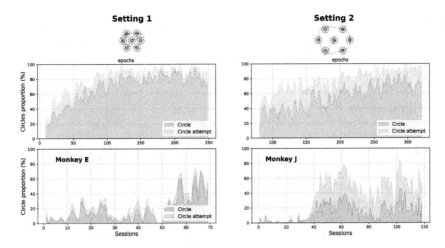

Fig. 4. Circle proportion along sessions. Proportion (percentile) of circular hand trajectories observed across sessions (monkey) and corresponding epochs (model). Top: examples of "circular" trajectories in both settings. Plain color : circles; shaded color : circle attempts. Left : Setting 1/Monkey E. Right : Setting 2/Monkey J.

However, beside the slight differences in variability between the models and the monkeys, an important difference was observed in the number of hand motions developed during learning. We measured the proportion of circular trajectories realized across trials during learning, both for the monkeys and the model (see Fig. 4). The circularity of trajectories was measured by the difference in indices between consecutively visited targets within a single trial. If the sequence of these differences (modulo 6) contains only 1's, the trajectory corresponds to a circular visit in counter clockwise order. Conversely, if the sequence of these differences (modulo 6) contains only 5's, the trajectory corresponds to a circular visit in clockwise order. As shown at the bottom of Fig. 4, both monkeys manifest a clear tendency to increase the number of circle trajectories during their learning process, but this trend is non-monotonic and characterized by abrupt transitions. Notably, in the initial phase of the training, Monkey J predominantly performed trajectories following an infinity shape (∞), shown in the second column of Fig. 3-B. This high diversity of behaviors is observed in both monkeys also at later stages of learning. Like the monkeys, the models provide an increase in the proportion of circular trajectories, with a vast majority of trajectories performed in an counter-clockwise fashion. However, for the models, this increase is constant and reaches close to maximal values, with little variability towards the end of the learning process. This highlights an important difference between the monkeys and the models. The higher movement variability in monkeys might have mixed origins: 1) muscle tiredness, boredom and decreased interest for rewards are only possible in monkeys, and might lead to a desire of varying motion, and 2) our learning algorithm, despite owning an incentive to promote variability, tends to converge to a unique solution with growing accuracy, leading to more stereotyped movements towards the end of the training. While the models capture the main aspects of monkey's learning, they do not entirely render the complexity of a macaque's cognition, and lack internal body states to drive behavior.

5 Discussion

Our study sheds light on two aspects often treated separately in motor learning: the learning of motor sequences to produce coordinated behavior in time or space [16,23,24,26], and the consideration of the long-term reward in the plasticity process [6,14]. In particular, our free-moving self-paced reaching task reveals the presence of a significant variability in both the execution of motor trajectories (motor variability) and in the *choice* of hand motor sequences (hand trajectory variations), that is absent in more constrained setups. This mixed variability indicates the existence of multiple levels of control in the learning process.

- The most elementary level (combination of chunks) appears to be subject to a conventional "model-free" optimization process, solving a temporal-credit assignment problem at the motor command level, and considering the energetic cost of the command, resembling the principles implemented in most deep reinforcement learning algorithms for continuous control [18,29,35]. The relatively inflexible nature of motor sequences observed in monkeys suggests a preference for a "memory-based" sequence of ballistic commands, with minimal visual feedback.
- The more integrated level of control (choice of a motor pattern among several possible patterns), which appears to be present in the monkeys only, would require the establishment of a hierarchical learning model (or meta-learning [22]), where different policies would compete. It is worth noting that the implementation of a meta-learning principle does not necessarily involve the presence of a forward model and/or the definition of goals.

From the machine learning perspective, our study confirms the relevance of considering kinematic features (and energetic penalties) in refining motor learning. However, it also highlights the significantly higher flexibility of motor learning in animals compared to mere value iteration in the Bellman equation. Thus, to develop more powerful learning algorithms, it is necessary to take into account the observed variability both at the level of motor execution and at the level of the selection of motor sequences to perform. Without the need to learn an inverse model, our observations rather suggest the implementation of flexible hierarchical mechanism that makes a choice among several motor programs to solve a task, and promotes adaptability to changing environmental conditions. These mechanisms should be able to account for the variability seen in motor learning, enabling the development of more versatile and robust algorithms for motor skill acquisition.

Acknowledgments. The project was funded by the International Research Project "Vision for Action" (CNRS/FZ Juelich), the Deutsche Forschungsgemeinschaft (GR

1753/4-2, DE 2175/2-1) Priority Program (SPP 1665); the Helmholtz Association through the Helmholtz Portfolio Theme Supercomputing and Modeling for the Human Brain; the European Union's Horizon 2020 Framework Programme for Research and Innovation (No. 720270, 785907). We are grateful to our colleagues, Frédéric Barthélémy, Shrabasti Jana, Alexander Kleinjohann for their scientific and technical support in carrying out this work. We would also like to thank Luc Renaud and Emilie Rapha for their technical support.

Disclosure of Interests. The authors have no competing interests to declare that are relevant to the content of this article.

References

1. Achiam, J.: Spinning up in deep reinforcement learning (2018). https://spinningup.openai.com/en/latest/
2. Alexander, R.M.: A minimum energy cost hypothesis for human arm trajectories. Biol. Cybern. **76**(2), 97–105 (1997)
3. Bellman, R.: Dynamic programming. Science **153**(3731), 34–37 (1966)
4. Bernstein, N.: he Co-Ordination and Regulation of Movements. Pergamon Press, Oxford (1967)
5. Condro, L.M., Barthélemy, F.V., Jana, S., De Haan, M., Riehle, A., Brochier, T.: Behavioral correlates of long-term motor skill learning in macaque monkeys. In: Federation of European Neuroscience Societies Forum (2022)
6. Daw, N.D., Tobler, P.N.: Value learning through reinforcement: the basics of dopamine and reinforcement learning. In: Neuroeconomics, pp. 283–298. Elsevier (2014)
7. Diamond, J.S., Wolpert, D.M., Flanagan, J.R.: Rapid target foraging with reach or gaze: the hand looks further ahead than the eye. PLoS Comput. Biol. **13**(7), e1005504 (2017)
8. Diedrichsen, J., Kornysheva, K.: Motor skill learning between selection and execution. Trends Cogn. Sci. **19**(4), 227–233 (2015)
9. Feldman, A.G.: Functional tuning of the nervous system with control of movement or maintenance of steady posture. iii. mechanographic analysis of the execution by man of the simplest motor tasks. Biofizika **11**, 766–775 (1966)
10. Fischer, F., Bachinski, M., Klar, M., Fleig, A., Müller, J.: Reinforcement learning control of a biomechanical model of the upper extremity. Sci. Rep. **11**(1), 14445 (2021)
11. Fitts, P.M.: The information capacity of the human motor system in controlling the amplitude of movement. J. Exp. Psychol. **47**(6), 381 (1954)
12. Fitts, P.M., Posner, M.I.: Human performance (1967)
13. Flash, T., Hochner, B.: Motor primitives in vertebrates and invertebrates. Curr. Opin. Neurobiol. **15**(6), 660–666 (2005)
14. Fu, W.T., Anderson, J.R.: Solving the credit assignment problem: explicit and implicit learning of action sequences with probabilistic outcomes. Psychol. Res. **72**, 321–330 (2008)
15. Gobet, F., Lane, P.C., Croker, S., Cheng, P.C., Jones, G., Oliver, I., Pine, J.M.: Chunking mechanisms in human learning. Trends Cogn. Sci. **5**(6), 236–243 (2001)
16. Graybiel, A.M.: The basal ganglia and chunking of action repertoires. Neurobiol. Learn. Mem. **70**(1–2), 119–136 (1998)

17. de Haan, M.J., Brochier, T., Grün, S., Riehle, A., Barthélemy, F.V.: Real-time visuomotor behavior and electrophysiology recording setup for use with humans and monkeys. J. Neurophysiol. **120**(2), 539–552 (2018)
18. Haarnoja, T., et al.: Soft actor-critic algorithms and applications. arXiv preprint arXiv:1812.05905 (2018)
19. Harris, C.M., Wolpert, D.M.: Signal-dependent noise determines motor planning. Nature **394**(6695), 780–784 (1998)
20. Heess, N., et al.: Emergence of locomotion behaviours in rich environments. arXiv preprint arXiv:1707.02286 (2017)
21. Hogan, N., Flash, T.: Moving gracefully: quantitative theories of motor coordination. Trends Neurosci. **10**(4), 170–174 (1987)
22. Jabri, A., Hsu, K., Gupta, A., Eysenbach, B., Levine, S., Finn, C.: Unsupervised curricula for visual meta-reinforcement learning. In: Advances in Neural Information Processing Systems, vol. 32 (2019)
23. Keele, S.W.: Movement control in skilled motor performance. Psychol. Bull. **70**(6p1), 387 (1968)
24. Keele, S.W., Ivry, R.I.: Modular analysis of timing in motor skill. In: Psychology of Learning and motivation, vol. 21, pp. 183–228. Elsevier (1988)
25. Kingma, D.P., Ba, J.: Adam: a method for stochastic optimization. arXiv preprint arXiv:1412.6980 (2014)
26. Koch, I., Hoffmann, J.: Patterns, chunks, and hierarchies in serial reaction-time tasks. Psychol. Res. **63**, 22–35 (2000)
27. Land, M., Mennie, N., Rusted, J.: The roles of vision and eye movements in the control of activities of daily living. Perception **28**(11), 1311–1328 (1999)
28. Latash, M.L., Scholz, J.P., Schöner, G.: Toward a new theory of motor synergies. Mot. Control **11**(3), 276–308 (2007)
29. Lillicrap, T.P., et al.: Continuous control with deep reinforcement learning. arXiv preprint arXiv:1509.02971 (2015)
30. Magill, R., Anderson, D.I.: Motor learning and control. McGraw-Hill Publishing, New York (2010)
31. Milner, D., Goodale, M.: The visual brain in action, vol. 27. OUP Oxford (2006)
32. Ramkumar, P., Acuna, D., Berniker, M., Grafton, S., Turner, R., Kording, K.: Chunking as the result of an efficiency computation trade-off. Nat Commun. **7**, 12176 (2016)
33. Rigoux, L., Guigon, E.: A model of reward-and effort-based optimal decision making and motor control (2012)
34. Schöner, G., Scholz, J.P.: Analyzing variance in multi-degree-of-freedom movements: uncovering structure versus extracting correlations. Mot. Control **11**(3), 259–275 (2007)
35. Schulman, J., Wolski, F., Dhariwal, P., Radford, A., Klimov, O.: Proximal policy optimization algorithms. arXiv preprint arXiv:1707.06347 (2017)
36. Sutton, R.S., Barto, A.G.: Reinforcement Learning: An Introduction. MIT Press, Cambridge (2018)
37. Todorov, E., Jordan, M.I.: Optimal feedback control as a theory of motor coordination. Nat. Neurosci. **5**(11), 1226–1235 (2002)

Memory-Feedback Controllers for Lifelong Sensorimotor Learning in Humanoid Robots

Magdalena Yordanova[✉] and Verena V. Hafner

Adaptive Systems, Department of Computer Science,
Humboldt University of Berlin, Berlin, Germany
{magdalena.yordanova,hafner}@informatik.hu-berlin.de

Abstract. The use of humanoid robots within the field of neuroscience has gained substantial interest in recent years, specifically as a means to implement and assess biological concepts. This conceptual paper addresses the vital challenge of uncertainty in the context of lifelong bio-inspired sensorimotor learning. Inspired by insights from developmental and neuroscientific studies, we examine the role of self-learning, exploration, and coordination dynamics. Building on principles derived from neural mechanisms and the concept of brain plasticity, we represent a robot's internal sensorimotor model and its synaptic-like reorganizational changes through dynamic self-organizing maps.

We propose a concept that builds on that and distinguishes itself by employing visuo-arm coordination not as an end goal with potential emergent behaviors, but as a feedback controller, emphasizing the integration of an explainable memory-embedded model for continuous sensorimotor self-learning. We illustrate the framework's potential in dynamic scenarios such as tool use, where enhanced adaptability and fast task resumption after motor perturbations or recurring tool changes provide significant benefits. Verifying a memory entry is significantly quicker than updating the visuo-motor model. Through the concept of a memory-embedded controller, we establish the groundwork for effective and lifelong learning of sensorimotor skills in humanoid robots.

Keywords: Sensorimotor coordination · Motor babbling · Self-monitoring · Self-exploration · Self-organizing maps · Continuous learning · Feedback control · Plasticity · Memory · Perturbation · Tool use · Humanoid robots

1 Introduction

In the field of cognitive robotics, the presence of uncertainty is significant. An adaptive system capable of learning and adapting to changes is essential. Humanoid

---

Supported by the German Research Foundation DFG under Germany's Excellence Strategy - EXC-2002/1 "Science of Intelligence" - Project number 390523135.

robots are social agents that engage in interactions not solely with external entities such as other agents, humans, or the environment but fundamentally with their own physical bodies. Uncertainty may arise not just externally but also from within the robot itself. Hence, a lifelong self-learning and self-monitoring process becomes essential to address this internal unpredictability.

As human beings, we are curious creatures [4,5], and already in infancy we engage in spontaneous exploration. These interactions exhibit coordination patterns, aligning with the principles of coordination dynamics [27]. Sloan et al. [26] explore how agency develops in infants through a coordination dynamics analysis using the Mobile Conjugate Reinforcement paradigm. The study reveals that conscious agency in infants can emerge as a self-organizing process from a dynamic system that includes both the infant and their surroundings.

Coordination approaches have also been implemented in humanoid robots, exploring the emergent behaviors resulting from sensorimotor coordination. Kuniyoshi et al. [11] discovered that through self-exploratory visuo-motor learning, the robot identifies and imitates human arm movements without prior human movement interpretation capabilities. Kajic et al. [9] and Hafner et al. [6] show how a humanoid robot acquires pointing skills through sensorimotor learning, specifically by coordinating hand-eye movements when reaching for an object beyond its immediate reach.

Luo et al. [13] applied a feedback control mechanism within their relative-location-based approximating strategy for reaching in humanoid robots, including two stages: an initial reaching phase followed by iterative adjustments according to the visual feedback of hand and target position. The reaching trajectory was subsequently refined through inner rehearsal to achieve smoother results [28]. Our approach lies in identifying distortions using a feedback controller, pausing the task for self-calibration, and then resuming. This ensures refinement occurs only when necessary, reducing unnecessary adjustments for each action.

Our research draws inspiration from brain plasticity, which undergoes continuous re-organization influenced by factors like hand activity. Surgical hand interventions, like those addressing median nerve injuries through nerve repair, lead to synaptic changes in the brain cortex. Sensory re-education is crucial after procedures, as the brain undergoes a re-learning process to enhance functional sensibility, involving touching and exploring items of increasing difficulty with eyes open or closed, using an alternative sense like vision to compensate for deficient sensation [12].

The human brain continues to be a mystery, and in a world saturated with artificial neural networks, the aspect of explainability also remains deficient. Transparency is acknowledged as an essential characteristic for comprehending and anticipating the behavior of robots, as demonstrated in human-robot interaction studies in the context of both mobile [29] and humanoid robots [16]. Providing transparency is a non-trivial task. Minimizing the complexity of the learning algorithm allows for the implementation of coordination dynamics using single-layer feed-forward neural networks, e.g. self-organizing maps.

Created in the 80 s, self-organizing maps (SOMs) [10] have the ability to intuitively represent intricate datasets by mapping high-dimensional data onto a lower-dimensional grid. Schillaci et al. [22] investigated the use of dynamic SOMs [18], allowing online learning and adaptation, focusing on the formation and connection via biologically inspired Hebbian links [1,15] of sensory (vision) and motor maps in a humanoid robot. The acquisition of data follows an unsupervised approach, achieved through a motor, goal, or hybrid babbling [20,25].

Apart from vision, in the context of humanoid robots, SOMs can also serve to represent touch or proprioception [3,7,8]. In [14] the researchers integrated MRF-SOMs for both purposes, exploring learning mechanisms through self-touch. The authors in [23] explore the essential role of sensorimotor control and learning in cognitive development for humans and animals. They suggest that the ability to mentally simulate action-perception loops, based on internal body representations [21,24], might be a key mechanism for higher cognitive functions.

In this work, we employed a model designed for lifelong sensorimotor self-learning having sensorimotor coordination not as an end goal [22] but as a feedback controller during the execution of particular tasks, such as pointing or reaching. For the sake of simplicity and as a demonstration of the concept, we exclusively relied on visual feedback. The model is capable of recognizing and addressing uncertainty while also retaining memory of states to facilitate quicker adaptation, like in scenarios involving motor perturbation or tool use. Dynamic self-organizing maps are employed to build the internal sensorimotor model of the robot needed for the coordination. We explore the range of tasks sensitive to distortions that can be effectively performed with the memory-embedded controller, without narrowing down and training the model for a specific task [9].

This paper is structured as follows. Section 2 serves as the primary segment, presenting the motivation, methodology along with its constraints, and assessment of the proposed method's design. Section 2.1 introduces the bio-inspired visuo-motor coordination learning algorithm proposed by Schillaci et al. [22]. Section 2.2 addresses the main approach of using and adapting the offline learned visio-motor coordination as a feedback controller for online continuous learning in the humanoid robot Nao. In Sect. 2.3, a potential uncertainty-responsive tool use scenario is outlined and subsequently examined in Sect. 2.4. The paper concludes with a discussion on future work and impact in Sect. 3.

2 Design

2.1 Related Work: Visuo-Motor Coordination

In this manuscript, we adopted the visuo-motor coordination learning introduced by Schillaci et al. [22]. The approach consists of two dynamic self-organizing maps (DSOMs): one for the head (housing vision - the robot's camera) and another for the arm (representing the selected motor group). Additionally, there are Hebbian links connecting neurons between both DSOMs. Each neuron is associated with a weight that corresponds to the joint angles of the respective body part. The neurons' weights in the arm DSOM include shoulder pitch and roll, and elbow yaw and roll, while in the head DSOM: head yaw and pitch.

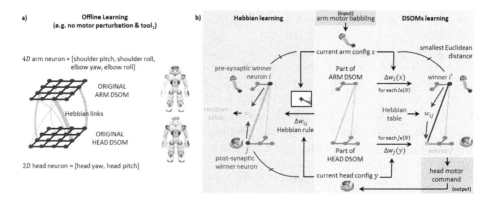

Fig. 1. Overview of the presented model - a) utilization of dynamic self-organizing maps in the offline learning phase, b) visualization of both learning algorithms.

An illustration of the visuo-motor coordination process is shown in Fig. 1a, where the arm and head neurons are arranged in two separate 2D lattices and fully interconnected by Hebbian connections, with link thickness representing their strength. The spatial arrangement was kept two-dimensional for simplicity and better comprehension. The 2D lattice organizes neurons to reflect their functional relationships during learning and adaptation rather than their weight dimensions. A notable characteristic of this coordination approach is its resilience to distortions, such as motor perturbations or tool changes, enabling real-time adaptation and ensuring effective coordination.

The learning process comprises two parallel activities, producing updated DSOMs and Hebbian table, as well as head motor commands (refer to Fig. 1b): training both DSOMs based on random arm movements (motor babbling) and a Hebbian learning paradigm reinforcing connections between the DSOMs, recording them in a Hebbian table. The self-organizing structure of the DSOMs is facilitated by a learning function, responsible for adjusting neuron weights according to input patterns (current random arm configurations), and a neighborhood function, which specifies the impact of neighboring neurons in this updating process. The Hebbian process is inspired by biological principles, reflecting the concept that neurons that fire together wire together, in the context of detecting the end-effector in the center of the field of view. Simulating synapse behavior, the inter-neuronal links are strengthened when the robot's end-effector is visible.

2.2 Proposed Approach: Feedback Controller

We leverage visuo-motor coordination as a feedback controller in our approach. The coordination, detailed in the preceding section, is initially trained offline and subsequently employed for online task performance evaluation. Figure 2 illustrates the steps of our methodology. We initiate the process by loading the offline pre-trained model and restructuring the two dynamic self-organizing maps (one

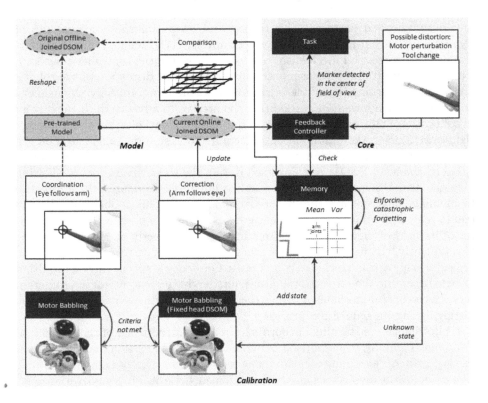

Fig. 2. Flow diagram of the outlined methodology. The layout is categorized into three primary sections: the creation and reshaping of the visuo-motor coordination model, the core program logic, and the calibration ensuring efficient adaptivity.

for the arm and one for the head) along with the Hebbian links into a unified DSOM. This transformation involves extending the neuron weights of the head DSOM by the arm configurations associated with their strongest connections. We refer to this patched dynamic self-organizing map as the original joined DSOM, serving as our template. It is worth noting that reverting the reshaping process to two distinct DSOMs will not result in a loss-free reconstruction. After creating a duplicate of this combined representation, denoted as the current joined DSOM, we proceed with the primary task. While we will delve into its constraints later, let's focus on the task at hand for now: detecting the color the robot is pointing at, requiring the robot to keep the green end-effector on its right arm or tool in the middle of its field of view. Initially, using the offline-trained model, the system functions as expected. However, suppose the system unexpectedly alters its behavior due to factors like motor perturbation or tool changes, disrupting the pointing task's success. In that case, the system detects this error and triggers a correction mechanism.

In the brain, neural circuits encode information about various states or experiences, and when a new state arises, these circuits assess whether it aligns with previously encoded patterns. This involves retrieving information from neural representations that indicate familiarity or novelty [17]. Similarly, in a simplified manner, our correction mechanism begins by reviewing the controller's memory, which stores vectors representing the online-calculated distortion of the robot's arm joints concerning the offline-learned original mapping. Initially, the memory is devoid of any information, prompting the process to advance to the next step: re-calibration. This entails initiating a motor babbling behavior of the robot's right arm, akin to the offline phase, to acquire coordination skills based on the current DSOMs and Hebbian links, which in this case are still the original ones. Notably, the head DSOM remains fixed during the re-learning process, allowing only the arm DSOM and the synaptic-like links to be updated. It is essential to highlight that despite the motor babbling in the offline phase, during the online re-organization, the robot learns to position its end-effector at the center of its field of view instead of directing its gaze directly at the end-effector. Consequently, the robot adjusts its arm position rather than its head position, typically employed in tracking objects tasks. Our focus is specifically on acquiring knowledge about the errors in the arm joints, and the visuo-motor coordination serves as a control mechanism rather than being the ultimate objective, thereby distinguishing its acquisition process.

Following the re-learning of coordination and restructuring of the model into a combined DSOM, a comparison between this updated version and the original one is possible. This online use of dynamic self-organizing maps is shown in Fig. 3. The comparison involves calculating an element-wise mean μ and variance σ^2 over the differences in their arm joint configurations, resulting in a 4×2 matrix representing the calculated distortion of the robot's 4-joined arm, which is then stored in the memory. Subsequently, the task execution resumes, and if another distortion occurs, the system checks the memory by adding each 4D mean vector to the original arm DSOM to determine if the task can be resumed. This facilitates quicker task resumption following recurring changes in the system.

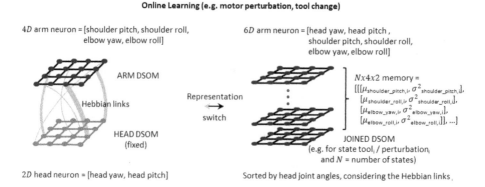

Fig. 3. Overview of the presented model - utilization of dynamic self-organizing maps in the online learning phase, incorporating a memory element to enhance adaptation.

The two primary interconnected tasks are determining the distortion and establishing criteria for when the coordination is considered learned or sufficiently suitable for the task. Criteria such as the frequency of end-effector detection within a defined time window or its average proximity to the center of the field of view can be employed. Given that the perturbation is calculated as the mean of the differences among the arm joint configurations, we also compute the variance and opt to utilize it as a threshold, serving as a criterion.

Since the memory integration of our method is a focal point, we must carefully consider how we manage the memory retention process for lifelong learning. Evolution favors organisms that use their resources efficiently, retaining only the most relevant information, and facilitating rapid adaptation and survival in dynamic settings. Neuroscientific studies show that the human brain undergoes synaptic pruning, eliminating less-used neural connections to strengthen more frequently used ones [19]. Enforcing catastrophic forgetting in robots aims to mirror this process, promoting efficient neural plasticity, as well as ensuring optimal memory management and processing efficiency. Our straightforward approach enables robots to prioritize new experiences over outdated ones in a first-in-first-out fashion with a fixed memory size.

2.3 Experimental Setup

We employ visuo-motor coordination as a mechanism to assess task accuracy. Specifically, we have selected the task of determining the color at which the robot is pointing, referring to the color directly behind the end-effector in relation to the orientation of the robot's head. This straightforward scenario is inspired by the developmental behavior observed in children exploring the world, as they draw their parents' attention [2] by e.g. pointing towards different objects, seeking to understand what those objects are, respectively what colors they have. We selected the simple task of generating random arm angles for pointing, akin to arm motor babbling, to avoid introducing additional mechanisms like object detection that might divert attention from the main point.

We utilize the offline learned and continuously online updated sensorimotor coordination model as a feedback controller. The assessment of coordination requires continuous monitoring through the robot's camera, particularly in uncertain or hazardous situations. In our case, the control mechanism remains responsive to alterations such as motor perturbation or tool changes. By placing a marker on the body part or tool to be evaluated, it can effectively undergo re-calibration, enabling a comprehensive performance check.

Figure 4 shows the humanoid robot Aldebaran Nao in a simulated colored room within the Cyberbotics Webots simulator, performing the task of recognizing the color it is pointing at. The available tools include a short tool, a long tool, an L-shaped tool, or no tool, where the robot uses its hand. A green marker is attached to the tip of each tool and in the absence of a tool encapsulates the entire hand for optimal visibility from each arm rotation. Throughout tool change and motor perturbation, involving shifting the arm motor command vector, the robot ensures that the marker stays positioned at the center of its

Fig. 4. Detection of the pointed-at color, utilizing four tool variations (short tool, long tool, L-shaped tool, or no tool - solely the hand), while maintaining the green marker on the arm or tool positioned at the center of the field of view and automatically adapting upon motor perturbation or tool change (instant tool swap in the simulation) using visuo-arm coordination through self-learning and memory. (Color figure online)

visual field. The color detection task starts with the known tool and undergoes unexpected changes. In this example scenario, the visuo-arm coordination is pre-trained offline exclusively with the short tool, leaving the robot unaware of the other tool options. In the event of a misalignment between the marker and the camera view, an automatic re-calibration is initiated, involving a memory check and, if necessary, engaging in random motor babbling until the task can resume properly.

The rationale behind the choice of tool use is to treat tools as extensions of the robot's hand, disregarding the tool's weight for now as negligible, allowing for examination of potential body modifications that may arise. Drawing a parallel to real-world scenarios like rehabilitation, this concept draws inspiration from the analogy to human joints, which undergo changes over time due to growth or accidents, sometimes requiring replacement with artificial joints that may differ in shape. Robots, similarly, may encounter challenges on difficult terrains, and the ability to continue performing their tasks, even e.g. with a missing or replaced joint at the last minute, without calibration due to time constraints, becomes essential for their functionality and adaptability.

An illustrative instance highlighting the efficacy of incorporating a memory component in our approach is depicted in Fig. 5. For visualization clarity, each state in the memory is identified not by the calculated angles' disparities from the original coordination model, but rather by a descriptive label, such as the name of the tool in use. However, in reality, the system can not infer the state label from the stored data. The depicted scenario involves the robot initially employing a familiar $tool_{short}$, successfully completing the task without issues, followed by an unforeseen motor disturbance that results in the use of a modified version of the tool, referred to as $tool_{short}'$. This prompts a re-calibration process through motor babbling. Upon learning and adapting the visuo-motor coordination, it resumes the task, saving the established differences between the updated and original joined DSOM in its memory. Subsequently, while the motor perturbation persists, it encounters another tool, $tool_{long}'$, not present in its memory, triggering another round of memory check and re-learning, the duration of which can differ. After returning to the main task, in a subsequent tool change back to $tool_{short}'$, the robot recognizes its presence in the memory, promptly adjusting

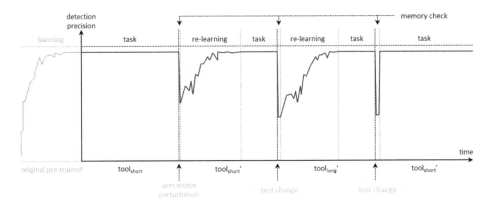

Fig. 5. Example pipeline of the decision precision in time. The conceptual non-data-driven sketch visually portrays various stages of the program, beginning with offline model training, progressing to task execution, addressing unexpected changes like motor perturbation and tool change, incorporating different re-calibration mechanisms, and, ultimately, ensuring the task's successful continuation.

its coordination without the need for extensive re-calibration through babbling, thus smoothly continuing the task.

2.4 Results and Limitations

The performance of the framework depends on the whole re-calibration process. However, the novelty of this approach lies in the utilization of a memory-embedded controller rather than the process of babbling and coordination model updating. It takes roughly two seconds to check a memory entry, whereas updating the coordination model through babbling again takes minutes. Solely the moving window over which the authors of [22] are computing the coordination performance is 2 min and they report a positive learning trade over a period of approximately 250 min but do not outline specific criteria for concluding the re-calibration process. We constrained the memory size to a maximum of 4 states, allowing for up to 4 tools or fewer depending on unexpected motor perturbations. Due to the manageable memory size, the time taken for memory checks also remains within reasonable limits and is much shorter than re-learning from scratch.

To compute the mean difference in arm configurations, it is necessary, visually speaking, to align the joined DSOMs (as shown on the right side of Fig. 3) while maintaining the sequence of head configurations. The creation of a joined DSOM involves adding the arm weight with the highest connection strength to the head weight. However, multiple connections may share the same strength, meaning that for the same head position, there could be several arm configurations, all leading to the end-effector being approximately at the center of the field of view. When reshaping both the original and updated DSOMs, a common method of selection is required, as a random choice would not yield accurate results. Assuming that both the pre-trained coordination and the online updated one are sufficient for the robot to keep the end-effector within a small error range around the center of the camera, one possible approach is to consistently select the arm configuration that has the smallest Euclidean distance from the one of the current joined DSOM. This method seeks to identify the minimal required adjustment in the arm configuration for each neuron pair to achieve re-calibration, thereby avoiding drastic changes caused by ambiguity. If none of the candidate configurations are sufficiently close according to a set threshold, the calculated correction should be excluded from the calculation of the mean and variance for the current state. In cases where ambiguity persists, utilizing the variance as a criterion, supported by a satisfactory amount of collected data, should sufficiently address any uncertainties. Maintaining minimal variance values might still result in a prolonged re-calibration process, necessitating additional measures to ensure variance reduction and avoid stagnation. However, once the re-learning is completed, it could, for instance, enable much swifter tool changes.

We employ sensorimotor coordination as a feedback controller, with the camera serving as the sensor, providing 2D images. Having a 2D controller limits our capability to perform tasks that do not require depth information, such as random pointing behavior or reaching without considering the depth with respect to the robot's orientation towards the end-effector. We incorporate both motor perturbation and tool change as sources of system uncertainty. Our focus is not on differentiating between tools in 3D space but rather on learning the motor arm translation errors associated with these changes. This implies that the effectiveness of the task relies significantly on well-learned coordination and optimal data distribution within the robot's accessible space. An insufficient coordination model could lead to errors in the model being mistakenly attributed to simulated perturbations.

3 Discussion on Future Development and Conclusion

We aimed to begin with a minimal and straightforward approach, allowing a subsequent study to build on this foundation and thoroughly address all the mentioned limitations. A forthcoming work could examine the use of alternative sensors for learning sensorimotor coordination, possibly 3D sensors to overcome the task limitations inherent in using a 2D camera image. Additionally, it could delve into examining neurogenesis within the coordination model, removing the necessity of predefining the lattice of dynamic self-organizing maps. More complex models could be tested by increasing the depth of the neural networks, as well as different spike-timing-dependent plasticity rules could be employed. Furthermore, scaling the problem and incorporating more advanced selective catastrophic forgetting within the controller's memory component is worth more extensive exploration. Employing adaptive learning algorithms that dynamically manage memory based on task relevance and performance feedback can improve the robot's continuous learning and adaptability while preserving essential knowledge. In addition, future research could investigate knowledge transfer between various humanoid robots with differing body structures. For instance, it would be valuable to examine if the Pepper robot, following initial calibration, can leverage the memory of the Nao robot to work directly with the tools previously encountered by Nao. Throughout these endeavors, a crucial consideration is to prioritize and incorporate principles of explainability and transparency whenever possible. Finally, conducting experiments in real-world scenarios may reveal new limitations and provide fresh insights.

We introduced an innovative approach aimed at tackling the challenges associated with lifelong sensorimotor learning in humanoid robots, particularly when unforeseen alterations occur in the robot's physical structure. This methodology draws inspiration from biological mechanisms, incorporating principles of self-learning, exploration, and coordination dynamics. Central to our approach is the utilization of dynamic self-organizing maps, subject to synaptic-like reorganizational changes and interconnected through Hebbian links. This interconnected structure is harnessed to learn visuo-arm coordination, subsequently employed as a feedback controller with an inherent reliance on memory embeddedness. The efficacy and adaptability of our proposed framework are highlighted in dynamic and uncertain scenarios, particularly those involving motor perturbation or tool changes. We contend that the utilization of self-learning as a bio-inspired memory-embedded controller holds significant promise within the realm of transparent open-ended learning for humanoid robots.

References

1. Attneave, F.B.M., Hebb, D.O.: The Organization of behavior; a Neuropsychological Theory. In: The American Journal of Psychology, vol. 63, p. 633 (1950)
2. Butterworth, G., C., E.: Towards a mechanism of joint visual attention in human infancy. Int. J. Behav. Dev. **3**(3), 253–272 (1980)
3. Gama, F., Hoffmann, M.: The homunculus for proprioception: Toward learning the representation of a humanoid robot's joint space using self-organizing maps. In: IEEE ICDL-EpiRob, pp. 113–114, September 2019
4. Gottlieb, J., Oudeyer, P.Y., Lopes, M., Baranes, A.: Information-seeking, curiosity, and attention: computational and neural mechanisms. Trends Cogn. Sci. **17**(11), 585–593 (2013)
5. Gureckis, T.M., Markant, D.B.: Self-directed learning: a cognitive and computational perspective. Perspect. Psychol. Sci. **7**(5), 464–481 (2012)
6. Hafner, V., Schillaci, G.: From field of view to field of reach-could pointing emerge from the development of grasping? Frontiers in Computational Neuroscience (2011)
7. Hoffmann, M., B., N.: The encoding of proprioceptive inputs in the brain: Knowns and unknowns from a robotic perspective. In: Vavrecka, M., Becev, O., Hoffmann, M., Stepanova, K. (eds.) Cognition and Artificial Life XVI, pp. 55–66, July 2016
8. Hoffmann, M., Straka, Z., Farkaš, I., Vavrečka, M., Metta, G.: Robotic Homunculus: learning of artificial skin representation in a humanoid robot motivated by primary somatosensory cortex. IEEE Trans. Cogn. Dev. Syst. **10**(2), 163–176 (2018)
9. Kajic, I., Schillaci, G., Bodiroža, S., Hafner, V.: A biologically inspired model for coding sensorimotor experience leading to the development of pointing behaviour

in a humanoid robot. In: Proceedings of the Workshop "HRI: a bridge between Robotics and Neuroscience", 9th ACM/IEEE HRI, March 2014
10. Kohonen, T.: Self-organized formation of topologically correct feature maps. Biol. Cybern. **43**(1), 59–69 (1982)
11. Kuniyoshi, Y., Yorozu, Y., Inaba, M., Inoue, H.: From visuo-motor self learning to early imitation-a neural architecture for humanoid learning. In: IEEE International Conference on Robotics and Automation, vol. 3, pp. 3132–3139 (2003)
12. Lundborg, G.: Brain plasticity and hand surgery: an overview. J. Hand Surgery Br. Eur. **25**(3), 242–252 (2000)
13. Luo, D., Nie, M., Zhang, T., Wu, X.: Developing robot reaching skill with relative-location based approximating. In: IEEE ICDL-EpiRob, pp. 147–154. IEEE (2018)
14. Malinovská, K., Farkaš, I., Harvanová, J., Hoffmann, M.: A connectionist model of associating proprioceptive and tactile modalities in a humanoid robot. In: IEEE ICDL, pp. 336–342, September 2022
15. Markram, H., Gerstner, W., Sjöström, P.J.: Spike-timing-dependent plasticity: a comprehensive overview. Front. Synaptic Neurosci. **4**, 2 (2012)
16. Mellmann, H., et al.: Effects of transparency in humanoid robots - a pilot study. In: Companion of ACM/IEEE HRI, pp. 750–754, March 2024
17. Reggev, N., Bein, O., Maril, A.: Distinct neural suppression and encoding effects for conceptual novelty and familiarity. J. Cogn. Neurosci. **28**(10), 1455–1470 (2016)
18. Rougier, N., Boniface, Y.: Dynamic self-organising map. Neurocomputing **74**(11), 1840–1847 (2011)
19. Sakai, J.: How synaptic pruning shapes neural wiring during development and possibly, in disease. Proc. Natl. Acad. Sci. **117**(28), 16096–16099 (2020)
20. Schillaci, G., Hafner, V.: Random movement strategies in self-exploration for a humanoid robot. In: 6th ACM/IEEE HRI, vol. 2011, pp. 245–246, March 2011
21. Schillaci, G., Hafner, V., Lara, B.: Coupled inverse-forward models for action execution leading to tool-use in a humanoid robot. In: ACM/IEEE HRI (2012)
22. Schillaci, G., Hafner, V.V., Lara, B.: Online learning of visuo-motor coordination in a humanoid robot. A biologically inspired model. In: IEEE ICDL-EpiRob, pp. 130–136, October 2014
23. Schillaci, G., Hafner, V.V., Lara, B.: Exploration behaviors, body representations, and simulation processes for the development of cognition in artificial agents. Front. Robot. AI **3**, 39 (2016)
24. Schillaci, G., Lara, B., Hafner, V.V.: Internal simulations for behaviour selection and recognition. In: Salah, A.A., Ruiz-del-Solar, J., Meriçli, Ç., Oudeyer, P.-Y. (eds.) HBU 2012. LNCS, vol. 7559, pp. 148–160. Springer, Heidelberg (2012). https://doi.org/10.1007/978-3-642-34014-7_13
25. Schmerling, M., Schillaci, G., Hafner, V.V.: Goal-directed learning of hand-eye coordination in a humanoid robot. In: IEEE ICDL-EpiRob, pp. 168–175 (2015)
26. Sloan, A.T., Jones, N.A., Kelso, J.A.S.: Meaning from movement and stillness: signatures of coordination dynamics reveal infant agency. Proc. Natl. Acad. Sci. **120**(39), e2306732120 (2023)
27. Tognoli, E., Zhang, M., Fuchs, A., Beetle, C., Kelso, J.A.S.: Coordination dynamics: a foundation for understanding social behavior. Front. Hum. Neurosci. **14**, 317 (2020)
28. Wang, J., Zou, Y., Wei, Y., Nie, M., Liu, T., Luo, D.: Robot arm reaching based on inner rehearsal. Biomimetics **8**(6), 491 (2023)
29. Wortham, R., Theodorou, A., Bryson, J.: Improving robot transparency: real-time visualisation of robot AI substantially improves understanding in naive observers. In: 26th IEEE International Symposium RO-MAN, pp. 1424–1431, August 2017

Problem Solving and Decision-Making

Extracting Principles of Exploration Strategies with a Complex Ecological Task

Oussama Zenkri[1,4](✉), Florian Bolenz[2,4], Thorsten Pachur[3,4], and Oliver Brock[1,4]

[1] Robotics and Biology Laboratory, Technische Universität Berlin, Berlin, Germany
{zenkri,oliver.brock}@tu-berlin.de
[2] Center for Adaptive Rationality, Max Planck Institute for Human Development, Berlin, Germany
{bolenz,pachur}@mpib-berlin.mpg.de
[3] School of Management, Technical University of Munich, Munich, Germany
[4] Science of Intelligence, Research Cluster of Excellence, Berlin, Germany

Abstract. Human exploration, a cornerstone of our ability to solve novel problems, is a complex process, posing significant research challenges. Most previous studies simplify tasks to isolate specific variables, creating artificial problems that do not align with those humans have evolved to solve, thus limiting the generalizability of findings. To address this gap, we introduce the Lockbox paradigm: a novel, ecologically valid, and challenging task that promotes active exploration and physical interaction. Data from 91 participants interacting with the Lockbox reveal a remarkable human ability to adapt and solve problems efficiently in complex scenarios. By comparing different interaction methods, we demonstrate the critical role of cost variations, such as physical and cognitive costs, in driving attentiveness and shaping exploration strategies. These findings provide valuable insights into human exploration strategies, with potential applications in fields such as robotics and artificial intelligence.

Keywords: Problem solving · Exploration · Physical costs

1 Introduction

How do humans develop strategies for tackling novel problems based on previous experience with similar situations? Answering this question is fundamental for understanding the mechanisms underlying human problem solving. Moreover, this knowledge can be leveraged to engineer synthetic agents that are more adept at solving real-world problems with human-like efficiency and robustness. To answer this question, it is important to investigate human exploration in complex environments that mimic those in which humans evolved [1,6,9]. These

---

Funded by the Deutsche Forschungsgemeinschaft (DFG, German Research Foundation) under Germany's Excellence Strategy - EXC 2002/1 "Science of Intelligence" - project number 390523135.

environments should promote physical interaction, active exploration, and inherent variability, which fosters the development of generalizable strategies.

To investigate human exploration in the context of problem solving, we developed the Lockbox task, a mechanical puzzle with a hidden mechanism. By manipulating the Lockbox's visual configuration, we can examine how humans approach different instantiations of problems that share a common underlying structure, mimicking the variability encountered in real-world problems. By hiding the mechanism interlocking the elements of the puzzle, we also impose the active exploration of the environment beyond passive observation [4]. We add a strong physical component to the task by introducing haptic devices as the interaction interface, providing a more realistic experience.

Here we introduce the Lockbox paradigm and illustrate how it can be used to investigate human exploration during problem solving. We collected data from 91 participants who interacted with Lockboxes of different complexities. By studying participants' interactions, we aim to gain insights into the fundamental principles behind human exploration of novel problems and the development of problem-solving strategies that are generalizable across problem instantiations. Our findings underscore the significant roles of attentiveness, physical interaction, and cost minimization in shaping exploration and strategy formation. The observed transition from exploratory to exploitative behavior as participants familiarized themselves with the task's intricacies reflects the adaptability and efficiency of human problem solving. Moreover, our comparative analysis between haptic device and mouse interactions sheds light on the impact of tangible, physical engagement on learning efficiency.

2 Background

Researchers have employed various paradigms to study exploration during problem solving, but these often face limitations in ecological validity. I.e. they may not reflect real-world settings. For instance, multiple research have shown that animals like cockatoos and primates can learn sequencial actions through exploration in physical puzzles [2,8,11]. However, these studies often involve repeated exposure to identical problems, limiting the need for generalizable strategies.

Traditional human problem-solving research has relied on either relatively simple tasks or semantically rich computer-simulated microworlds [3]. Simple tasks often require minimal exploration [5], while semantically-rich tasks might introduce biases by prior knowledge about the real-world context they encapsulate. Shrager & Klahr's study [10] showed that people can learn the functioning of a complex toy-car merely by exploratory interaction with the object under investigation. The study concluded that people engage in a systematic formation and testing of hypotheses to acquire a functional understanding of an unknown system, highlighting the role of physical exploration and information seeking in human problem-solving behavior. However, this task did not involve learning a complex sequence of actions for reaching a specific goal.

More recently, human exploration behavior has predominantly been investigated in foraging situations, often using variants of multi-arm bandit tasks (e.g. [14,15]) where exploration is commonly understood as forgoing a known reward for the sake of gaining information about the environment. This exploration-exploitation trade-off does not exist during problem solving, as any reward can only be achieved through the successful completion of the task.

This research addresses these gaps by investigating human exploration in a physically interactive environment that promotes active exploration and generalization, a scenario closer to real-world problem solving.

3 Methodology

We begin by outlining the Lockbox paradigm. Next, we introduce the haptic device and highlight how it integrates a strong physical component into the Lockbox interaction model. We then describe the experimental conduct and data collection processes for an experiment with the Lockbox paradigm. Finally, we detail the metrics used to analyze the collected data.

3.1 The Lockbox Paradigm

Our Lockbox paradigm is designed to investigate human ability to develop generalizable strategies for solving novel problems. It taps into the complex interplay between exploration and strategy development by presenting a challenging long-horizon task. The Lockbox environment consists of a 3×3 grid with six linear

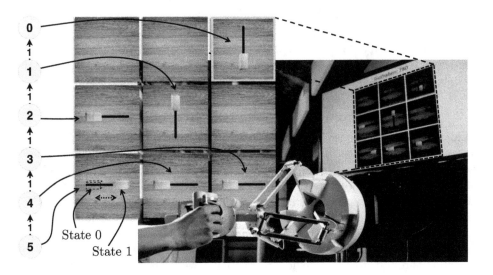

Fig. 1. Sample Lockbox instance and corresponding dependency structure. The sliders' visual features (state, position, orientation) are independent of the hidden mechanism governing the Lockbox (Graph on the left). All slider have binary states. Users interact naturally with the Lockbox using the haptic device.

sliders. Participants must move a specific slider (*target slider*), always located in the highlighted upper right corner of the grid (Fig. 1), to unlock the Lockbox.

Initially, the target slider is locked by other sliders, which in turn might be locked by other ones, requiring participants to explore and discover these dependencies. While slider locations and orientations vary across trials (except the target slider location), the underlying dependency structure, which determines how moving one slider affects others, remains constant. This offers visual variability without altering the core problem. The dependency structure is hidden from participants to encourage active exploration and reveal how individuals integrate feedback into decision-making during exploration. By hiding these underlying dependencies, we also minimize the potential influence of prior knowledge that could otherwise bias the exploratory behavior we aim to study. The dependency structure, represented as a *directed acyclic graph* (DAG) as visualized in Fig. 2, can be learned by participants through observing the relationships between sliders over time. This learning is crucial for developing generalizable strategies applicable to future problems with similar underlying structures.

To increase complexity and promote the development of generalizable strategies, we introduce *bistable-locking* (BL) sliders (Fig. 2). These sliders are visually identical to the normal ones but can lock or unlock others depending on their current state. BL sliders significantly expand the solution space, requiring more interactions and pushing participants' cognitive processes. This challenging aspect allows us to investigate how humans develop generalizable strategies at various difficulties. We attribute binary state (two possible positions) to all sliders. This simplification allows us to isolate the core cognitive processes involved in problem solving, mitigating the potential role of fine motor skills.

In summary, our Lockbox paradigm leverages **long-horizon** challenges, **visual variability**, **active exploration**, **variable difficulty**, and **simple manipulation** to create a challenging and engaging environment that taps into people's exploration and strategy formation.

3.2 Leveraging Haptic Technology for Physical Interaction

While a computer mouse is a common interface for psychological experiments, it lacks the natural feel of real-world manipulation. Humans evolved problem-solving skills through physically interacting with the environment, relying heavily on sensory and motor feedback. Richer interaction is thus crucial for leveraging these complex processes in exploring tasks and generalizing solutions. To create a more immersive and realistic experience, we employed a haptic device. Unlike traditional interfaces, haptic devices introduce a strong physical component to exploration and provide richer sensory feedback. By interacting with the Lockbox through a haptic device, participants were able to employ real-world actions like reaching, grasping, and pushing. We hypothesize that physical interaction impacts exploratory behavior and strategy development when solving the Lockbox. Section 4.4 presents findings that support this hypothesis.

Our experiment utilized the *Sigma 7*, a seven degree-of-freedom haptic device from *Force Dimension*. Participants interacted with the virtual Lockbox by maneuvering a virtual pointer within the simulated environment. To interact with elements of the Lockbox, they guided the pointer by maneuvering the haptic device's handle towards the desired slider. Once aligned, they then used the handle's gripper to secure the slider and applied force to push it. Throughout these interactions, the Sigma 7 accurately simulated forces and frictions encountered, mimicking the experience of manipulating a physical Lockbox. This haptic feedback provided a realistic sensation of movement and resistance. Additionally, the device consistently operated in a gravity compensation mode, neutralizing the handle's weight and enhancing ease of use.

3.3 Procedure and Data Collection

The experiment adhered to ethical guidelines with written informed consent obtained from all participants. The Max Planck Institute for Human Development's institutional review board approved the study. Before engaging with the main experiment, participants completed training phases to ensure a proper understanding of the Lockbox's functionality, the rules of the experiment, and the operation of the haptic device. These training phases involved simpler 2×2 Lockboxes and incorporated true/false questionnaires to assess comprehension. Participants were randomly assigned one of five underlying dependency structures (Fig. 2) that differed in complexity. They solved multiple Lockboxes that shared the same underlying structure, but differed in their visual appearance (layout $\mathscr{L}$), which were randomized across trials. To maintain a consistent experiment duration (approx. 1 h), the number of trials per participant was adjusted based on the structure's complexity.

Table 1. Participant Count per Underlying Structure

Underl. Struc.	C_0	C_1	C_2	C_3	C_4
Difficulty Level	Easy	Medium	Medium	Hard	Very Hard
# Participants	10	27	29	21	4
# Trials/Part.	30	30	30	25	20

Throughout each trial, we tracked participants' interactions with the Lockbox. An interaction, defined as an attempt to move a slider, comprises two steps: establishing contact between the pointer and the slider's edge, followed by pressing the gripper to confirm manipulation intent. For each interaction, successful or not, 10 points were deducted from a starting balance of 1,000 points displayed above the grid structure. The trial ended when either the target slider was moved or the balance reached zero. A total of 91 participants between 18 and 35 years old participated in the experiment. Participants were randomly distributed across the underlying structures as listed in Table 1. Data collection for Structures C_0 and C_4 was halted prematurely. Participants perceived C_0 as too easy, converged to the optimal solution after only a few trials, and required significantly less time than those, who engaged with other structures. We included

C_0 data in our analysis as a reference point. Conversely, C_4 presented excessive difficulty for participants. Due to the low participant turnout for C_4, we excluded the data from this condition from the analysis.

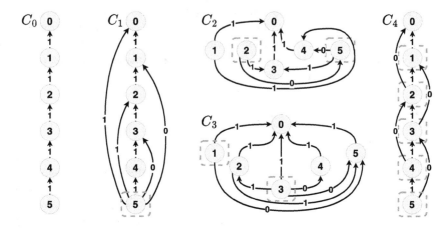

Fig. 2. The five underlying dependency structures (C_0 to C_4) used in our experiment, visualized as directed acyclic graphs (DAGs). Each node corresponds to a slider in the Lockbox. Directed edges point from locking sliders to the sliders they lock. The edge value shows the specific state (position) the locking slider needs to be in to unlock the connected slider. A slider is unlocked only if all its locking sliders are positioned according to the corresponding edge values. Nodes with multiple outgoing edges and opposing edge values (marked by orange frames) correspond to bistable-locking (BL) sliders. The number of BL sliders in a Lockbox is one factor contributing to its complexity. (Color figure online)

3.4 Metrics

To unveil the exploration principles participants employed within the Lockbox task, this section introduces three key metrics used to analyze their interactions.

Target Slider Selection. Analyzing the timing and frequency of attempts to unlock the Lockbox (manipulating the target slider) provides insights into participants' decision-making during exploration. Participants are likely to attempt the target slider only when they believe that the current configuration (mainly reached through their interactions) has a good chance of success. By studying the frequency of these attempts, along with the steps leading up to them, we can infer the mental processes behind these actions. This analysis allows us to identify patterns in human exploration, revealing how participants form and test hypotheses, ultimately refining their strategies for solving the Lockbox.

State Revisits. To assess how effectively participants explored the Lockbox solution space, we analyze state revisits. A state revisit occurs when a participant attempts the target slider in a configuration (state of all sliders at that moment) they have already tried and found unsuccessful. Ideally, efficient exploration minimizes revisiting these dead ends. Therefore, the state revisit rate, calculated as the number of revisits divided by the total number of target slider attempts within a trial, serves as an indicator of a participants' attentiveness during exploration. A lower revisit rate suggests a more focused exploration strategy, while higher rates can result from cognitively less demanding strategies [7]. To distinguish intentional exploration from potential manipulation errors, we do not consider revisits of the same state if they are consecutive.

Randomness. Measuring the randomness of the taken steps reveals whether certain patterns are followed when exploring the Lockbox. The change in this randomness measure across trials is indicative of the refinement of the employed exploration strategies. To quantify the randomness of a sequence of attempted sliders **s** of length N, we first calculate the autocorrelation vector $\mathbf{r}_{\tilde{s}\tilde{s}}$. Here, $\tilde{\mathbf{s}}$ of length $N - 1$, is the sequence of L1 distances between every two consecutively attempted sliders $s_{i-1}, s_i \in \mathbf{s}, \forall i = 2, \ldots, N$. The average of $\mathbf{r}_{\tilde{s}\tilde{s}}$ across all lags, excluding lag 0, is then used to compute a randomness score R_s:

$$R_s = 1 - \frac{2}{N-2} \cdot \sum_{\forall k \neq 0} \mathbf{r}_{\tilde{s}\tilde{s}}[k], \quad \tilde{s}_i = |\mathscr{L}_x(s_i) - \mathscr{L}_x(s_{i-1})| + |\mathscr{L}_y(s_i) - \mathscr{L}_y(s_{i-1})|, \quad (1)$$

with $\tilde{s}_i \in \tilde{\mathbf{s}}$, and the coordinates $\mathscr{L}_x(s_i)$ and $\mathscr{L}_y(s_i)$ corresponding to the x-, and y-coordinates of slider s_i within the Lockbox-layout $\mathscr{L}$ respectively.

4 Results

This section presents key insights from our ongoing data analysis. We begin by analyzing the transitioning from initial random exploration to focused exploitation. We then investigate how cost minimization influences exploration behavior. Next, we evaluate the impact of attentiveness on exploration efficiency. Finally, we explore the relationship between physical costs and learning efficiency.

4.1 Initial Exploration Transitions to Exploitation

Our data analysis reveals a clear shift in participants' strategies, transitioning from exploration to exploitation. Two key factors highlight this shift.

Target Slider Attempts. A decrease in the frequency of attempts to manipulate the target slider signifies convergence towards a successful solution. Conversely, persistently high attempt frequencies indicate difficulty in unlocking the Lockbox. Figure 3**A** visualizes a pattern, with a steady decrease in target slider attempts suggesting participants honing in on efficient solution approaches.

Action Sequence Randomness. Early interactions exhibit high randomness (metric defined in Sect. 3.4), reflecting initial uncertainty and lack of a systematic

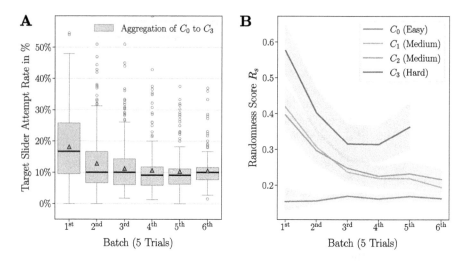

Fig. 3. As expertise with the Lockbox increases (increasing batch order), participants' strategies become more efficient (lower target slider attempts and randomness). Figure **A** shows a decreasing rate of target slider attempts, indicating a refined approach. Similarly, Figure **B** shows a decreasing randomness in slider selection as participants gain proficiency. Solid lines represent the mean randomness scores, and the surrounding shaded area indicates the confidence intervals.

exploration pattern. However, randomness decreases as participants progress in the experiment. This is depicted in Fig. 3B, where the decline in randomness with accumulated knowledge implies a shift toward more focused and efficient behavior, indicative of exploitation.

The observed learning curve can be interpreted as a transition from exploratory behavior, where participants try various options, to exploitative behavior, where they leverage their acquired knowledge to solve the Lockbox efficiently.

4.2 Cost Minimization Shapes Exploration

Human decision-making during exploration appears to be shaped by the principle of cost minimization. This is evident in several ways throughout our experiment:

Local Exploration after Failure: When encountering a locked slider, participants primarily explored neighboring sliders, as shown in the transition probability graph (Fig. 4A). This figure depicts the two most probable transitions from every unsuccessful slider attempt, with the majority of transitions pointing to direct neighbors. This suggests a focus on local exploration to avoid effortful motions. This behavior can also be motivated by an inductive bias: objects in close proximity are often more likely to be interconnected than distant ones [13].

Prioritizing the Target Slider after Success: Conversely, when a slider was moved successfully (Fig. 4B), participants tended to directly attempt the target slider. This highlights a preference for quicker solutions that minimize overall effort invested in exploration. Interestingly, participants seem to be willing to incur a higher immediate physical cost (moving a distant slider) to achieve a lower overall cost (balance and effort) in the long run, as reflected by shorter solution sequences. This tendency to attempt the target slider can be further understood as leveraging prior knowledge-target slider unlocks the Lockbox.

In Sect. 4.4, we will delve deeper into how physical interaction costs introduced by the interaction-interface further influence exploration strategies.

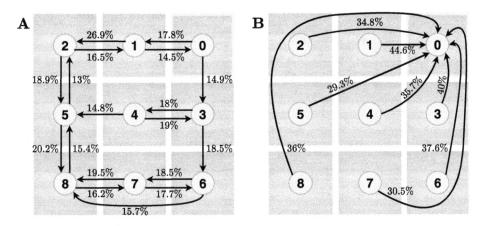

Fig. 4. These graphs depict the most likely sliders to be attempted by participants after a failed slider attempt (**A**) and after a successful one (**B**). Arrows indicate the most probable transitions, with the probability labeled on the edge. Following a failed attempt, participants tend to explore direct neighboring sliders (**A**). Conversely, after a successful move, the most likely next attempt is the target slider (**B**). The results suggest a cost-minimizing exploration strategy.

4.3 Attentiveness Drives Efficiency in Exploration

In this section we examine how attentiveness, measured by a participants' ability to avoid revisiting unsuccessful configurations (see Sect. 3.4), influences learning efficiency in solving the Lockbox task. Lower revisit rates indicate more focused exploration and potentially higher attentiveness. Figure 5A illustrates the connection between state revisits and learning efficiency. Participants with fewer revisits in the first five trials showed better performance in solving the Lockbox in the remaining trials. This suggests that excessive revisiting, characteristic of inattentive exploration, might hinder learning efficiency. We further investigated how revisits relate to game complexity and player experience. Figure 5B reveals two key trends. First, there is a **complexity effect**, such that more complex Lockboxes (indicated by higher lines) tend to have higher state revisit rates. While the increased exploration demand for more challenging configurations seems to be a plausible cause, this trend holds even when the revisit rate is normalized by the trial length (number of interactions per trial), suggesting a potential influence of complexity on attentiveness as well. Second, there is a **learning effect**, such that regardless of complexity, the revisit rate decreases across trials as participants develop more efficient exploration strategies with increasing experience and the overall decreasing need for exploration.

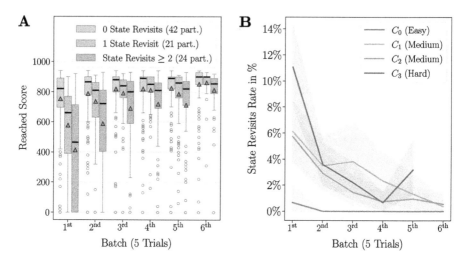

Fig. 5. Attentiveness and Learning Efficiency. Figure **A** shows the evolution of the Lockbox-solving performance based on the average number of state revisits in the first five trials. Participants with fewer revisits exhibit higher learning efficiency, highlighting the importance of attentiveness during exploration. Figure **B** examines how revisits relate to game complexity and player expertise. Game complexity triggers an increase in state revisits, while experience leads to a decrease, suggesting a transition towards more efficient exploration strategies.

4.4 Physical Interaction Alters Learning

Building on the principle of cost minimization, we also investigated how physical costs, introduced by the haptic device's interaction demands, influence exploration and strategy development. To that end, we repeated the experiment using a computer mouse as the interface instead of the haptic device. Eighty two new participants (27 in C_1, 29 in C_2, and 26 in C_3) interacted with the Lockbox using the mouse. Figure 6 compares the aggregated data of the three underlying structures for both interaction interfaces. Participants, who used the haptic device completed the Lockbox in fewer steps and converged faster on more efficient strategies compared to those, who used the mouse (Fig. 6**A**). This suggests that physical interaction with the Lockbox promotes more efficient exploration. Furthermore, Fig. 6**B** shows that participants, who used the mouse, revisited previously explored states at a higher rate compared to those, who used the haptic device. This indicates that physical interaction, facilitated by the haptic device, encourages more thoughtful exploration, potentially due to the increased investment (physical cost) associated with each action.

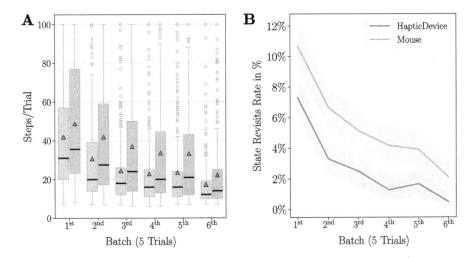

Fig. 6. Participants using the haptic device to interact with the Lockbox solved it in fewer steps across all trial batches compared to those using the mouse (**A**). Additionally, they exhibited a significantly lower revisiting of already explored configurations (**B**), indicating more focused exploration. This suggests that the physical interaction provided by the haptic device contributes to a higher exploration efficiency and the development of more optimized strategies.

This finding aligns with Thompson's observations in a maze experiment with rats [12]. When navigating a maze with levers requiring greater pressing force, rats learned the maze faster. This suggests that physical costs, might have a distinct impact on how users explore problems and develop strategies, compared to the costs that a diminishing balance introduces. However, further research is needed to investigate how other factors, beyond physical cost, such as interaction duration or sensory feedback, contribute to this behavior.

5 Conclusion

Our study introduces the Lockbox, a complex and ecologically valid paradigm designed to deepen our understanding of human exploration strategies during problem solving. Utilizing a haptic device, we highlighted the potential influence of physical costs on cognitive processes, revealing how these costs enhance attentiveness and exploration efficiency. Additionally, the observed transition from exploratory to exploitative behavior exemplifies the adaptability and efficiency of human cognition in complex problem-solving scenarios. These findings motivate research on transferring these dynamics into robotic systems to potentially improve their learning efficiency and robustness in real-world tasks.

In the future, collecting more data using an extended Lockbox paradigm that allows for adjustable levels of physical demand and other cost factors, such as time, will further illuminate the impact of these elements on exploration in problem solving. By bridging the gap between controlled laboratory experiments and the complexity of real-world environments, we pave the way for a deeper understanding of human problem solving and the development of more robust synthetic agents.

References

1. Allen, K.R., et al.: Using games to understand the mind (2023)
2. Auersperg, A., Kacelnik, A., von Bayern, A.: Explorative learning and functional inferences on a five-step means-means-end problem in Goffin's cockatoos (Cacatua goffini). PLoS ONE **8**(7), e68979 (2013)
3. Funke, J.: Problem solving. In: Sternberg, R., Funke, J. (eds.) The Psychology of Human Thought: An Introduction, pp. 155–176. Heidelberg University Publishing (2019)
4. Gibson, J.J.: The Ecological Approach to Visual Perception. Houghton Mifflin (1979)
5. Huber, O.: Beyond gambles and lotteries: naturalistic risky decisions. In: Decision Making, pp. 159–176. Routledge (2002)
6. Huber, O., Wider, R., Huber, O.W.: Active information search and complete information presentation in naturalistic risky decision tasks. Acta Physiol. (Oxf) **95**(1), 15–29 (1997)
7. Kahneman, D.: Thinking, fast and slow. Farrar, Straus and Giroux (2011)
8. Lang, B., et al.: Challenges and advanced concepts for the assessment of learning and memory function in mice. Front. Behav. Neurosci. **17**, 1230082 (2023)
9. Lejarraga, T., Hertwig, R.: How experimental methods shaped views on human competence and rationality. Psychol. Bull. **147**(6), 535–564 (2021)
10. Shrager, J., Klahr, D.: Instructionless learning about a complex device: the paradigm and observations. Int. J. Man Mach. Stud. **25**(2), 153–189 (1986)
11. Tecwyn, E.C., Thorpe, S.K., Chappell, J.: A novel test of planning ability: great apes can plan step-by-step but not in advance of action. Behav. Proc. **100**, 174–184 (2013)
12. Thompson, M.E.: Learning as a function of the absolute and relative amounts of work. J. Exp. Psychol. **34**(6), 506 (1944)
13. Tobler, W.R.: A computer movie simulating urban growth in the detroit region. Econ. Geogr. **46**(sup1), 234–240 (1970)
14. Wilson, R.C., Geana, A., White, J.M., Ludvig, E.A., Cohen, J.D.: Humans use directed and random exploration to solve the explore-exploit dilemma. J. Exp. Psychol. Gen. **143**(6), 2074 (2014)
15. Wu, C.M., Schulz, E., Speekenbrink, M., Nelson, J.D., Meder, B.: Generalization guides human exploration in vast decision spaces. Nat. Hum. Behav. **2**(12), 915–924 (2018)

"Value" Emerges from Imperfect Memory

Jorge Ramírez-Ruiz[✉] and R. Becket Ebitz

Département de Neurosciences, Faculté de Medicine, Université de Montréal,
Montréal, Canada
jorgeerrz@gmail.com

Abstract. Whereas computational models of value-based decision-making generally assume that past rewards are perfectly remembered, biological brains regularly forget, fail to encode, or misremember past events. Here, we ask how realistic memory retrieval errors would affect decision-making. We build a simple decision-making model that systematically misremembers the timing of past rewards but performs no other value computations. We call these agents "Imperfect Memory Programs" (IMPs) and their single free parameter optimizes the trade-off between the magnitude of error and the complexity of imperfect recall. Surprisingly, we found that IMPs perform similarly or better than a simple agent with perfect memory in multiple classic decision-making tasks. IMPs also generated multiple behavioral signatures of value-based decision-making without ever calculating value. These results suggest that mnemonic errors (1) can improve, rather than impair decision-making, and (2) provide a plausible alternative explanation for some behavioral correlates of "value".

Keywords: memory errors · probabilistic choice · value integration

1 Introduction

The brain is an imperfect computer. There are fundamental constraints on its ability to accurately represent, retain, and recall the information we have encountered in the world [14,16,21]. While neural processing constraints can cause errors that have no obvious utility [17], others may generate "errors" that are computationally useful [8,21,34]. Especially in uncertain environments, there is both theoretical [32] and empirical [8,27] evidence that noise is critical for exploratory discovery and learning. By contrast, in cognitive models, noise is generally only added as an error term, with limited but important exceptions [11,12,21]. Developing cognitive models that meaningfully incorporate noise could be the key to determining when the constraints on the brain are truly a limitation and when they serve a computational purpose.

This paper offers a thought experiment: What if neural processing errors were sufficient for making good decisions in uncertain environments? Sequential decision-making in changing and only partially observable environments requires continual learning and discovery. This is exactly the kind of task where computational noise is most likely to be useful [12,27]. Yet realistic errors in representing, retaining, or recalling past rewards are not a feature of cognitive models of sequential decision-making. Instead, the standard approach assumes that previous rewards are perfectly encoded and integrated into a value signal that is then used to guide decisions [31,37,38]. While some recent work has considered alternative architectures, including sampling from episodic memory [2,24], these studies have not permitted memory errors. As a result, we know little about how errors would impact decision making, nor how an agent imbued with this noise would differ from other models or real decision-makers.

To address this omission, we developed a novel and minimal decision-making agent that we call "Imperfect Memory Programs" (IMPs). IMPs systematically misremember the timing of past rewards in a biologically realistic way, but perform none of the value computations thought to be important for sequential decision-making in the cognitive sciences. We find that IMPs are capable of performing well above chance in multiple decision-making environments and generate behavior that is typically interpreted as evidence of value integration. These results have implications for understanding sequential-decision making, but also for designing new algorithms for decision-making in uncertain environments.

2 Imperfect Memory Programs (IMPs)

IMPs (Fig. 1A) use a 2-step process to make decisions. First, in the sampling stage, IMPs "remember" a past outcome via a noisy sampling process. Second, in the choice stage, IMPs decide whether to exploit the previous action or to explore the action space.

In the **sampling stage**, IMPs sample a reward outcome from a memory store that contains the set of previous rewards associated with the current action. That is, they remember some prior reward R from a previous time point $\Delta t \in (-\infty, 0]$. Inspired by previous theoretical work in the domain of structure learning [21], we assume that this memory store is organized such that it balances 2 competing objectives (Fig. 1B). First, the memory store should maximize the likelihood of recalling relevant past outcomes. In changing environments, the relevance of past information decreases as a function of time, such that recalling outcomes distant in time (Δt with respect to the present), incurs a high error. The sampling distribution Q that would maximize timeliness is the one that minimizes the average temporal error in remembering, $E(Q) = \sum_{\Delta t} Q(\Delta t) \Delta t$. Second, the recall process should be as minimally complex as possible (i.e., efficient to implement), while also maximizing the information that can be sampled from memory. This fact implies that we should choose the maximum entropy sampling distribution Q: the one that minimizes the negative entropy, $-H(Q) = \sum_{\Delta t} Q(\Delta t) \log Q(\Delta t)$. We can balance these two competing objectives by calculating the total cost, or free energy of Q, $F(Q) = \beta E(Q) - H(Q)$, and minimizing $F(Q)$ with respect to Q. In doing so, we find that the sampling distribution that would simultaneously minimize the magnitude of mnemonic errors and maximize information storage is the Boltzman distribution,

$$p(R_{\Delta t}) = \frac{1}{Z} e^{-\beta \Delta t}, \tag{1}$$

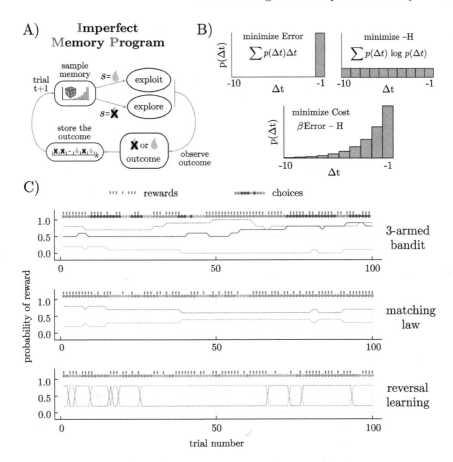

Fig. 1. Agents and environments. A) Schematic illustrating the structure of the decision-making process in Imperfect Memory Programs (IMPs). B) IMPs sample previous rewards from memory in a way that optimizes the trade-off between (1) minimizing the average temporal error of the recalled samples (top left) and (2) minimizing the complexity of the sampling process (top right). The Boltzman distribution (bottom) minimizes the total cost (or free energy) of the sampling distribution, thereby naturally balancing these two objectives. Here β is both the inverse temperature of the sampling process and the single free parameter of the IMPs. C) Examples of the reward schedules from 3 testbeds: a restless 3-armed bandit (top), a matching law task (middle), and a probabilistic reversal learning task (bottom). Delivered rewards and the choices generated by one example IMP are overlaid.

where Z is the partition function and β is the model's single free parameter, which controls the trade-off between error magnitude and storage capacity. Although this choice is theoretically motivated [21,23], there is also empirical evidence that memory retrieval tends to be exponentially recency-weighted [16,21,30], suggesting that the brain may also have struck a balance between error magnitude and complexity. In instances where IMPs are unable to sample a past outcome from memory (for example, during the first choice to an option), an outcome (reward omission or delivery) is chosen at random as the sample.

In the **choice stage**, IMPs make a deterministic choice based on the sampled outcome. Inspired by an ϵ-greedy decision rule, IMPs choose whether to explore or exploit. If the remembered outcome, $R_{\Delta t}$ was not positive (not rewarded), the agent explores through choosing a new action policy at random,

$$p(\text{choice}_{t+1} = i | R_{\Delta t} = 0) = \frac{1}{k}, \quad (2)$$

where k is the total number of options. If the remembered outcome was positive (rewarded), the agent continues to exploit its current action policy,

$$p(\text{choice}_{t+1} = i | R_{\Delta t} = 1) = \begin{cases} 1, \text{if choice}_t = i \\ 0, \text{otherwise} \end{cases}. \quad (3)$$

This decision policy was chosen for its simplicity, but there is also evidence that biological decision-makers produce distinct explore and exploit choices in similar tasks, where the exploration resembles random decision-making and exploitation resembles directed, reward-dependent decision-making [5,7,19].

3 Testbeds

We simulated behavior from the IMP in 3 sequential value-based decision-making tasks that are common in the neuroscience and psychology literature (Fig. 1C). These included a restless bandit task [5–8,26], a matching law task [36], and a probabilistic reversal learning task [3,4]. Unless otherwise noted, simulations involved 500 sessions of 500 trials. All agents experienced identical environments.

In each task, choices are made between a set of k options, each of which is associated with some probability of reward. Reward probabilities can only be inferred by choosing each option and combining information over multiple samples. The tasks are all uncertain because the reward probabilities are not fixed, but instead evolve over time. This encourages decision-makers to exploit valuables option when they are discovered while also occasionally exploring alternative options that have the potential to become more rewarding at any time.

In the **restless bandit task** (Fig. 1C, top), the reward probabilities of each option i are independently updated at each trial t according to

$$p(\text{reward}_{i,t+1}) = p(\text{reward}_{i,t}) \pm \begin{cases} \text{step}, & \text{if } \mathcal{U}(0,1) < \text{hazard} \\ 0, & \text{otherwise} \end{cases} \quad (4)$$

where "hazard" is a fixed rate of change $\in [0,1]$, $\mathcal{U}(0,1)$ is a draw from a uniform random distribution, and the sign of the step is chosen independently at random for each option on each trial. The hazard rate and step size were both fixed at 0.1, and the number of options was fixed at 3 except as otherwise noted, after [4,8,19,34].

In the **matching law task** (Fig. 1C, middle), reward probabilities are updated according to the same function, but not independently because

$$\sum_{i=1}^{k} p(\text{reward}_{i,t}) := 1 \tag{5}$$

for each option i. The matching law task is often used to illustrate that biological decision-makers tend to be imperfect reward maximizers: more likely to allocate their choices in proportion to the rate of reward than to chose the best option. Matching law tasks are typically 2-alternative; we followed that convention here.

The **probabilistic reversal learning task** (Fig. 1C, bottom) is another 2-alternative decision-making task that is very common in rodents. As in the matching law task, reward probabilities are symmetrical such that one option is high-value and the other is low-value. However, here the high and low values are fixed, often at $p(\text{reward}|\text{high}) = 80\%$ and $p(\text{reward}|\text{low})$ 20%, with the identity of the high and low values swapping at specific reversal points.

4 Results

4.1 IMPs Performed Similarly to Agents with Perfect Memory

Simulations in a restless k-armed bandit environment revealed how IMPs behavior and performance depends on the inverse temperature parameter β. When $\beta \gg 1$, the cost of recall error takes over the total error, and IMP agents become perfect memory agents with a Win-Stay, Lose-Shift (WSLS) strategy (Fig. 2A-B, right). In contrast, when $\beta \ll 1$, the complexity cost dominates and IMPs stochastically recall reward episodes far in the past; hence, performance degrades in this unpredictable environment (Fig. 2A-B, left). Crucially, there is an intermediate value of β that not only optimizes the performance of the IMPs (Fig. 2A-B), but does so while switching less frequently (Fig. 2C). Given that switch decisions take longer than stay decisions [39], IMP's tendency to repeat choices suggests that a biological agent would be able to achieve a higher rate of reward via an IMP-like algorithm than a WSLS-algorithm.

To benchmark IMPs' performance, we also simulated an oracle (that knows the probability of each arm and always selects the best), a random agent (which chooses an arm uniformly randomly), and a reinforcement learning algorithm (SARSA, [37]). SARSA updates the value $Q(a)$ of an arm a after receiving reward R at time t, and sampling another arm a' from its policy,

$$Q_{t+1}(a) = Q_t(a) + \alpha \left(R_t + \gamma Q_t(a') - Q_t(a)\right), \tag{6}$$

where α is a learning rate and γ is a discount factor. Then, the SARSA agent defines a probability $\pi(a)$ of choosing an arm a using the action value $Q(a)$, as a softmax distribution

$$\pi(a) = \frac{1}{Z} \exp(\beta_{\text{SARSA}} Q(a)), \tag{7}$$

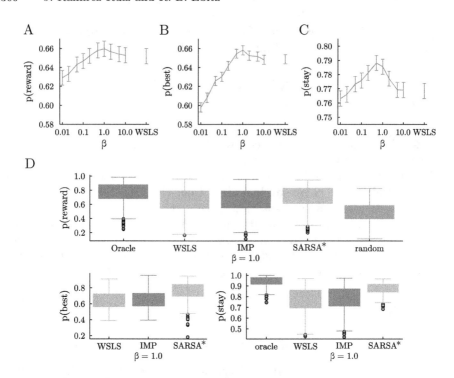

Fig. 2. IMPs perform a restless multi-armed bandit task without perfect memory. Probability of (A) obtaining reward, (B) choosing the best arm, and (C) sticking with the same arm across sessions, as a function of the inverse temperature β that trades off complexity and error. WSLS is a special case of an IMP with perfect memory, corresponding to infinite β. (D) Probability of obtaining reward, choosing the best arm (or any of the best arms) and of repeating a choice for various agents, with their free parameters optimized for this task.

where β_{SARSA} is an inverse temperature parameter that controls the noise for the SARSA agent, and Z is a normalizing function.

The optimal IMPs did not significantly perform better or worse than WSLS agents in the restless arm bandit task, in either probability of obtaining reward or choosing the best arm (logit-transformed Welch's t-test, two-sided $p > 0.5$ for both metrics). Both of these agents underperformed SARSA ($p < 10^{-6}$ for both metrics), with optimized parameters $\alpha = 0.8, \gamma = 0.9$, and fixed $\beta_{SARSA} = 10$ (Fig. 2D). All of these agents perform better than chance ($p < 10^{-50}$). For the parameters of this task, both the oracle and the optimized SARSA agent show that staying persistently is a good strategy to obtain more rewards (Fig. 2D, probability of staying is higher for these agents), which the IMPs tend to do slightly more than perfect memory WSLS agents ($p < 0.02$).

4.2 IMPs Generate Reward History Kernels

Sequential decision-making algorithms like SARSA, generally assume that agents calculate a quantity known as "value" by integrating experienced rewards. By setting the discount factor γ to 0 and rearranging Eq. 6, we see that SARSA is one example of the broad class of delta-rule learning models where

$$Q_{t+1}(a) = (1-\alpha)Q_t(a) + \alpha R_t \,, \tag{8}$$

which makes explicit that action values $Q(a)$ are the α-weighted average of value at the previous time step and the newest reward. This implies that the weight of past rewards falls off exponentially in these models, mirroring the pattern that is commonly seen in biological decision-makers [18].

IMPs do not calculate value. Nonetheless, it seems plausible that they might generate value-like reward history kernels because of their memory swaps. To measure their reward history kernels, we simulated IMPs in a 2-armed bandit and fit a logistic regression model,

$$\log \frac{p(c_t = 1)}{p(c_t = -1)} = \beta_0 + \sum_{i=1}^{N} \alpha_i c_{t-i} + \sum_{i=1}^{N} \phi_i c_{t-i} r_{t-i} + \eta \,, \tag{9}$$

where c_{t-i} is 1 if the first option is chosen on trial $t-i$ (-1 if the second is chosen), and r_{t-i} is 1 if they were rewarded on that trial (0 otherwise). Together, the $\phi_{1:N}$ parameters represent the unique effect of previous rewards on the log odds of choice, beyond the contribution of choice history ($\alpha_{1:N}$) and bias (β_0). Models were fit via ridge-regularized maximum likelihood ($\lambda = 1$). To determine if the influence of previous rewards decayed exponentially quickly, we fit a 3-parameter exponential curve, $Ae^{-Bx} - C$, to $\phi_{1:N}$. Here, A represents a scaling parameter, B is the decay rate of the influence of previous rewards, and C is an offset.

IMPs reliably generated reward history kernels that were well-described by exponential decay (Fig. 3A-C; median $R^2 = 0.9999$). The decay in the reward history kernel largely increased as a function of β before saturating at values > 5 (Fig. 3D). In short, IMPs had exponentially decaying reward history kernels because of their imperfect memory, rather than any value calculations.

If IMPs imperfect memory accomplishes something like reward integration, then in less volatile environments, where longer reward history integration offers an advantage, (1) IMPs should have a greater advantage over WSLS, and (2) the optimal β should decrease. To test these predictions, we simulated IMPs in the restless bandit with varying hazard rates. We found that the optimal β and the performance of the optimal agent both scaled with volatility (Fig. 3E-H). This observation suggests that the imperfect memory process in IMPs functioned like the reward history integration in delta-rule learning agents.

4.3 IMPs Engage in Behaviors Between Matching and Maximizing

Under some circumstances, biological decision-makers tend to match the relative rate of reward of their options rather than maximize their reward by consistently

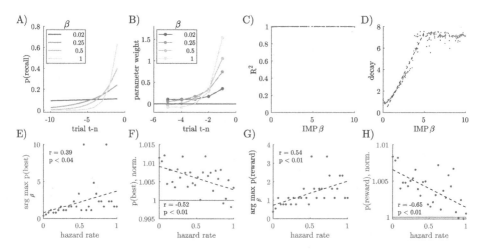

Fig. 3. IMPs integrate reward history via imperfect memory. A) Probability of recall as a function of previous trial for 4 example IMPs with different β. B) Reward history kernels for the 4 example IMPs, with exponential fits overlaid. C) R^2 as a function of β for 200 simulated IMPs with $\beta \sim U(10^{-2}, 10^1)$. D) Exponential decay parameter for the simulated IMPs in panel C. Gray line is the least squares linear fit for $0.5 \geq \beta \leq 5$ (slope = 1.54, offset = −0.23). E) The β that maximizes the probability of choosing the best option, plotted as a function of volatility (hazard), identified via grid search (20 log-distributed bins $\in [10^{-2}, 10^1]$). F) Probability of choosing the best option for the optimal IMP, plotted as a function of volatility. P(best) is normalized to WSLS performance at each volatility level (WSLS at 1). G-H) Same as panels E-F, for p(reward).

choosing the best option [10,33,35,36]. In order to determine whether IMPs tended to match or maximize, we simulated IMPs in a matching law testbed. We found that IMPs tended to match: allocating their choices in proportion to the rate of reward associated with each option (Fig. 4A). However, whereas a WSLS agent matched perfectly, some IMPs had a slight tendency towards maximizing (Fig. 4A, inset). In order to compare IMPs against a true value-integrating agent, we also simulated matching law behavior from an optimized SARSA agent ($\alpha = 0.8, \gamma = 0.9$). SARSA had a substantially stronger maximizing effect. Thus, although IMPs did have a slight tendency to maximize, they did not maximize to the full extent possible in this task.

4.4 IMPs Trade-Off Flexibility and Stability

In biological brains, there is a natural trade-off between the ability to persist in stable environments and the ability to adapt to change [8,20]. In order to determine if IMPs resolve this trade-off, we simulated their behavior in a probabilistic reversal learning task in which stable periods (where one option is clearly more valuable than the other) are interspersed with "reversals" (where the values flip).

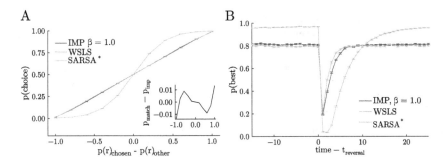

Fig. 4. IMPs tend to be matching agents with a slight tendency to maximize reward. (A) Probability of choosing a particular arm as a function of that arm's probability of reward minus the probability of reward of the unchosen arm. Inset: probability of choice of a perfect matching strategy (a straight line) minus the probability of choice for the IMP agent as a function of the arm's probability of reward minus the probability of reward of the unchosen arm. (B) Probability of choosing the best arm as a function of time relative to the onset of a reversal event. Tasks described in Sect. 3, with a hazard rate of 0.02. All agents received the same random walks in each session.

During stable periods, some IMPs were able to outperform WSLS agents (Fig. 4B) because they were better able to persist in choosing the high-value option despite the noisy reward. By contrast, at reversals, these same IMPs learned less quickly than the WSLS agents. To estimate what a value-integrating agent would do, we again simulated behavior from the optimal SARSA agent ($\alpha = 0.8, \gamma = 0.4$). Although the IMPs were less capable than SARSA during the stable periods state, IMPs adapted faster at reversals—already beginning to reverse after the first omitted reward, whereas SARSA required at least 2 omitted rewards. In sum, IMPs were again able to solve a classic sequential decision-making task and their performance levels were in between the extremes of a memory-less agent and a full reinforcement learning agent.

5 Discussion

The ability to robustly store and retrieve information about past interactions with the world is crucial for decision-making. In this paper, we have explored a simple, yet surprisingly competent model for decision-making that incorporates a "faulty" memory. We showed that Imperfect Memory Programs (IMPs) have a memory system that provably trades off the cost of retrieval errors and the cost of high complexity. We found that IMPs do well in classic decision-making tasks under uncertainty, performing similarly or slightly better than agents with a similar strategy but perfect memory. At the same time, IMPs showed characteristics that are more aligned to certain natural and normative decision-makers, such as an exponential reward history kernel, an increased persistence with chosen targets, and a slight tendency to maximize (versus match) rewards.

A memory retrieval system that trades off error and complexity costs also improves performance in structure learning [21]. In that context, imperfect memory can enhance generalization in hierarchically organized environments, task spaces, and other networks because it permits smoothing in learned associations in time. Here, we have applied a similar idea to the notion of reward learning and found that the same errors in retrieval help agents perform a rudimentary form of value integration. Given that human behavior is consistent with hierarchical reinforcement learning [9], imperfect memory may be a good candidate mechanism for smoothing reward associations in such hierarchically organized spaces.

This paper complements recent work on resource-limited decision-making. A particularly relevant line of work is solving a *reward-complexity* trade-off, where a highly complex policy can achieve high rewards, but is cognitively costly [1]. Although the policies derived also involve the Boltzmann distribution, our work aims specifically to solve a error-complexity trade-off in *memory recall*. Combining a resource-limited decision-making system and a resource-limited memory could be an opportunity to offer a parsimonious account of resource-limited cognitive systems more broadly [15, 29].

Memory limitations have also been explored in the field of reinforcement learning (RL). Because the deep neural networks used in RL are sample inefficient, several groups have begun using episodic memory-based systems to improve learning in data-limited settings [13, 28]. One approach leverages an external memory system to retrieve experiences of high relevance to the agent, in order to update the agent's policy and value function [22, 25]. In contrast, our approach does not start with the assumption that value is calculated at all. Instead, IMPs demonstrate that samples from episodic memory can be sufficient to imply that value integration is occurring even when it is not.

Although IMPs are simplistic, additional work here could have implications for artificial intelligence research. First, the stochastic retrieval of episodic memories could be extended to actions and states and not just rewards. Agents with stochastic retrieval in all 3 domains would presumably generalize better [21]. Second, retrieving multiple episodic memories might make IMPs even more capable, especially if samples are weighted according to time-relevance or state/action similarity. Overall, given that memory errors are consistent with realistic computations, showing how they can enhance generalization and performance is a promising avenue of research.

Acknowledgement. Support was provided by the Natural Sciences and Engineering Research Council of Canada (RGPIN-2020-05577), the Research Corporation for Science Advancement & Frederick Gardner Cottrell Foundation (Project 29087), the Canada Research Chair Dynamics of Cognition (FD507106) and a Research Fellowship from the Jacobs Foundation.

References

1. Bhui, R., Lai, L., Gershman, S.J.: Resource-rational decision making. Curr. Opin. Behav. Sci. **41**, 15–21 (2021). https://doi.org/10.1016/j.cobeha.2021.02.015
2. Bornstein, A.M., Khaw, M.W., Shohamy, D., Daw, N.D.: Reminders of past choices bias decisions for reward in humans. Nat. Commun. **8**(1), 15958 (2017). https://doi.org/10.1038/ncomms15958. https://www.nature.com/articles/ncomms15958
3. Butter, C.M.: Perseveration in extinction and in discrimination reversal tasks following selective frontal ablations in Macaca mulatta. Physiol. Behav. **4**(2), 163–171 (1969)
4. Chen, C.S., Ebitz, R.B., Bindas, S.R., Redish, A.D., Hayden, B.Y., Grissom, N.M.: Divergent strategies for learning in males and females. Curr. Biol. **31**(1), 39–50 (2021)
5. Chen, C.S., Knep, E., Han, A., Ebitz, R.B., Grissom, N.M.: Sex differences in learning from exploration. eLife **10**, e69748 (2021). https://doi.org/10.7554/eLife.69748
6. Daw, N.D., O'Doherty, J.P., Dayan, P., Seymour, B., Dolan, R.J.: Cortical substrates for exploratory decisions in humans. Nature **441**(7095), 876–879 (2006). https://doi.org/10.1038/nature04766
7. Ebitz, R.B., Albarran, E., Moore, T.: Exploration disrupts choice-predictive signals and alters dynamics in prefrontal cortex. Neuron **97**(2), 450–461.e9 (2018). https://doi.org/10.1016/j.neuron.2017.12.007
8. Ebitz, R.B., Sleezer, B.J., Jedema, H.P., Bradberry, C.W., Hayden, B.Y.: Tonic exploration governs both flexibility and lapses. PLoS Comput. Biol. **15**(11), e1007475 (2019). https://doi.org/10.1371/journal.pcbi.1007475
9. Eckstein, M.K., Collins, A.G.E.: Computational evidence for hierarchically structured reinforcement learning in humans. Proc. Natl. Acad. Sci. **117**(47), 29381–29389 (2020). https://doi.org/10.1073/pnas.1912330117. https://www.pnas.org/doi/10.1073/pnas.1912330117
10. Fantino, E.J.: Is maximization theory general, and is it refutable? Behav. Brain Sci. **4**(3), 390–391 (1981). https://doi.org/10.1017/S0140525X00009444. https://www.cambridge.org/core/journals/behavioral-and-brain-sciences/article/abs/is-maximization-theory-general-and-is-it-refutable/816AF98716906B780F8AD21BEC12B232
11. Findling, C., Chopin, N., Koechlin, E.: Imprecise neural computations as a source of adaptive behaviour in volatile environments. Nat. Hum. Behav. **5**(1), 99–112 (2021). https://doi.org/10.1038/s41562-020-00971-z. https://www.nature.com/articles/s41562-020-00971-z
12. Findling, C., Wyart, V.: Computation noise in human learning and decision-making: origin, impact, function. Curr. Opin. Behav. Sci. **38**, 124–132 (2021). https://doi.org/10.1016/j.cobeha.2021.02.018. https://www.sciencedirect.com/science/article/pii/S2352154621000401
13. Gershman, S.J., Daw, N.D.: Reinforcement learning and episodic memory in humans and animals: an integrative framework. Annu. Rev. Psychol. **68**, 101–128 (2017). https://doi.org/10.1146/annurev-psych-122414-033625. https://www.annualreviews.org/content/journals/10.1146/annurev-psych-122414-033625
14. Gregory, R.L.: Perceptions as Hypotheses. Philos. Trans. Roy. Soc. London Ser. B Biol. Sci. **290**(1038), 181–197 (1980). https://www.jstor.org/stable/2395424
15. Griffiths, T.L., Lieder, F., Goodman, N.D.: Rational use of cognitive resources: levels of analysis between the computational and the algorithmic. Top. Cogn. Sci.

7(2), 217–229 (2015). https://doi.org/10.1111/tops.12142. mAG ID: 2141467654 S2ID: 50485a11fc03e14031b08960370358c26553d7e5

16. Howard, M.W., Kahana, M.J.: A distributed representation of temporal context. J. Math. Psychol. **46**(3), 269–299 (2002). https://doi.org/10.1006/jmps.2001.1388. https://www.sciencedirect.com/science/article/pii/S0022249601913884
17. Jurewicz, K., Sleezer, B.J., Mehta, P.S., Hayden, B.Y., Ebitz, R.B.: Irrational choices via a curvilinear representational geometry for value. bioRxiv (2022). https://www.biorxiv.org/content/10.1101/2022.03.31.486635.abstract
18. Lau, B., Glimcher, P.W.: DYnamic Response-by-response Models Of Matching Behavior In Rhesus Monkeys. J. Exp. Anal. Behav. **84**(3), 555–579 (2005). https://doi.org/10.1901/jeab.2005.110-04. http://doi.wiley.com/10.1901/jeab.2005.110-04
19. Laurie, V.J., Shourkeshti, A., Chen, C.S., Herman, A.B., Grissom, N.M., Ebitz, R.B.: Persistent Decision-Making in Mice, Monkeys, and Humans. bioRxiv (2024). https://www.biorxiv.org/content/10.1101/2024.05.07.592970.abstract
20. Liljenström, H.: Neural stability and flexibility: a computational approach. Neuropsychopharmacology **28**(1), S64–S73 (2003). https://www.nature.com/articles/1300137
21. Lynn, C.W., Kahn, A.E., Nyema, N., Bassett, D.S.: Abstract representations of events arise from mental errors in learning and memory. Nat. Commun. **11**(1), 2313 (2020). https://doi.org/10.1038/s41467-020-15146-7. https://www.nature.com/articles/s41467-020-15146-7
22. Mattar, M.G., Daw, N.D.: Prioritized memory access explains planning and hippocampal replay. Nat. Neurosci. **21**(11), 1609–1617 (2018). https://doi.org/10.1038/s41593-018-0232-z. https://www.nature.com/articles/s41593-018-0232-z
23. McNamara, J.M., Houston, A.I.: Memory and the efficient use of information. J. Theor. Biol. **125**(4), 385–395 (1987). https://doi.org/10.1016/S0022-5193(87)80209-6. https://www.sciencedirect.com/science/article/pii/S0022519387802096
24. Nicholas, J., Daw, N.D., Shohamy, D.: Uncertainty alters the balance between incremental learning and episodic memory. eLife **11**, e81679 (2022). https://doi.org/10.7554/eLife.81679
25. Patel, N., Acerbi, L., Pouget, A.: Dynamic allocation of limited memory resources in reinforcement learning. In: Advances in Neural Information Processing Systems, vol. 33, pp. 16948–16960. Curran Associates, Inc. (2020). https://proceedings.neurips.cc/paper/2020/hash/c4fac8fb3c9e17a2f4553a001f631975-Abstract.html
26. Pearson, J.M., Hayden, B.Y., Raghavachari, S., Platt, M.L.: Neurons in posterior cingulate cortex signal exploratory decisions in a dynamic multioption choice task. Curr. Biol. **19**(18), 1532–1537 (2009). https://doi.org/10.1016/j.cub.2009.07.048. https://linkinghub.elsevier.com/retrieve/pii/S0960982209014742
27. Pisupati, S., Chartarifsky-Lynn, L., Khanal, A., Churchland, A.K.: Lapses in perceptual decisions reflect exploration. Elife **10**, e55490 (2021). https://elifesciences.org/articles/55490
28. Ramani, D.: A Short Survey On Memory Based Reinforcement Learning (2019). http://arxiv.org/abs/1904.06736. arXiv:1904.06736
29. Ramírez-Ruiz, J., Moreno-Bote, R.: Optimal allocation of finite sampling capacity in accumulator models of multialternative decision making. Cogn. Sci. **46**(5), e13143 (2022). https://doi.org/10.1111/cogs.13143. https://onlinelibrary.wiley.com/doi/10.1111/cogs.13143
30. Ranc, N., Moorcroft, P.R., Ossi, F., Cagnacci, F.: Experimental evidence of memory-based foraging decisions in a large wild mammal. Proc. Natl. Acad. Sci.

118(15), e2014856118 (2021). https://doi.org/10.1073/pnas.2014856118. https://www.pnas.org/doi/abs/10.1073/pnas.2014856118
31. Rescorla, R.A., Wagner, A.R.: A theory of Pavlovian conditioning: variations on the effectiveness of reinforcement and non-reinforcement. In: Black, A.H., Prokasy, W.F. (eds.) Classical conditioning II: Current research and theory, pp. 64–99. Appleton-Century-Crofts, New York (1972)
32. Robbins, H., Monro, S.: A stochastic approximation method. Ann. Math. Stat. **22**(3), 400–407 (1951). https://doi.org/10.1214/aoms/1177729586. https://projecteuclid.org/journals/annals-of-mathematical-statistics/volume-22/issue-3/A-Stochastic-Approximation-Method/10.1214/aoms/1177729586.full
33. Sakai, Y., Fukai, T.: When does reward maximization lead to matching law? PLoS ONE **3**(11), e3795 (2008). https://doi.org/10.1371/journal.pone.0003795. https://journals.plos.org/plosone/article?id=10.1371/journal.pone.0003795
34. Shourkeshti, A., Marrocco, G., Jurewicz, K., Moore, T., Ebitz, R.B.: Pupil size predicts the onset of exploration in brain and behavior. bioRxiv (2023). https://www.ncbi.nlm.nih.gov/pmc/articles/PMC10245915/
35. Soltani, A., Wang, X.J.: A biophysically based neural model of matching law behavior: melioration by stochastic synapses. J. Neurosci. **26**(14), 3731–3744 (2006). https://doi.org/10.1523/JNEUROSCI.5159-05.2006. https://www.jneurosci.org/content/26/14/3731
36. Sugrue, L.P., Corrado, G.S., Newsome, W.T.: Matching behavior and the representation of value in the parietal cortex. Science **304**(5678), 1782–1787 (2004). https://doi.org/10.1126/science.1094765. https://www.science.org/doi/full/10.1126/science.1094765
37. Sutton, R.S., Barto, A.G., et al.: Introduction to Reinforcement Learning. MIT Press, Cambridge (1998)
38. Wilson, R.C., Collins, A.G.: Ten simple rules for the computational modeling of behavioral data. eLife **8**, e49547 (2019). https://doi.org/10.7554/eLife.49547. https://doi.org/10.7554/eLife.49547
39. Wylie, G., Allport, A.: Task switching and the measurement of "switch costs." Psychol. Res. **63**(3), 212–233 (2000). https://doi.org/10.1007/s004269900003. https://doi.org/10.1007/s004269900003

The Role of Theory of Mind in Finding Predator-Prey Nash Equilibria

Tiffany Hwu[1]()), Chase McDonald[1,2], Simon Haxby[1], Flávio Teixeira[1], Israel Knight[1], and Albert Wang[1]

[1] Riot Games, Santa Monica, CA 90064, USA
thwu@riotgames.com
[2] Carnegie Mellon University, Pittsburgh, PA 15213, USA

Abstract. When a predator chases its prey, a mind game ensues, requiring both predator and prey to predict what the other will do next. These elements of uncertainty and opponency are also seen in analyses of real-world tasks and games. For instance, one way to define an optimal solution of a non-cooperative game is to find the Nash equilibrium, a state in which each agent in a game has optimized its strategy given the strategies of others. The Regularized Nash Dynamics (R-NaD) algorithm guarantees that policies will converge to the Nash equilibrium, creating AIs that beat top human players in tasks with hidden information. Our research compares the performance of deep reinforcement learning agents trained with and without R-NaD in a simple hide-and-seek game, aiming to see how well the agents process unknowns in the environment. We then apply explainable AI (XAI) techniques to the trained model to examine the kinds of information that trained policies encode about opponent strategies. We find that policies trained with R-NaD outperform policies trained in regular self-play when there is hidden information. Furthermore, R-NaD policies use their opponent's past positions to decide which actions to take, more so than regular self-play. These findings yield insights on how animals and artificial agents operate under spatial uncertainty.

Keywords: Nash Equilibrium · Explainable AI

1 Introduction

When animals compete in an uncertain environment, such as in the predator-prey scenario, they must account for where the enemy may travel next and optimize over all possibilities. As seen in empirical experiments [2], as well as in simulation [4], this sort of theory of mind may play a large role in projecting possible outcomes. To help us understand the predator-prey scenario, we turn to the game-theoretic concept of the Nash equilibrium. The Nash equilibrium is a set of strategies over all players where no player can do better by unilaterally changing their own strategy. Put another way, the Nash equilibrium is the optimal strategy for each player, given that all other players are also playing optimal strategies. Finding the Nash equilibrium is particularly useful in games of hidden information, where there is often no deterministic best action, and agents must choose actions stochastically in order to not be exploited by their adversary. In the Regularized Nash Dynamics (R-NaD) paper [6], the authors introduced a method of training a reinforcement learning model that guaranteed convergence to the Nash equilibrium. With their method, they were able to train an AI to beat top human players at Stratego, a game that has an abundance of hidden information. While the effectiveness of R-NaD was clearly demonstrated in this work, it is still unintuitive to understand why and how the trained policies are so responsive to hidden information. In the process of finding the Nash equilibrium, one may think of the agents as imagining the different possibilities of what their opponent may do. While R-NaD does not require any explicit structure to model the opponent's actions, are there still traceable elements of theory of mind embedded within the model?

In this paper, we hope to gain a better understanding of what kinds of information are encoded by models trained on the predator-prey scenario. Specifically, we hypothesize that agents trained with R-NaD will emphasize information about the opponent's moves. We explore this hypothesis by training a deep reinforcement learning policy with and without R-NaD on a hide-and-seek game within simple grid-world mazes. The mazes differ in amounts of hidden information, revealing insights on how the agents handle spatial unknowns. This paper makes the following novel contributions: 1) The R-NaD algorithm, which was previously demonstrated in the game of Stratego, is extended to a predator-prey game involving spatial uncertainty, and 2) Explainable AI techniques are applied to the trained models to see how a trained R-NaD model may differ from a regular self-play model.

2 Background

2.1 R-NaD

In multi-agent deep reinforcement learning, one common method for training agents is self-play. In self-play, each agent is controlled by a copy of a single policy. By playing copies of itself, the policy continually learns how to improve upon its current performance. While self-play can be effective at finding good deterministic strategies, it fails to converge when there are strategy cycles in the game and where the Nash equilibrium policy must be stochastic. The game Rock Paper Scissors is an example of this. As seen in Fig. 1, self-play leads to a continuous cycling of rock, then paper, then scissors as the dominant strategy. The policy never converges upon the optimal strategy of selecting the actions at equal probability.

In R-NaD, a reward regularizer is applied to the reward component of self-play:

$$r^i(\pi^i, a^i) = r^i(a^i) - \eta log(\frac{\pi^i(a^i)}{\pi^i_{reg}(a^i)}). \tag{1}$$

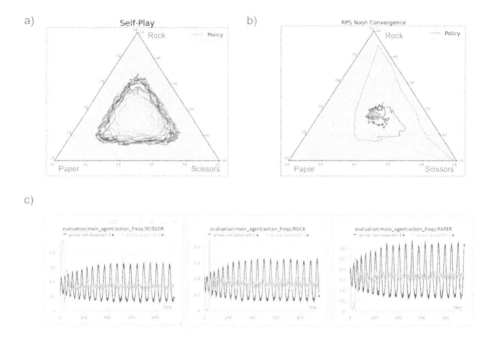

Fig. 1. Comparing regular self-play and R-NaD on the game of Rock Paper Scissors. a) Simplex diagram of a policy as it is trained in self-play. The strategy continuously cycles around the equilibrium. b) Simplex diagram of a policy as it is trained with R-NaD. The policy quickly converges to the Nash equilibrium. c) Action frequencies of rock, paper, and scissors across training steps. The action frequencies oscillate continuously in regular self-play whereas they quickly converge in R-NaD.

1. **Reward Transformation**: A regularization policy $\pi_{n,reg}$ is initialized. It can initially be set to an arbitrary policy.
2. **Dynamics**: The policy is trained as in regular self-play. By applying the regularized reward function, the policy follows replicator dynamics, eventually converging to $\pi_{n,fix}$.
3. **Update**: The regularization policy is set: $\pi_{n+1,reg} = \pi_{n,fix}$.

Applying these steps to a regular self-play algorithm, which may be any existing reinforcement learning algorithm such as PPO, the policy is guaranteed to converge to a Nash equilibrium. As regular self-play has no such guarantees, R-NaD would be much more likely to yield better policies.

2.2 Explainable AI

In R-NaD, there are no extra restrictions on the training of model parameters, requiring only the reward regularization. It would therefore be interesting to inspect the parameters of models trained under regular self-play and under R-NaD to see if there are distinguishable differences in the types of information encoded. Explainable AI (XAI) techniques enable us to examine the parameters of trained neural networks. One such technique is called layer-wise relevance propagation, or LRP [1,5]. This technique can be used on deep neural networks, or in fact any multi-layered algorithm in which each layer is calculated via the weighted combination of the previous layer. In LRP, relevance is a mass that is conserved as it propagates backwards from output to input. For a given output neuron k and input neuron j, the relevance of neuron j is calculated as the following:

$$R_j = \sum_k \frac{a_j w_{jk}}{\epsilon + \sum_{0,j} a_j w_{jk}} R_k, \qquad (2)$$

where R_j is the relevance of neuron j, a_j is the activation of neuron j, w_{jk} is the weight from neuron j to neuron k, R_k is the relevance of neuron k, and ϵ is a small error term to absorb some relevance from weaker inputs, leaving only the stronger relevance signals. There are other variants of this LRP rule as well, some emphasizing positive relevance scores over negative ones. As shown in Fig. 2, LRP can be used to create a visual heatmap of relevance for a given input example. In this paper, we use LRP to gain insights on how the agents are encoding information about their opponents.

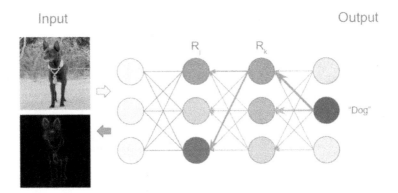

Fig. 2. Overview of LRP. An input, which in this case is an image, is fed through a deep neural network to produce a classification of the input. A relevance score for the output is first assigned to be the output values themselves. Then, this relevance score is treated as a conserved mass, dividing itself amongst the neurons of the previous layer such that the sum of relevance in both layers remains the same. The relevance is propagated backwards until it reaches the input. If the input is an image, a relevance heatmap may be visualized.

3 Methods

3.1 Training Environment

The gridworld environment is based on an existing open source reinforcement learning environment [3]. For this experiment, grids are set to 9 rows and 9 columns. Terrain types are restricted to two: traversable and non-traversable. The visibility of each square can be configured for each player and is fixed for each game. By default, each player has visibility of their neighboring squares in eight directions. The possible actions are limited to NONE, UP, RIGHT, DOWN, and LEFT. There are two players, a hider and a seeker, which are represented by mouse and cat icons, respectively. Both hider and seeker step through the environment simultaneously, acting at the same time. The objective of the hider is to avoid getting caught, and it receives a small reward of 5e-5 for each timestep it is not caught and -1 when it gets caught. Getting caught is defined as the seeker occupying the same square as the hider. The objective of the seeker is to catch the hider, and it receives a negative reward of -5e-5 for each timestep it has not caught the hider, and 1 when it catches the hider. Thus, the rewards are zero-sum, compatible with R-NaD. To encourage exploration, the starting positions of the hider and seeker are randomized at the beginning of each episode.

3.2 Neural Network Architecture

The PPO algorithm [7] is used to train neural network policies to play the game. The neural network architecture consists of a simple convolutional network. The layers of the network are described in Table 1 and the hyperparameters are described in Table 2.

Table 1. Neural Network Architecture

Layer	Dimensions
Convolutional Layer	$3 \times 3 \times 10$
Fully Connected Layer	812×256
Rectified Linear Unit	-
Fully Connected Layer	256×5

A single neural network policy of this architecture is used to represent both players in the game. The inputs to the neural network consist of a stack of two 9 by 9 grids, one grid representing the player's own location trace and one grid representing the opponent's location trace. The opponent's location trace only shows the visible portion of locations from the player's perspective, whereas the player's own location trace shows all locations, regardless of visibility. Both grids start as zero values at the beginning of a game, with a value of 1 set at

Table 2. Hyperparameters

Hyperparameter	Value
Train batch size	128
Learning rate	4e-5
Gamma	0.995
Entropy coefficient	3e-3
Clipping	40.0
Eta (R-NaD)	0.2
Regularizer update interval (R-NaD)	100 iterations

the player's location in the first grid and 1 set at the opponent's location in the second grid. At each time step, a trace of current and past states is encoded by setting the value of the players' locations to 1 on the corresponding grid location in the input and decaying all previous values by multiplying them by .9. Two additional inputs include the number of timesteps remaining out of a maximum of 50, and a binary input of whether the player is the seeker or not. These additional inputs are concatenated to the flattened output of the first convolutional layer.

3.3 Experimental Conditions

With this neural network architecture, two conditions are compared: regular self-play and self-play with R-NaD. In regular self-play, the PPO algorithm is applied. In R-NaD, the training steps are the same, but the regularizer is applied to the reward function. Figure 3 describes the mazes used in the experiment. Maze 1 is in the shape of a plus sign, with all squares of the plus sign visible to both players. The optimal behavior of this maze would be for the seeker to immediately find the hider. Maze 2 is also a plus sign, but the intersection squares are visible to the hider only. The lack of cyclical paths and the chokepoint of the centermost square make both mazes advantageous for the seeker. However, in Maze 2, the hider can find opportunities to escape into different arms of the maze, taking advantage of the seeker's lack of visibility. By escaping to different arms, the hider can delay its eventual capture.

4 Results

4.1 Training Results

The policies were trained for approximately 4000 training iterations (network updates) each. Since the rewards were zero-sum, the reward could not be used to monitor performance when training. Since our mazes were simple and the optimal solutions were known, we used the metric of episode length to approximate the convergence and performance of the policies. Figure 4 shows episode

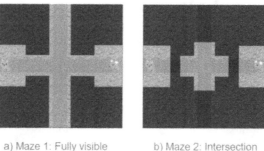

a) Maze 1: Fully visible
b) Maze 2: Intersection visible to hider only

Fig. 3. Mazes used in experiment, where green squares are traversable, brown squares are nontraversable, and lit squares are visible to at least one player. In both mazes, players have visibility of immediately neighboring squares. a) A plus sign maze. Every traversable square is visible to both the hider and the seeker. b) The same plus sign maze but with the intersection squares visible to the hider only. (Color figure online)

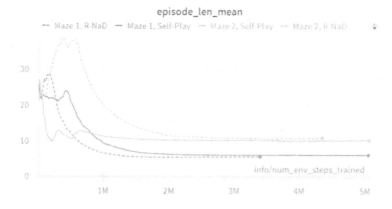

Fig. 4. Episode length of training episodes over time. For Maze 1, the episode length of R-NaD decreased faster than regular self-play, indicating that the policy was learning the optimal solution faster. For Maze 2, episode length initially increased for R-NaD. This may be an indication of extra spatial exploration from the R-NaD policy, enabling seekers to learn to escape to different arms of the maze.

length over the course of training. For Maze 1, we expected episode length to decrease over time as the seeker should learn to move directly towards the hider, giving the hider no time for escape. We saw that the episode length indeed decreased and converged, decreasing faster in R-NaD than in regular self-play. For Maze 2, the episode length was not as clear of a signal for performance, as an optimal policy would show some increase in episode length if the hider learned to escape into different arms of the maze rather than giving up and staying in place. We saw that episode length increased far more in R-NaD and stayed a bit above the episode length of self-play as it decreased and converged.

This matched our empirical observation of hiders finding some opportunities to escape. While the opportunities were rare, the acts of escaping increased the episode length noticeably. Table 3 shows the number of squares covered when evaluating the trained policy on 25 games. The results show that the R-NaD hider traversed more squares than the self-play hider. In Maze 1, both policies had the seeker covering just enough squares to reach the hider, while the hider covered few squares as there was no way for it to escape. In Maze 2, we ideally would have liked to see the R-NaD hider take advantage of every single opportunity for escape, always moving into an arm of the maze that the seeker explored. We did not reach this perfect behavior, likely due to the difficulty of finding the exact time windows for escape. However, the R-NaD policy still covered more squares due to its ability to find at least a few escape opportunities. Overall, we saw that both self-play and R-NaD could learn the optimal solution for Maze 1. For Maze 2, where there was hidden spatial information, the R-NaD policy attempted to escape into other arms of the maze, whereas the regular self-play policy tended to stay in place. These training results show an advantage of the R-NaD algorithm when there is hidden spatial information.

Table 3. Number of squares covered per game over 25 games, SP = self-play, RN = R-NaD. This serves as a rough estimation of the amount of exploration of a policy. In Maze 1, the optimal solution was for the seeker to cover 8 squares to catch the hider quickly, while the hider had no extra reward for moving, as it was not possible to escape. In Maze 2, the R-NaD hider explored more squares. This accounts for the instances in which the hider learned to escape into a different arm of the maze.

	SP, Seeker	RN, Seeker	SP, Hider	RN, Hider
Maze 1	7.64 (SE:0.10)	7.68 (SE:0.09)	1.76 (SE:0.17)	1.12 (SE:0.06)
Maze 2	10.72 (SE:0.57)	11.44 (SE:0.47)	1.60 (SE:0.13)	2.12 (SE:0.38)

4.2 XAI Results

It was clear that R-NaD policies had some quantitative behavioral advantages over regular self-play policies, as expected. However, it was less clear how the trained parameters contributed to this advantage. To examine this, we applied LRP to a specific test scenario of the game. With the trained policies, we set up a series of actions for the hider and seeker to perform, creating a position in which the hider had to use some theory of mind to decide on the next optimal action. Figure 5 shows a series of setup moves performed in Maze 2. The seeker traversed three arms of the maze, while the hider remained hidden, advancing to the centermost square only when the seeker was out of sight. We expected the optimal action of the hider to be moving left or right. Since the seeker had no vision beyond its immediate vicinity, its best strategy was to explore all arms of the maze. At the end of the setup sequence, while the seeker was away from the intersection, we expected the hider to try to escape into an arm that the seeker had already explored.

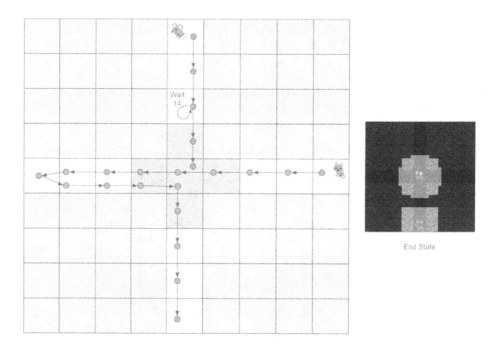

Fig. 5. Setup moves taken before using XAI techniques to examine the next action. The seeker explored three arms of the maze, while the hider remained out of sight, using the visible intersection squares to figure out where the seeker had visited. The hider then moved into the centermost square, setting up a decision opportunity. The optimal solution for the hider was to travel into a part of the maze that had already been explored by the seeker.

Figure 6 shows heatmaps from the perspectives of the hider and seeker after the setup moves had been made. Since the seeker had not seen the hider at all during the setup sequence, it could only rely on its own movement trace to make decisions. A notable finding is seen in the R-Nad hider of Maze 2, where the hider placed much relevance on the seeker's last seen locations to decide to go left, where the seeker had already visited. In Maze 2, the R-NaD hider placed much more relevance on the opponent than the self-play hider did.

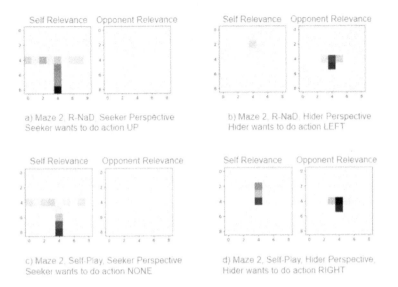

Fig. 6. LRP heatmaps from perspectives of the hider and seeker. Red values indicate positive values and blue values indicate negative values. The color bar ranges from $-1 * max(abs(relevance_{max}), abs(relevance_{min}))$ to $max(abs(relevance_{max}), abs(relevance_{min}))$. The heatmap only includes position traces of the hider and seeker, not the extra input information about steps remaining or player identity. a) The R-NaD seeker relies on its own position trace, with an interesting striped pattern in the downward arm. b) The R-NaD hider places great relevance on the seeker's past history of locations seen in the intersection. c) The self-play seeker seems to have a heatmap that matches closely with its own position trace. d) The self-play hider places relevance on its own trace, with some mixed relevance placed on the opponent. (Color figure online)

Figure 7 shows the heatmaps of the top 2 most probable actions after the setup sequence, from the hider's perspective only. While both R-NaD and self-play highly favor the top action at 99%, the top two actions from R-NaD represent the optimal choices, with left and right actions both being previously visited locations of the seeker, and the left arm being the most recently visited arm. The regular self-play policy's actions are not as optimal, and there are more negative relevance values in the opponent's trace. This may mean that regular self-play policies tend to use the opponent's absence to decide where to move. Table 4 compares the relevance of a player's own trace compared to the opponent's trace, a measure of theory of mind. In Maze 2, where there is hidden information and a need for theory of mind, R-NaD policies placed more relevance on their opponents' traces than regular self-play policies did. This was particularly important for the hider, as timing and planning an escape to a different arm requires great opponent awareness.

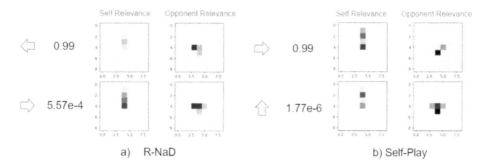

Fig. 7. LRP heatmap explanations for the top 2 most probable next actions of the hider and seeker, from the hider's perspective only. In both R-NaD and self-play, the top action is strongly preferred, at .99 probability. a) In R-NaD, there is much more emphasis placed on the opponent's trace. The top two actions represent the most optimal choices, as both left and right are places that the seeker has already visited. b) In self-play, there is emphasis placed on its own trace and mixed relevance on the opponent's trace. It does not make the most optimal choices.

Table 4. Proportion of LRP input relevance on opponent's position trace over 25 games. Given the LRP heatmaps, the proportion of relevance of the opponent's trace over the total relevance of both traces was calculated. The proportion was calculated over positive relevance scores only. This serves as an indication of how much a policy regards its opponents moves when selecting actions. In Maze 2, R-NaD policies have more relevance on the opponent's position trace than regular self-play.

	SP, Seeker	RN, Seeker	SP, Hider	RN, Hider
Maze 1	0.454 (SE:0.06)	0.276 (SE:0.04)	0.143 (SE:0.02)	0.182 (SE:0.02)
Maze 2	0.001 (SE:0.00)	0.021 (SE:0.00)	0.136 (SE:0.03)	0.307 (SE:0.02)

5 Conclusions and Future Directions

In this paper, we have examined the R-NaD algorithm, applying it to a simple environment with spatial uncertainty. We saw that when there was hidden spatial information in the maze, policies trained with R-NaD accounted for their opponent's position traces more often than policies trained with regular self-play. Empirically, R-NaD hiders were more likely to escape into different arms of the maze. These findings may indicate that R-NaD policies use some elements of theory of mind. For a relatively simple game such as the one used in this experiment, it may be quite possible to adjust the settings of regular self-play to improve performance. However, the convergence guarantees of R-NaD lessen the need for such adjustments. Expanding the mazes to more complex environments may yield more insights into the long term planning behaviors of the agents. For instance, a maze with multiple intersections will require more creative strategies. More agency can be given in controlling the visibility of the maze by allowing players to place down visibility wards anywhere they choose. It would also be interesting to try this analysis on the trained Stratego model, where the hidden information lies in the identity of the pieces rather than spatial location.

The opponency of predator-prey relationships in uncertain spatial environments extends to many real-world applications, for instance, the multi-agent motion planning used in autonomous driving and swarm coordination. Many video games have spatial hidden information as well, and the design of game AI bots would benefit from being able to navigate in such environments. In existing game AI navigation algorithms, cost-based path planners lead to deterministic, predictable pathing. A planning algorithm utilizing R-NaD can find the optimal probabilities of path-taking, leading to more interesting bots in-game. Furthermore, the application of XAI to trained R-NaD policies can create powerful explanation tools for coaching. Tasks requiring a human to learn probabilistic actions can be quite unintuitive. A technique that ties these probabilities to the behaviors of other agents can help humans attend to relevant information.

Acknowledgments. We credit Riot Games for funding this research.

Disclosure of Interests. We declare that the authors have no competing interests.

References

1. Bach, S., Binder, A., Montavon, G., Klauschen, F., Müller, K.R., Samek, W.: On pixel-wise explanations for non-linear classifier decisions by layer-wise relevance propagation. PLoS ONE **10**(7), e0130140 (2015)
2. Ellis, B.J., Jordan, A.C., Grotuss, J., Csinady, A., Keenan, T., Bjorklund, D.F.: The predator-avoidance effect: an evolved constraint on emerging theory of mind. Evol. Hum. Behav. **35**(3), 245–256 (2014)
3. Knight, I.: So you want to make AI bots? A gentle intro into reinforcement learning. Github (2023). https://github.com/isknight/rl_intro
4. Krichmar, J.L., Hwu, T., Zou, X., Hylton, T.: Advantage of prediction and mental imagery for goal-directed behaviour in agents and robots. Cogn. Comput. Syst. **1**(1), 12–19 (2019)
5. Montavon, G., Binder, A., Lapuschkin, S., Samek, W., Müller, K.R.: Layer-wise relevance propagation: an overview. In: Explainable AI: Interpreting, Explaining and Visualizing Deep Learning, pp. 193–209 (2019)
6. Perolat, J., et al.: Mastering the game of stratego with model-free multiagent reinforcement learning. Science **378**(6623), 990–996 (2022)
7. Schulman, J., Wolski, F., Dhariwal, P., Radford, A., Klimov, O.: Proximal policy optimization algorithms. arXiv preprint arXiv:1707.06347 (2017)

Nonverbal Immediacy Analysis in Education: A Multimodal Computational Model

Uroš Petković[1,3](✉), Jonas Frenkel[2,3], Olaf Hellwich[1,3], and Rebecca Lazarides[2,3]

[1] Computer Vision and Remote Sensing, Technische Universität Berlin, Berlin, Germany
uros.petkovic@tu-berlin.de
[2] Department of Educational Sciences, University of Potsdam, Potsdam, Germany
[3] Science of Intelligence (SCIoI), Cluster of Excellence, Berlin, Germany

Abstract. This paper introduces a novel computational approach for analyzing nonverbal social behavior in educational settings. Integrating multimodal behavioral cues, including facial expressions, gesture intensity, and spatial dynamics, the model assesses the nonverbal immediacy (NVI) of teachers from RGB classroom videos. A dataset of 400 30-s video segments from German classrooms was constructed for model training and validation. The gesture intensity regressor achieved a correlation of 0.84, the perceived distance regressor 0.55, and the NVI model 0.44 with median human ratings. The model demonstrates the potential to provide a valuable support in nonverbal behavior assessment, approximating the accuracy of individual human raters. Validated against both questionnaire data and trained observer ratings, our models show moderate to strong correlations with relevant educational outcomes, indicating their efficacy in reflecting effective teaching behaviors. This research advances the objective assessment of nonverbal communication behaviors, opening new pathways for educational research.

Keywords: Automatic Classroom Observation · Nonverbal Communication · Gesture Analysis · Facial Expression Recognition

1 Introduction

The way in which instructors communicate is crucial for successful knowledge transfer. While traditionally, research has often focused on verbal communication as the primary mode of interaction in educational settings, empirical evidence highlights the critical role of instructors' nonverbal behaviors in promoting cognitive, affective, and motivational learning [27].

The field of computer science, particularly within the domains of Social Signal Processing [26], Affective Computing [21], Educational Technology [13], and Human-Robot Interaction (HRI) [9], has increasingly recognized the substantial influence of nonverbal communication in educational contexts. Advanced computational models and machine learning techniques offers novel avenues, that may hold distinct advantages over traditional observer or questionnaire assessments of nonverbal behaviors. Computer vision-based approaches, for instance, could provide increased objectivity and consistency by applying uniform criteria across diverse instances, mitigating subjective human biases [26]. Additionally, these techniques allow for unobtrusive observations, preserving the authenticity of the educational setting by minimizing observer effects [6]. The scalability and efficiency of such systems facilitate the processing of large volumes of video data, while supporting cost-effective longitudinal studies, thereby aiding the evaluation of educational interventions [15]. Yet, despite the considerable advances, the challenge of capturing the complex and interactive dynamics of nonverbal signals remains significant. Existing research has predominantly adopted a reductive approach, focusing on specific nonverbal cues such as pointing or gaze, often analyzing them in isolation from each other [1,3]. The nature of nonverbal communication, however, is characterized by the intricate interaction of multiple cues that function collectively. The effects of individual cues are not merely additive; rather, the combination of different nonverbal cues can alter their individual intensities and inherent meanings [17]. As social cues represent a unified percept, they need be assessed holistically [9,29], necessitating an integrated approach to studying the collective effects of nonverbal behaviors alongside individual cues.

A construct that effectively describes nonverbal behavior in such a manner is that of nonverbal immediacy (NVI) [14]. Closely related to the construct of enthusiasm, NVI comprises behaviors that transmit positive signals, such as sympathy and warmth, effectively reducing the psychological distance between instructors and students. Common behaviors associated with NVI include gestures, eye contact, a relaxed posture, and smiling [12]. Meta-analyses have consistently demonstrated positive correlations between NVI and various cognitive and affective-motivational learning outcomes [12,27].

Against this background, this paper presents a novel computer vision-based approach for estimating NVI scores of teachers by integrating multiple dimensions of nonverbal cues. Our research aims to develop computational models that analyze facial expressions, gesture intensity, and spatial dynamics collectively. By leveraging machine learning techniques, we aim to provide a more comprehensive understanding of nonverbal communication in an educational context.

Our main contributions in this paper are as follows: a) We present the first known effort to estimate NVI computationally from video recordings. b) We develop a gesture intensity measurement model that captures a continuous spectrum of gestures, moving beyond binary recognition to more subtly distinguish variations in gesture intensity. c) Our research introduces the first model for assessing perceived distance as a key aspect of spatial dynamics in educational settings. d) We have developed a specialized dataset, labeled by trained raters, for the training and validation of models in NVI, perceived distance, and gesture intensity.

2 Related Work

This section presents a review of studies on teachers' behaviors in educational settings, categorized into two main areas: 1) individual behavior recognition identifies specific actions and movements, such as hand raises and body poses, and 2) overall teaching behavior assessment evaluates teachers' performance by analyzing broader behavioral patterns and interactions, providing insights into teaching quality and classroom dynamics.

Individual Behavior Analysis. In [1], depth cameras extract skeletons for classroom monitoring, enhancing teacher development by analyzing activities. The models use these skeletons to identify behaviors like hand raises, body poses, and speech patterns via machine learning. The dataset includes 30 participants (5 instructors, 25 students), with 1545 body instances and 60 speech/silence audio instances. Challenges include the need for high-resolution cameras for precise skeleton extraction and adapting to diverse classroom layouts.

Similarly, TeachLivE [2] tracks teachers' interactions with virtual 3D students using skeleton extraction, offering posture feedback. In a study with 34 trainee teachers, initial skeletal posture feedback led to improved body language later, showcasing real-time behavioral assessment's efficacy. However, TeachLivE's need for clear views for accurate tracking and extraction limits its use in typical classrooms.

The study presented in [3], a video-based motion estimation model evaluates teachers' non-verbal behaviors in classrooms, including gesturing and walking. Analyzing nine lectures from a Canadian university, it utilizes motion detection, camera pan, and zoom techniques on 5415 30-s video segments. While effective for sustained activities like walking, accuracy for brief gestures is constrained by the 30-s analysis window and the limited dataset of nine teachers. Shorter time frames and a diverse dataset are necessary to enhance detection of transient movements.

In the research documented in [28], videos of distinguished educators are analyzed for behaviors like blackboard-writing, questioning, displaying, instructing, describing, and non-gesture behavior. Using 3-s clips, the study combines RGB video and skeleton data, exploring early and late fusion methods. It stresses the necessity of detailed dataset information to improve model generalizability, highlighting the limited effectiveness of late fusion techniques here.

These single-behavior approaches focus primarily on action recognition using skeletal data. While this approach significantly reduces visual information and improves robustness, it also lacks crucial contextual details needed to estimate behaviors such as gesture intensity and perceived distance. To address these gaps, our proposed models retain visual information, allowing for a more comprehensive analysis. Furthermore, unlike previous studies that rely on depth cameras, our approach does not require such specialized equipment, thereby increasing its practicality for diverse educational settings.

Overall Teaching Behavior Assessment. In the study [25], a novel framework is proposed for analyzing teacher perceptions in classroom settings, utilizing a combination of egocentric video recordings and mobile eye tracker data. This method employs face detection and tracking technologies to meticulously monitor the teacher's focus on individual students, offering insights into attention patterns in relation to student characteristics such as gender. While this approach represents a significant advancement in educational research, enabling detailed attention maps and a deeper understanding of teacher-student interactions, it is constrained by the small size of the dataset and the challenges posed by variable recording conditions, such as motion and lighting variations.

In [4], a model evaluates university instructors' teaching enthusiasm through sound feature extraction, facial expression recognition, and pose estimation. A cascading feature fusion method enhances predictive accuracy, with estimations analyzed using a BP neural network. Challenges arise in processing missing values, like substituting data when faces or body parts are obscured, potentially impacting prediction accuracy. The dataset includes 1004 10-s videos of 46 university teachers, offering a broad foundation for analyzing teaching behaviors across diverse course settings.

The study in [19] used the ACORN system to assess the 'positive climate' in classrooms by analyzing 15-min videos, longer than typical datasets. This included the UVA Toddler dataset with 300 videos and the Measures of Effective Teaching (MET) dataset [8], with 5574 CLASS-coded segments [18]. The MET dataset and the related experiment focused exclusively on auditory data. The system utilized auditory features, such as low-level acoustic signals, key phrase detection, and speech recognition, which made up the larger proportion of the input, along with visual features like facial expressions. Unlike our focus on predicting the NVI of teachers, their estimation of positive and negative climate included both teachers and students.

While there is a lack of studies specifically focused on measuring NVI, the broader field of automated teaching behavior assessment, which includes nonverbal elements, has seen some exploration. This field shows a significant variation in observation time windows, ranging from brief 10-s clips in [4] to more extended 15-min sessions in [19]. These differences in approaches illustrate the challenges related to the duration of video analysis, as well as the diversity of sensors utilized, ranging from standard RGB cameras to advanced eye-tracking devices.

3 Dataset and Labeling

The study utilized the TALIS dataset [11], which includes 135 videos from German schools and represents the German educational system. Each 90-min video focuses on teaching quadratic equations to 9th and 10th graders. Our study used video data from 46 teachers in the TALIS dataset.

From these videos, three 30-s segments with frontal presentations were extracted, forming the training and validation sets. Only segments without excessive camera movement, where the teacher remained visible, were included. Audio was available for human raters. Segments from 9 teachers were used for validation, and those from 37 teachers for training. To prevent data leakage and memorization, no segments from the same teacher were used in both sets.

For the perceived distance and gesture intensity regressors, 3056 frames were labeled, divided into 2451 for training and 605 for validation. These frames were

(a) Hand gesture (b) No gesture

Fig. 1. Examples of different gesture intensities with similar extracted skeletons. In 1a, the person is pointing, representing a specific hand gesture. In 1b, the person is leaning on the whiteboard, which is not considered gesturing.

rated by three trained raters, resulting in an ICC(2,3) of 0.683 for proximity and 0.885 for gesture intensity.

In the context of this study, a 'hand gesture' refers to deliberate movements or signals made by the hand that are intended to communicate specific messages or emotions, as shown in Fig. 1a. In contrast, a 'non-hand gesture' involves hand movements that are more incidental or casual, not aimed at communication, as illustrated in Fig. 1b. The Fig. 1 demonstrates the difference in gesture intensity even with similar extracted skeletons.

Perceived distance is evaluated based on the physical space and intervening objects between the teacher and students. Objects like desks placed between the teacher and students can increase the perceived distance, creating a sense of separation. Conversely, the absence of such barriers can lead to a perception of a smaller, more engaging distance.

In a similar approach, 400 30-s video segments (321 for training and 79 for validation) were used for the NVI model. The ICC(2,3) for the NVI score was 0.684, reflecting a substantial level of rater agreement.

Histograms of all datasets are presented in Fig. 2.

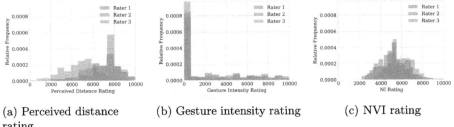

(a) Perceived distance rating (b) Gesture intensity rating (c) NVI rating

Fig. 2. Histograms showing the distribution of ratings for perceived distance, gesture intensity, and nonverbal immediacy.

4 Methodology

In this section, we describe the methods used to measure NVI from teachers in video recordings. NVI estimation is approached as a regression task involving facial expressions, gesture intensity, and perceived distance. We extract facial expressions and gesture intensity from the teacher's segmentation masks. To assess perceived distance, we use estimated depth images and segmentation masks of the teacher and students. These elements are combined, as shown in Fig. 3, to calculate the NI Score from RGB video inputs.

Tracking and Segmenting. In each video, the teacher's initial position is manually identified in the first frame. The subsequent tracking of the teacher throughout the video is conducted using Segment and Track Anything [5]. This involves estimating the semantic segmentation of each frame with the prompt "human" [10]. This approach not only segments the teacher but also the students. The segmented data of both teachers and students are later utilized in the gesture intensity and proximity regressors.

Gesture Intensity Regressor Model. For the gesture intensity analysis, we developed a specialized model based on a fine-tuned ResNet architecture [7]. It extends the standard ResNet18 model, pretrained on ImageNet [23], to suit our specific task. The model's input is a masked RGB image of the teacher, obtained from semantic segmentation. It processes this input through ResNet18's layers, followed by additional fully connected layers that adapt the output to our dataset's requirements.

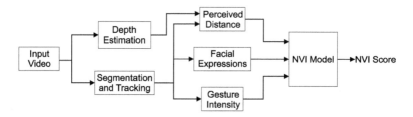

Fig. 3. Pipeline overview: The model takes RGB videos as input, performing tracking, segmentation, and depth estimation. We extract facial expressions and gesture intensity from the teacher's segmentation masks. To estimate perceived distance, we use estimated depth images and segmentation masks from both the teacher and students. These features - facial expressions, gesture intensity, and perceived distance - converge in the NVI model to calculate the NVI score.

Perceived Distance Regressor Model. The perceived distance regressor model mirrors the architecture used in the gesture intensity regressor, based on a fine-tuned ResNet18 architecture [7]. The input for this model, consists of a three-channel input: depth image, a mask of the teacher, and a combined mask of all students in a single channel. The depth image was estimated from the RGB frame using [16]. This depth channel provides spatial context, while the masks isolate the subjects of interest, enabling accurate estimation of physical distances in the classroom setting. This format enables the model to recognize spatial relationships between the teacher and students.

Facial Expression Analysis. For analyzing facial expressions, we utilized the state-of-the-art HSEmotion library [24], designed for high accuracy and speed in recognizing and quantifying facial expressions. The output from HSEmotion includes confidence scores for eight emotions: anger, contempt, disgust, fear, happiness, neutral, sadness, and surprise, providing a comprehensive emotional profile for each frame. Additionally, in cases where the face is not visible in a frame, we store a distinct output to account for the absence of facial expression data.

Nonverbal Immediacy Model. The NVI model in our study, while straightforward in its approach, aligns with methods commonly used by educational psychologists for analyzing NVI [20]. It integrates outputs from gesture intensity and distance regressors, along with facial expression analysis, into a 10-dimensional feature vector. This vector thus combines gesture intensity, distance, and eight emotion confidence ratings.

The integration process weights the feature vector based on the sum of frames with visible faces. The resulting vector feeds into a multilayer perceptron model [22] with three fully connected layers, designed to predict the NVI score. This model reflects the practical approaches in educational psychology, offering a quantifiable measure of various nonverbal aspects.

5 Experiments

Regressor Models. To train the gesture intensity regressor model, we employed the Adam optimizer with a 0.001 learning rate and resized input images to 360 × 360 pixels. The model was trained with mean squared error (MSE) loss. To ensure the quality of the training subset, we excluded samples with high rater disagreement, defined by a standard deviation of 1600 among ratings. This exclusion reduced the training samples from 2451 to 2121. The median of the three ratings was used as the target for training. For validation, all samples were included, regardless of rater agreement.

Similarly, the perceived distance regressor model was trained using the Adam optimizer with a 0.001 learning rate and 360 × 360 pixel input images, employing MSE loss. Due to higher rater disagreement, indicated by a standard deviation of 1600, we retained 1604 out of 2451 samples for training. The median of the three ratings was used as the target. For validation, all samples were included for comprehensive model evaluation.

Nonverbal Immediacy Model. The NVI model was trained using the Adam optimizer with a learning rate of 0.001 and MSE loss. The model's architecture included three fully connected layers with sizes 300, 100, and 10, respectively. For training, we utilized a majority of the dataset, specifically 316 out of the 321 available samples, selecting based on data quality and consistency. For validation, all samples were included to comprehensively assess the model's performance in various scenarios.

This model being the first, to our knowledge, to analyze NVI using computer vision and given the lack of additional labeled datasets, the model was further validated through an assessment of ecological validity. An external validation was conducted, to examine the correlation between model-estimated NVI scores and relevant questionnaire data and observer ratings from the TALIS study [11].

To this end, 220 additional 30-s video clips were extracted from the available video data, applying the same selection criteria as for the original clips. Despite efforts to extract the same number of additional clips per video, some teachers had to be excluded, due to the limited availability of sufficient additional video material.

We examined the hypothesized relationships using both the entire dataset (a) and only the additionally identified clips (b). Based on existing NVI-literature [12,27], we anticipated positive correlations between the model-estimated NVI scores and student-reported interest in mathematics (H1a + b), cognitive activation (H2a + b), and perceived teacher enthusiasm (H3a + b). Additionally, we hypothesized a positive correlation between NVI scores and teachers' socio-emotional support behaviors, focusing on the domain of encouragement and warmth (H4a + b), as rated in regular intervals by trained observers during the TALIS-study. For detailed information regarding the rating process and the scales employed in the TALIS-study see [11]. To test the hypotheses, a series of Pearson correlations was computed (with false discovery rate adjustments for multiple testing). Questionnaire data and NVI scores were aggregated at the teacher level. For correlations with socio-emotional support ratings, mean aggregates were calculated at the video level.

6 Results

Regressor Models. The gesture intensity regressor model showed a Pearson correlation of 0.84, indicating a high level of predictive accuracy. When used as a binary classifier through thresholding, it achieved an accuracy of 84.9%.

The perceived distance regressor model demonstrated a Pearson correlation of 0.55, indicating a moderate positive relationship between predicted values and ground truth.

Table 1. ICC comparison between human raters and the model on the validation dataset. The ICC values are different from those for the entire dataset as they are estimated solely based on the validation subset.

Rater Combination	ICC
Rater0, Rater1, Rater2	0.648
Rater0, Rater1, Model	0.611
Rater0, Model, Rater2	0.621
Model, Rater1, Rater2	0.613
Rater0, Rater1, Rater2, Model	0.690

Nonverbal Immediacy Model. The NVI model exhibited a Pearson correlation of 0.44 with the ground truth based on human raters, indicating a moderate correlation.

As the first to measure NVI computationally, direct comparisons are challenging. To further evaluate the model, we integrated the model-based NVI ratings those of the human raters by calculating median values. This approach allowed for the assessment of bivariate correlations between individual ratings and this new ground truth. Significant and strong positive correlations were found between the median rating and each individual rater, as well as the model: the correlations were 0.74 ($p < .01$) for the first rater, 0.68 ($p < .01$) for the second rater, 0.66 ($p < .01$) for the third rater, and 0.69 ($p < .01$) for the model.

Additionally, as shown in Table 1, replacing one human rater with our model does not significantly drop the ICC, and including the model alongside three human raters slightly improves the ICC. It is important to note that these ICC values differ from those calculated for the entire dataset, as they are estimated solely based on the validation subset.

The results of the correlation analyses in the external validation, considering the full available data set, revealed positive and significant moderate correlations between NVI scores and students' interest in mathematics ($r = .33, p = .03$)

(H1a), students' cognitive activation ($r = .32, p = .03$) (H2a), perceived teacher enthusiasm ($r = .31, p = 04$) (H3a), and social-emotional support ($r = .34, p = < .01$) (H4a). Using only the additional video data clips, a positive and significant moderate correlation was found between NVI scores and students' interest in mathematics ($r = ..45, p < .01$) (H1b) as well as students' cognitive activation ($r = .32, p = .03$) (H2b). No significant correlation was observed between NVI scores and perceived teacher enthusiasm ($r = .23, p = .13$) or social-emotional support ($r = .02, p = .86$) using only the additional video data clips.

7 Discussion and Conclusion

In this study, we demonstrated the potential of a multimodal computational approach using only RGB cameras to estimate NVI from 30-s video segments in educational settings. We developed a gesture intensity regressor, a perceived distance regressor, and an NVI model integrating these outputs with facial expression analysis. Using our labeled data for training and validation, these models were further evaluated with available data from the TALIS study [11] to correlate our model's outputs with real-world educational outcomes.

The gesture intensity regressor model showed strong predictive accuracy, effectively processing gesture intensity from RGB images. However, limitations in processing the broader context of scenes remain, particularly when faced with complex or ambiguous scenarios. The perceived distance regressor model exhibited moderate performance, highlighting the complexity of measuring perceived distance compared to geometrical distance. The NVI model demonstrated a moderate correlation with median human ratings. Comparing bivariate correlations while including the models' scores alongside those of the human raters, demonstrated that the model performs comparable to two of the three observers. ICC analyses showed, that interrater reliabilities remain stable, when replacing one human rater with the model, while including the model alongside three human raters slightly improves ICC values. These results indicate, that the model can effectively be leveraged, to support human observers, while fully replacing human raters remains more challenging, due to the complexity and context specificity of nonverbal communication.

The external validation showed significant moderate correlations with students' interest in mathematics, cognitive activation, perceived teacher enthusiasm, and socio-emotional support. These findings indicate that the NVI model effectively captures key aspects of nonverbal communication in educational settings, supporting its potential as a tool for enhancing teacher-student interactions. The discrepancies observed between the outcomes of the two analyses, one utilizing the complete data set and the other employing only the additionally extracted video clips, may be attributed to the reduced quantity of available data in the latter case. The reduced amount of data may limit the representativeness of the analyzed clips for the overall teaching behavior. Moreover, the corresponding clips were not missing at random, with some teachers being excluded completely.

However, we also recognize the challenges encountered. The sensitivity of the data limited the size of our datasets and restricted public availability, making direct comparisons with other studies challenging. Additionally, the variability in video recording conditions across different classrooms, such as camera placement and angles, posed substantial challenges, requiring our models to be adaptable and robust.

Future work will incorporate additional modalities like gaze tracking to enrich understanding of nonverbal communication. While our current study integrated multiple nonverbal modalities, future research will focus on developing new architectures to enhance and support these modalities. Additionally, we will investigate the comprehensive integration of the NVI model and its modalities with scores and relevant questionnaire data and observer ratings from the TALIS study [11].

In conclusion, our study provides a foundational framework for future research in enhancing teacher-student interactions through a deeper understanding of NVI. It opens promising pathways for educational research and practice, offering a more comprehensive understanding of nonverbal communication between teachers and students.

Acknowledgments. We gratefully acknowledge funding by the Deutsche Forschungsgemeinschaft (DFG, German Research Foundation) under Germany's Excellence Strategy - EXC 2002/1 "Science of Intelligence" - project number 390523135.

References

1. Ahuja, K., et al.: Edusense: practical classroom sensing at scale. Proc. ACM Interact. Mob. Wearable Ubiquitous Technol. **3**(3), 1–26 (2019)
2. Barmaki, R., Hughes, C.E.: Providing real-time feedback for student teachers in a virtual rehearsal environment. In: Proceedings of the 2015 ACM on International Conference on Multimodal Interaction, pp. 531–537 (2015)
3. Bosch, N., Mills, C., Wammes, J.D., Smilek, D.: Quantifying classroom instructor dynamics with computer vision. In: Penstein Rosé, C., et al. (eds.) AIED 2018. LNCS (LNAI), vol. 10947, pp. 30–42. Springer, Cham (2018). https://doi.org/10.1007/978-3-319-93843-1_3
4. Chen, Y., Wang, C., Jian, Z.: Research on evaluation algorithm of teacher's teaching enthusiasm based on video. In: Proceedings of the 6th International Conference on Robotics and Artificial Intelligence, pp. 184–191 (2020)
5. Cheng, Y., et al.: Segment and track anything. arXiv preprint arXiv:2305.06558 (2023)
6. Gupta, S.K., Ashwin, T., Guddeti, R.M.R.: Cvucams: computer vision based unobtrusive classroom attendance management system. In: 2018 IEEE 18th International Conference on advanced learning technologies (ICALT), pp. 101–102. IEEE (2018)
7. He, K., Zhang, X., Ren, S., Sun, J.: Deep residual learning for image recognition. In: Proceedings of the IEEE Conference on Computer Vision and Pattern Recognition, pp. 770–778 (2016)

8. Kane, T.J., McCaffrey, D.F., Miller, T., Staiger, D.O.: Have we identified effective teachers? Validating measures of effective teaching using random assignment. Research Paper. MET Project. Bill & Melinda Gates Foundation (2013)
9. Kennedy, J., Baxter, P., Belpaeme, T.: Nonverbal immediacy as a characterisation of social behaviour for human-robot interaction. Int. J. Soc. Robot. **9**, 109–128 (2017)
10. Kirillov, A., et al.: Segment anything. arXiv preprint arXiv:2304.02643 (2023)
11. Klieme, E., Grünkorn, J., Praetorius, A.K., Schreyer, P., Herbert, B., Käfer, J.: Talis-videostudie deutschland - unterrichtsbeobachtung (tvd). Forschungsdatenzentrum Bildung am DIPF (2019). https://doi.org/10.7477/352:1:0. datenerhebung 2017-2018
12. Liu, W.: Does teacher immediacy affect students? A systematic review of the association between teacher verbal and non-verbal immediacy and student motivation. Front. Psychol. **12**, 713978 (2021)
13. Mangal, S., Mangal, U.: Essentials of educational technology. PHI Learning Pvt, Ltd (2019)
14. Mehrabian, A.: Some referents and measures of nonverbal behavior. Behav. Res. Methods Instrum. **1**(6), 203–207 (1968)
15. Murphy, K.P.: Machine Learning: A Probabilistic Perspective. MIT Press, Cambridge (2012)
16. Oquab, M., et al.: DINOv2: Learning Robust Visual Features without Supervision (2023)
17. Patterson, M.L.: Reflections on historical trends and prospects in contemporary nonverbal research. J. Nonverbal Behav. **38**, 171–180 (2014)
18. Pianta, R.C., La Paro, K.M., Hamre, B.K.: Classroom Assessment Scoring System^{TM}: Manual K-3. Paul H Brookes Publishing (2008)
19. Ramakrishnan, A., Zylich, B., Ottmar, E., LoCasale-Crouch, J., Whitehill, J.: Toward automated classroom observation: multimodal machine learning to estimate class positive climate and negative climate. IEEE Trans. Affect. Comput. **14**(1), 664–679 (2021)
20. Richmond, V.P., McCroskey, J.C., Johnson, A.D.: Development of the nonverbal immediacy scale (NIS): measures of self-and other-perceived nonverbal immediacy. Commun. Q. **51**(4), 504–517 (2003)
21. Rouast, P.V., Adam, M.T., Chiong, R.: Deep learning for human affect recognition: insights and new developments. IEEE Trans. Affect. Comput. **12**(2), 524–543 (2019)
22. Rumelhart, D.E., Hinton, G.E., Williams, R.J.: Learning representations by back-propagating errors. Nature **323**(6088), 533–536 (1986)
23. Russakovsky, O., et al.: ImageNet large scale visual recognition challenge. Int. J. Comput. Vis. (IJCV) **115**(3), 211–252 (2015)
24. Savchenko, A.V.: Facial expression and attributes recognition based on multi-task learning of lightweight neural networks. In: Proceedings of the 19th International Symposium on Intelligent Systems and Informatics (SISY), pp. 119–124. IEEE (2021)
25. Sumer, O., et al.: Teachers' perception in the classroom. In: Proceedings of the IEEE Conference on Computer Vision and Pattern Recognition Workshops, pp. 2315–2324 (2018)
26. Vinciarelli, A., Pantic, M., Bourlard, H.: Social signal processing: survey of an emerging domain. Image Vis. Comput. **27**(12), 1743–1759 (2009)

27. Witt, P.L., Wheeless, L.R., Allen, M.: A meta-analytical review of the relationship between teacher immediacy and student learning. Commun. Monogr. **71**(2), 184–207 (2004)
28. Wu, D., Chen, J., Deng, W., Wei, Y., Luo, H., Wei, Y.: The recognition of teacher behavior based on multimodal information fusion. Math. Probl. Eng. **2020**, 1–8 (2020)
29. Zaki, J.: Cue integration: a common framework for social cognition and physical perception. Perspect. Psychol. Sci. **8**(3), 296–312 (2013)

Author Index

A
Abdelwahed, Mehdi 106
Amin, Amany 3
Anand, Christopher 211
Arias-Rodriguez, Lenin 133

B
Bal, Baljinder Singh 106
Bartashevich, Palina 157
Bennewitz, Maren 169
Bergoin, Raphaël 106
Bierbach, David 133
Bolenz, Florian 289
Brass, Marcel 145
Brochier, Thomas 263
Brock, Oliver 289

C
Cañamero, Lola 106
Carminatti, Laurène 263
Cheney, Nick 81
Cheng, Sen 39
Choe, Yoonsuck 211
Chrastil, Elizabeth R. 63
Condro, Lucio 263

D
Daneshi, Asieh 145
Daucé, Emmanuel 263

E
Ebitz, R. Becket 301
Espino, Harrison 27

F
Frenkel, Jonas 326

G
Ghazinouri, Behnam 39
Ghosh, Anindya 3

Graham, Paul 3
Grün, Sonja 263

H
Hafner, Verena V. 275
Haxby, Simon 314
Hegarty, Mary 63
Hellwich, Olaf 326
Hemelrijk, Charlotte K. 194
Hildenbrandt, Hanno 194
Hinnen, Zachary 121
Hsu, Katherine 93
Hwu, Tiffany 314

J
Jaiton, Vatsanai 251

K
Kagioulis, Efstathios 3
Kandimalla, Sriskandha 93
Kang, William 211
Karimian, Maryam 145
Kirshner, Eleanore J. 93
Knight, Israel 314
Knight, James C. 3, 15
Knopf, Lars 157
Krause, Jens 133
Krichmar, Jeffrey L. 27, 63, 93, 236

L
L'Haridon, Louis 106
Lazarides, Rebecca 326
Liu, Jiahe 93
Lui, Hin Wai 93

M
Manoonpong, Poramate 251
Marcantonio, Matteo 223
McDonald, Chase 314
Mertan, Alican 81

Misiek, Thomas   15
Mohaddesi, Seyed A.   63

N
Nguyen, Thinh H.   51
Nowotny, Thomas   3

P
Pacher, Korbinian   133
Pachur, Thorsten   289
Pae, Hongju   236
Papadopoulou, Marina   194
Petković, Uroš   326
Pfeiffer, Michael A.   93
Philippides, Andrew   3, 15

R
Ramírez-Ruiz, Jorge   301
Reeh, Fabio   145
Riehle, Alexa   263
Rockbach, Jonas D.   169
Romanczuk, Pawel   133, 145, 157, 182

S
Sevinchan, Yunus   133
Sion, Antoine   223

T
Teixeira, Flávio   314
Tuci, Elio   223

V
Vanderelst, Dieter   51
Vollmoeller, Carla   133

W
Wang, Albert   314
Weitzenfeld, Alfredo   121

Y
Yordanova, Magdalena   275
Yuan, Alina   93

Z
Zenkri, Oussama   289
Zheng, Yating   182

Printed in the USA
CPSIA information can be obtained
at www.ICGtesting.com
CBHW060329230924
14771CB00005B/141